METHODS
OF
ORBIT DETERMINATION
FOR THE
MICROCOMPUTER

Dan L. Boulet

Willmann-Bell, Inc.
P.O. Box 35025
Richmond, Virginia 23235
United States of America

Published by Willmann Bell, Inc.
P.O. Box 35025, Richmond, Virginia 23235

Copyright © 1991 by Dan L. Boulet, Jr.
First English Edition

All rights reserved. Except for brief passages quoted in a review, no part of this book may be reproduced by any mechanical, photographic, or electronic process, nor may it be stored in any information retrieval system, transmitted, or otherwise copied for public or private use, without the written permission of the copyright owner. Requests for permission or further information should be addressed to the Permissions Department, Willmann-Bell, Inc. P.O. Box 35025, Richmond, VA 23235.

Printed in the United States of America

Library of Congress Cataloging-in-Publication Data

Boulet, Dan L.
    Methods of orbit determination for the microcomputer / Dan L. Boulet.
       p.   cm.
    Includes bibliographical references and index.
    ISBN 0-943396-34-4
    1. Orbits–Measurement–Data processing. 2. Mechanics, Celestial.
3. Microcomputers. 4. BASIC (Computer program language) I. Title.
QB355.B68   1991
521'.3–dc20   91-11837
                                                                               CIP

...for

*Jewel,
David,
Jonathan,
and
Rebekah.*

Preface

This book is a popular work which describes how the principles of celestial mechanics may be applied to determine the orbits of planets, comets, and Earth satellites. More specifically, my objective is to enable a dedicated novice to learn, by first-hand experience, the manner in which orbital motion conforms to Newtonian physics, how a set of orbital elements can be translated into quantities which can be compared with observations, and how a record of observed motion can be used to determine an orbit from scratch or improve a preliminary orbit. Until fairly recently, this exciting adventure with nature was beyond the reach of nearly all non-specialists. However, the present widespread use of microcomputers has swept away the drudgery of days, or weeks, of tedious calculations fraught with endless opportunities for careless error. With the aid of a computer, an enthusiast may have the satisfaction of conquering problems which preoccupied astronomy for hundreds of years, and, in the process, gain a fresh appreciation for the genius and industry of the great mathematicians of the seventeenth, eighteenth, and nineteenth centuries.

The emphasis throughout is practical rather than theoretical even though the derivations of many important relationships are described in some detail. It is only essential that the reader accept the validity of the key equations and understand their symbology in order to use the computer programs to explore the power of the mathematical models. However, to make extensive use of the material presented in the text, one should be familiar with the basic physics of motion and be well grounded in algebra, geometry, and trigonometry. Ideally, one should have an acquaintance with the operations of elementary calculus and vectors. If this is not the case, the desired familiarity can be gained by studying the appropriate chapters in almost any introductory calculus text. Indeed, my secondary purpose for this effort was to provide a source of material which could be used to enrich the teaching of basic calculus, physics, and astronomy. For convenience, the appropriate definitions and rules of vector algebra and calculus are summarized in Appendices A and B.

Thus, this is a how-to-do-it book. All the important principles have been reduced to complete computer programs written in simple BASIC. Each program is illustrated by at least one numerical example, and the output is shown in the format produced by the computer routine. The programs do not claim to represent the most widely used numerical methods or the most efficient ways to implement the various mathematical algorithms. Rather, they are intended to provide a starting point for developing individualized computer routines tailored to the reader's particular interests. Also, to a certain extent, the computational steps in these programs are displayed in a manner which makes it easier to show how they relate to the equations derived in the text and to maintain a little standardization among the various listings.

All the computer programs were originally written in BASIC for execution on a Macintosh SE. However, they have also been successfully run on an IBM Model 50Z, where the only adjustment which had to be made in some programs was the addition of a statement as line 1005 to reserve some extra space in memory. Although the programs listed here display their output on the computer screen, they can be easily modified to output to a printer or data file. This latter method was used to incorporate the numerical examples into the text.

I am deeply indebted to a number of people whose special contributions helped me to finally bring to completion this long project. I thank Mr. Perry Remaklus of Willmann-Bell for much good advice and kind patience from beginning to end. Dr. J. M. A. Danby and Dr. B. G. Marsden graciously consented to read the initial manuscript and provided some very helpful comments. Of course, any errors or shortcomings which remain are completely my responsibility. Finally, I would like to thank my wife, Jewel, and my family for the loving support which permitted me to continue this labor through some especially busy times.

D.L.B.

Wilmington, Delaware
February 1991

Contents

	Preface	v
1	**Fundamentals of Orbital Motion**	1
1.1	Introduction	1
1.2	The Laws of Motion	1
	1.2.1 The Law of Inertia	2
	1.2.2 The Law of Acceleration	5
	1.2.3 The Law of Action and Reaction	5
1.3	The Law of Gravitation	6
1.4	Equations of Motion	7
	1.4.1 The Equation of Inertial Motion	7
	1.4.2 The Equation of Relative Motion	9
1.5	Working Units and Constants	10
	1.5.1 The Heliocentric System	10
	1.5.2 The Geocentric System	11
1.6	The Working Equation of Motion	11
1.7	Numerical Example	13
2	**Time and Position**	15
2.1	Introduction	15
2.2	The Fundamental References	15
2.3	The Empirical Frame of Reference	16
2.4	Time Scales	18
	2.4.1 Universal Time	19
	2.4.2 Julian Date	19
	2.4.3 Sidereal Time	20
	2.4.4 Atomic Time	23
	2.4.5 Dynamical Time	24
2.5	Coordinate Systems	25
	2.5.1 Celestial Equatorial Systems	26
	2.5.2 Terrestrial Equatorial Systems	28

	2.5.3 Celestial Ecliptic Systems	30
2.6	Ecliptic-Equatorial Transformations	30
2.7	The Fundamental Vector Triangle	31
2.8	Reduction of Astronomical Coordinates	33
	2.8.1 Planetary Aberration	33
	2.8.2 The Instantaneous and Fixed Equator and Equinox	35
	2.8.3 Astrometric Positions	35
	2.8.4 Reductions for Aberration and Nutation	36
	2.8.5 Reductions for Precession	37
	2.8.6 Reductions for Geocentric Parallax	39
2.9	Computer Programs	41
	2.9.1 Program LMST	41
	2.9.2 Program XYZ	45
	2.9.3 Program RAD	47
	2.9.4 Program CQTRAN	51
	2.9.5 Program ADAPP	54
	2.9.6 Program ADCES	58
	2.9.7 Program XYZCES	62
	2.9.8 Program ADLAX	65
2.10	Numerical Examples	70
	2.10.1 Computing Local Mean Sidereal Time	70
	2.10.2 Converting Spherical to Rectangular Coordinates	71
	2.10.3 Converting Rectangular to Spherical Coordinates	72
	2.10.4 Converting Equatorial to Ecliptic Coordinates	73
	2.10.5 Reducing Apparent Place to Astrometric Place	74
	2.10.6 Reducing RA and DEC from Jxxxx.x to J2000.0	76
	2.10.7 Reducing Rectangular Coordinates from J2000.0 to Jxxxx.x	77
	2.10.8 Reducing Geocentric Place to Topocentric Place	78

3 The Two-Body Problem — 81

3.1	Introduction	81
3.2	The Two-Body Equation of Motion	81
3.3	The Orbital and Radial Rates	83
3.4	The Laws of Two-Body Motion	85
	3.4.1 The Conic Section Law	85
	3.4.2 The Law of Areas	91
	3.4.3 The Harmonic Law	92
	3.4.4 The Vis-viva Law	93
3.5	Two-Body Motion by Numerical Integration	95
	3.5.1 The f and g Series	95
	3.5.2 Taylor Series	100
	3.5.3 Runge-Kutta Five	102

	3.5.4	Numerical Error 104
3.6	Computer Programs 106	
	3.6.1	Program FANDG 106
	3.6.2	Program TAYLOR 111
	3.6.3	Program RUNGE 117
3.7	Numerical Examples 124	
	3.7.1	Two-Body Motion by f and g Series 124
	3.7.2	Two-Body Motion by Taylor Series 127
	3.7.3	Two-Body Motion by Runge-Kutta Five 129

4 Orbit Geometry   135

4.1	Introduction 135	
4.2	General Relationships 135	
	4.2.1	Angular Momentum and Angular Speed 137
	4.2.2	Radial Speed and True Anomaly 137
	4.2.3	True Anomaly and D 138
	4.2.4	Eccentricity, Semiparameter, and D 138
4.3	Relationships between Geometry and Time 139	
	4.3.1	Elliptic Formulation 140
	4.3.2	Hyperbolic Formulation 142
	4.3.3	Parabolic Formulation 146
4.4	The Classical Elements from Position and Velocity 149	
	4.4.1	Three Fundamental Vectors 150
	4.4.2	The Conic Parameters 153
	4.4.3	The Orientation Angles 153
	4.4.4	The Mean Anomaly 155
	4.4.5	The Time of Perifocal Passage 157
4.5	Position and Velocity from the Classical Elements 157	
	4.5.1	The Scalar Components of Elliptic Motion 158
	4.5.2	The Scalar Components of Hyperbolic Motion 159
	4.5.3	The Scalar Components of Parabolic Motion 160
	4.5.4	The Unit Vector Components of Motion 161
4.6	Computer Programs 165	
	4.6.1	Program CLASSEL 165
	4.6.2	Program POSVEL 174
4.7	Numerical Examples 184	
	4.7.1	Classical Elements for Mars 184
	4.7.2	Classical Elements for Comet X 186
	4.7.3	Classical Elements for Comet Y 188
	4.7.4	Classical Elements for GEOS 190
	4.7.5	Position and Velocity Elements for Pallas 192
	4.7.6	Position and Velocity Elements for Recon 1 194

	4.7.7	Position and Velocity Elements for Recon 2 196
	4.7.8	Position and Velocity Elements for Recon 3 198

5 Ephemeris Generation   201
   5.1 Introduction 201
   5.2 The Differenced Kepler Equations 201
       5.2.1 Elliptic Formulation 201
       5.2.2 Hyperbolic Formulation 204
       5.2.3 Parabolic Formulation 206
   5.3 The Closed f and g Expressions 207
       5.3.1 Elliptic Motion 208
       5.3.2 Hyperbolic Motion 212
       5.3.3 Parabolic Motion 214
   5.4 The Universal Formulation 217
       5.4.1 The Coefficients C, S, and U 217
       5.4.2 The Equations of Motion 219
   5.5 The Ephemeris 221
   5.6 Computer Programs 224
       5.6.1 Program SEARCH 224
       5.6.2 Program RADEC 234
   5.7 Numerical Examples 243
       5.7.1 Ephemeris for GEOS 243
       5.7.2 Ephemeris for Pallas 245
       5.7.3 Ephemeris for Comet X 248
       5.7.4 Right Ascension and Declination of Comet X 250

6 Special Perturbations   253
   6.1 Introduction 253
   6.2 Direct and Indirect Attractions 254
   6.3 The Method of Cowell 257
   6.4 The Method of Encke 258
   6.5 A Perturbed Ephemeris 262
   6.6 Computer Programs 263
       6.6.1 Program ATTRACT 263
       6.6.2 Program COWELL 272
       6.6.3 Program ENCKE 287
   6.7 Numerical Examples 304
       6.7.1 Solar and Planetary Attractions 304
       6.7.2 The Motion of Mars 307
       6.7.3 The Motion of Uranus 311

7 Applied Numerical Methods 317
- 7.1 Introduction 317
- 7.2 Finding the Root of an Equation 317
  - 7.2.1 The Bisection Method 319
  - 7.2.2 The Newton-Raphson Method 321
- 7.3 Solving a System of Linear Equations 323
  - 7.3.1 Naive Gauss Elimination 323
  - 7.3.2 Partial Pivoting 325
- 7.4 Polynomial Interpolation 326
- 7.5 Polynomial Regression 329
- 7.6 Multiple Linear Regression 333
- 7.7 Numerical Differentiation 335
  - 7.7.1 The Interpolating Polynomial 335
  - 7.7.2 The Regression Polynomial 337
- 7.8 Computer Programs 339
  - 7.8.1 Program PTERP 339
  - 7.8.2 Program PGRESS 341
  - 7.8.3 Program MGRESS 345
- 7.9 Numerical Examples 348
  - 7.9.1 Polynomial Interpolation and Differentiation 348
  - 7.9.2 Polynomial Regression and Differentiation 349
  - 7.9.3 Multiple Linear Regression 351

8 Preliminary Orbit Data 353
- 8.1 Introduction 353
- 8.2 Principal Constraints 354
- 8.3 The Topocentric Vector L 355
- 8.4 The Topocentric Vector R 357
  - 8.4.1 Vector R for Geocentric Orbits 357
  - 8.4.2 Vector R for Heliocentric Orbits 361
- 8.5 Computer Programs 364
  - 8.5.1 Program ADGRESS 364
  - 8.5.2 Program GEO 371
  - 8.5.3 Program HELO 376
- 8.6 Numerical Examples 382
  - 8.6.1 Regression of Angular Data for Satellite GEOS . . . 382
  - 8.6.2 Regression of Angular Data for Comet Rebek-Jewel . . . 384
  - 8.6.3 Topocentric Vector to the Geocenter 386
  - 8.6.4 Topocentric Vector to the Heliocenter 387

9 The Method of Laplace — 389
- 9.1 Introduction 389
- 9.2 Solution by Successive Differentiation 389
- 9.3 The Scalar Equations for the Range and Rate 390
- 9.4 The Scalar Equation for the Radial Distance 391
- 9.5 The Scalar Equation of Lagrange 392
- 9.6 The Vector Orbital Elements 392
- 9.7 Program LAPLACE 394
- 9.8 Numerical Examples 403
  - 9.8.1 The Orbit of Satellite GEOS 403
  - 9.8.2 The Orbit of Comet Rebek-Jewel 407

10 The Method of Gauss — 413
- 10.1 Introduction 413
- 10.2 Solution by f and g Expressions 413
- 10.3 The Scalar Equations for the Ranges 415
- 10.4 The First Approximation 415
- 10.5 The Scalar Equations Relating ρ and r at Epoch 417
- 10.6 The Scalar Equation of Lagrange 418
- 10.7 The Vector Orbital Elements 419
  - 10.7.1 Initial Position Vector 419
  - 10.7.2 Initial Velocity Vector 419
  - 10.7.3 Refinement of the Elements 420
- 10.8 Program GAUSS 422
- 10.9 Numerical Examples 435
  - 10.9.1 The Orbit of Pallas 435
  - 10.9.2 The Orbit of Comet Rebek-Jewel 442

11 The Method of Olbers — 451
- 11.1 Introduction 451
- 11.2 Solution by Euler's Equation 451
- 11.3 The Scalar Equations for the Range 453
- 11.4 The Vector Orbital Elements 454
  - 11.4.1 Three Radius Vectors 454
  - 11.4.2 The Velocity Vector 456
- 11.5 Program OLBERS 459
- 11.6 Numerical Example 469
  - 11.6.1 The Orbit of Comet Z 469
  - 11.6.2 The Orbit of Comet Rebek-Jewel 475

12 Orbit Improvement — 483
  12.1 Introduction 483
  12.2 The Differential Equations of Condition 483
  12.3 Numerical Evaluation of the Partial Derivatives 485
  12.4 Comparing Observation with Theory 486
  12.5 Computer Programs 487
      12.5.1 Program IMPROVE 487
      12.5.2 Program CORRECT 499
  12.6 Numerical Examples 509
      12.6.1 Improved Orbit for GEOS 509
      12.6.2 Improved Orbit for Rebek-Jewel 516
      12.6.3 Improved Orbit for Pallas 525

A Vectors — 535
  A.1 Basic Vector Operations 535
  A.2 The Dot and Cross Products 541

B Elementary Calculus — 545
  B.1 Differentiation 545
  B.2 Integration 551

C Astronomical Constants — 557
  C.1 Constants Related to Units 557
  C.2 Masses of the Planets 558

Index — 559

Chapter 1

Fundamentals of Orbital Motion

1.1 Introduction

Suppose we carefully follow the apparent motion of a celestial body as it travels across the heavens. Its track against the background of fixed stars might turn out to be an interesting curve such as that depicted in Figure 1.1. If we wish to determine the orbit of this celestial body, we must make accurate measurements of its position at a series of convenient times and use certain clever procedures to disentangle its orbital motion from the motion of our observing station on the surface of the moving Earth. Although such observational data contain a complex mixture of several independent motions, the orbit of the celestial body can, in principle, be determined by employing a theory of celestial mechanics developed from three general laws of motion and one law which accounts for the acceleration caused by gravity. Armed with these four fundamental principles, the orbits of planets, comets, and satellites can be computed using only elementary physics and simple calculus.

This chapter introduces the basic physics of celestial mechanics. The methods by which these principles are applied and the numerical techniques used to solve the orbital equations are taken up later in the text. Appendices A and B briefly review the necessary terminology and rules of vector algebra and calculus.

1.2 The Laws of Motion

Three laws of motion received explicit formulation by Isaac Newton in the Seventeenth Century. The importance of these principles to the development of celestial mechanics can hardly be overestimated. They may be stated as follows:

2 CHAPTER 1. FUNDAMENTALS OF ORBITAL MOTION

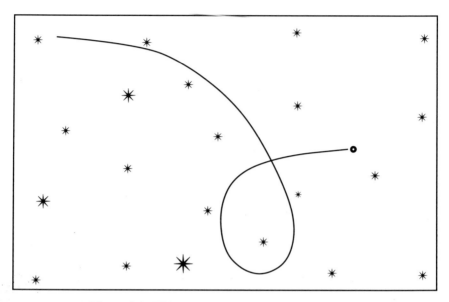

Figure 1.1: The apparent path of a celestial body.

Law 1 *A body continues in a state of rest or uniform motion in a straight line unless compelled to change its state by forces impressed upon it.*

Law 2 *The acceleration of a body is directly proportional to the net force impressed upon the body, inversely proportional to the mass of the body, and in the same direction as the net force.*

Law 3 *When one body exerts a force on a second body, the second body exerts a force of equal magnitude, but opposite direction, upon the first body.*

1.2.1 The Law of Inertia

Newton's first law is also known as the *law of inertia* because it expresses the sluggishness which matter exhibits when an attempt is made to move it or change any motion it may already possess. Assume that the origin O of the rectangular coordinate system shown in Figure 1.2 is either at rest or in uniform rectilinear motion and that its axes are not rotating with respect to the distant stars. Such a coordinate system is an *inertial* or *Newtonian frame of reference* and is necessary for a valid application of Newton's laws of motion.

The location of a celestial body can be defined by a *radius vector* **r** from the origin as shown in Figure 1.2. Any vector may be expressed in terms of three

1.2. THE LAWS OF MOTION

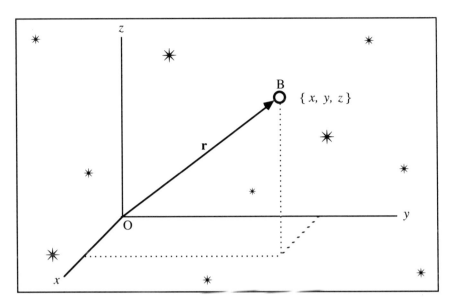

Figure 1.2: The instantaneous position of a body with respect to an inertial origin.

components which are projections of the vector onto the three coordinate axes. Thus, we may simply write

$$\mathbf{r} = \{x, y, z\}, \tag{1.1}$$

where x, y, and z are the rectangular coordinates of the given location. The distance r of B from the origin is equal to the magnitude of $\mathbf{r}$. Therefore,

$$r = |\mathbf{r}| = \sqrt{x^2 + y^2 + z^2}. \tag{1.2}$$

If B maintains constant direction and distance with respect to the inertial origin, it is said to be at rest.

Consider the situation shown in Figure 1.3, where a celestial body is moving in an inertial coordinate system. The rate at which its position changes with respect to time is a vector $\mathbf{v}$ called *velocity*. Thus, at a given instant of time

$$\mathbf{v} = \frac{d\mathbf{r}}{dt} = \left\{ \frac{dx}{dt}, \frac{dy}{dt}, \frac{dz}{dt} \right\}, \tag{1.3}$$

where $d\mathbf{r}$ is the incremental change in $\mathbf{r}$ during an infinitesimal time interval dt. In the terminology of calculus, velocity is the first derivative of position with respect to time. The *speed* v is a scalar quantity equal to the magnitude of the velocity. Therefore,

$$v = |\mathbf{v}| = \sqrt{v_x^2 + v_y^2 + v_z^2}, \tag{1.4}$$

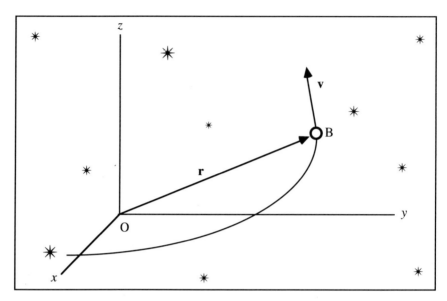

Figure 1.3: The instantaneous velocity of a body with respect to an inertial origin.

where v_x, v_y, and v_z are the rectangular components of the velocity vector in the x, y, and z directions, respectively.

If a body changes its speed or direction of motion, its velocity changes. The rate of change of velocity with respect to time is a vector called *acceleration* **a**. Thus, if the velocity changes by an increment $d\mathbf{v}$ during the infinitesimal time interval dt, then

$$\mathbf{a} = \frac{d\mathbf{v}}{dt} = \left\{ \frac{dv_x}{dt}, \frac{dv_y}{dt}, \frac{dv_z}{dt} \right\}. \tag{1.5}$$

Accordingly, acceleration is the first derivative of velocity with respect to time.

Another expression for acceleration follows immediately from substituting Equation 1.3 for **v** in Equation 1.5. Then

$$\mathbf{a} = \frac{d}{dt}\left(\frac{d\mathbf{r}}{dt}\right) = \frac{d^2\mathbf{r}}{dt^2}. \tag{1.6}$$

In other words, acceleration is equal to the second derivative of position with respect to time.

1.2. THE LAWS OF MOTION

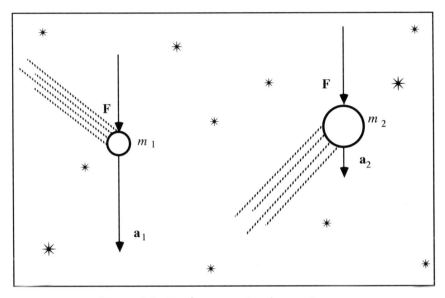

Figure 1.4: Net force, acceleration, and mass.

1.2.2 The Law of Acceleration

Newton's second law of motion is expressed by the following vector equation, where **F** represents the net force acting on a body of mass m:

$$\mathbf{a} = \frac{\mathbf{F}}{m}. \tag{1.7}$$

We assume that the physical units of length, mass, and time have been so defined that the proportional relationship between **a**, **F**, and m can be replaced by an exact equation.

This law is illustrated in the example of Figure 1.4 where a net force **F** acts on two unequal masses $m_1 = 1$ and $m_2 = 3$, causing both to accelerate. In accordance with Newton's second law, the acceleration of the lesser mass is three times that of the greater, and the directions of the accelerations are the same as that of the net force.

1.2.3 The Law of Action and Reaction

Newton's third law is also called the *law of action and reaction*. In Figure 1.5 two unequal masses are experiencing forces of mutual attraction. Thus, according to the third law of motion,

$$\mathbf{F}_1 = -\mathbf{F}_2. \tag{1.8}$$

6   CHAPTER 1. FUNDAMENTALS OF ORBITAL MOTION

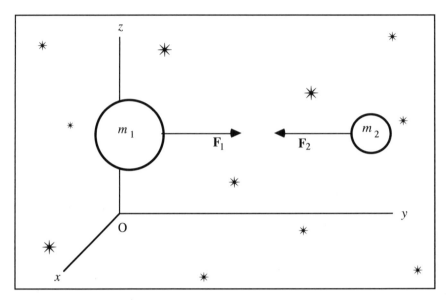

Figure 1.5: Forces of action and reaction.

However, in accordance with the second law, the accelerations of the two masses will be different. The law of action and reaction implies that in all orbital situations the central body is also accelerated, even though its mass may be very great in comparison to its satellite.

1.3 The Law of Gravitation

Newton's law of universal gravitation is the fourth fundamental principle of orbital motion. It may be stated as follows:

Law 4 *Every particle of matter in the universe attracts every other particle of matter with a force that is directly proportional to the product of their masses and inversely proportional to the square of the distance between them.*

The gravitational law is summarized by the following expression:

$$F = \frac{k^2 m_1 m_2}{r^2}, \qquad (1.9)$$

where F is the magnitude of the force, k^2 is the *gravitational constant*, m_1 and m_2 are the masses, and r is the distance. Although the law is stated for particles, Equation 1.9 can be applied to large accumulations of matter if we assume that

1.4. EQUATIONS OF MOTION

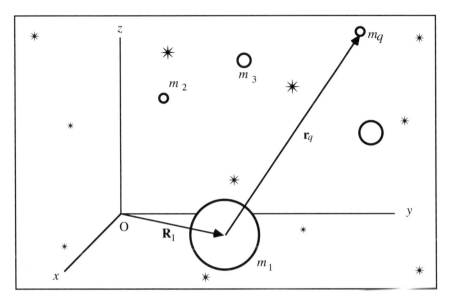

Figure 1.6: Mass m_1 subject to the gravitational attractions of other bodies.

the mass distribution of each body is spherically symmetric about its center or that all bodies are separated by distances which are very great in comparison to their sizes.

1.4 Equations of Motion

The orbital motion of a celestial body must be described by an equation which expresses its instantaneous acceleration in terms of all gravitational and nongravitational forces. We shall reduce the problem considerably by dealing only with spherically symmetric gravitational force fields and ignoring all nongravitational influences. Applying the fundamental principles already discussed in the context of these simplifying assumptions, the result is an *equation of motion* which can be used to compute the movement of a planet, comet, or satellite, taking into account the effects of any number of perturbing masses.

1.4.1 The Equation of Inertial Motion

Figure 1.6 depicts a mass m_1 subject to the gravitational attractions of several other masses $m_2, m_3, \ldots, m_N$. The position vector $\mathbf{R}_1$ defines the location of m_1 with respect to the inertial origin O. The vector $\mathbf{r}_q$ defines the position of the qth mass with respect to the center of m_1. According to Equation 1.9, the magnitude of the force exerted on m_1 by any other mass m_q is as follows:

$$F_q = \frac{k^2 m_1 m_q}{r_q^2}. \tag{1.10}$$

By defining a unit vector in the direction of m_q having the form

$$\textbf{unit vector} = \frac{\mathbf{r}_q}{r_q},$$

we can write the following expression for the force vector radiating from m_1 toward any of the other bodies:

$$\mathbf{F}_q = \frac{k^2 m_1 m_q}{r_q^2} \left(\frac{\mathbf{r}_q}{r_q} \right).$$

Carrying out the multiplication, we obtain

$$\mathbf{F}_q = \frac{k^2 m_1 m_q \mathbf{r}_q}{r_q^3}. \tag{1.11}$$

Thus, the net force $\mathbf{F}$ acting on m_1 is the vector sum

$$\mathbf{F} = \mathbf{F}_2 + \mathbf{F}_3 + \cdots + \mathbf{F}_N,$$

or, in sigma notation,

$$\mathbf{F} = \sum_{q=2}^{N} \mathbf{F}_q. \tag{1.12}$$

According to Newton's second law, the inertial acceleration $\mathbf{A}_1$ of the mass m_1 is given by

$$\mathbf{A}_1 = \frac{\mathbf{F}}{m_1}.$$

Dividing $\mathbf{F}$ by m_1 is equivalent to dividing each term on the right side of Equation 1.12 by m_1. Therefore, by Equation 1.11,

$$\mathbf{A}_1 = \sum_{q=2}^{N} \frac{k^2 m_q \mathbf{r}_q}{r_q^3}, \tag{1.13}$$

which is the *equation of inertial motion* for mass m_1 with respect to the Newtonian frame of reference.

1.4. EQUATIONS OF MOTION

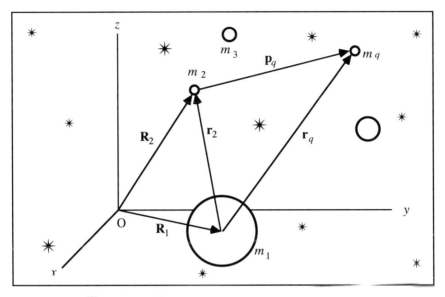

Figure 1.7: The many-body gravitational problem.

1.4.2 The Equation of Relative Motion

Figure 1.7 illustrates the gravitational problem we must solve in order to compute an orbit. Several masses $m_1, m_2, \ldots, m_N$ are moving under the influence of their mutual attractions. Vectors $\mathbf{R}_1$ and $\mathbf{R}_2$ define the locations of m_1 and m_2 with respect to the inertial origin O, and $\mathbf{r}_q$ and $\mathbf{p}_q$ define the position of any other mass m_q relative to m_1 and m_2, respectively. Thus, from the depiction,

$$\mathbf{r}_2 = \mathbf{R}_2 - \mathbf{R}_1 . \tag{1.14}$$

According to the second law of motion, the forces of gravitational attraction will cause m_1 and m_2 to accelerate with respect to the inertial origin. The acceleration $\mathbf{a}_2$ of m_2 with respect to m_1 can by found by differentiating Equation 1.14 twice with respect to time. When this is accomplished, Equation 1.6 can be used to obtain

$$\mathbf{a}_2 = \mathbf{A}_2 - \mathbf{A}_1 , \tag{1.15}$$

where $\mathbf{A}_1$ and $\mathbf{A}_2$ represent the inertial accelerations of m_1 and m_2, respectively. Now, let Equation 1.13 be rewritten as follows:

$$\mathbf{A}_1 = \frac{k^2 m_2 \mathbf{r}_2}{r_2^3} + \sum_{q=3}^{N} \frac{k^2 m_q \mathbf{r}_q}{r_q^3} . \tag{1.16}$$

By analogy with Equation 1.16, we can also write an expression for $\mathbf{A}_2$ in terms of m_1, $\mathbf{r}_2$, m_q, and $\mathbf{p}_q$. The result is

$$\mathbf{A}_2 = -\frac{k^2 m_1 \mathbf{r}_2}{r_2^3} + \sum_{q=3}^{N} \frac{k^2 m_q \mathbf{p}_q}{p_q^3}, \qquad (1.17)$$

where $\mathbf{p}_q = \mathbf{r}_q - \mathbf{r}_2$, $p_q = |\mathbf{p}_q|$, and the first term on the right side is negative because the acceleration of m_2 due to m_1 is in a direction opposite to vector $\mathbf{r}_2$. Finally, substituting Equations 1.16 and 1.17 into Equation 1.15, we obtain

$$\mathbf{a}_2 = -\frac{k^2(m_1 + m_2)\mathbf{r}_2}{r_2^3} + \sum_{q=3}^{N} k^2 m_q \left(\frac{\mathbf{p}_q}{p_q^3} - \frac{\mathbf{r}_q}{r_q^3} \right). \qquad (1.18)$$

This is the *equation of relative motion* for mass m_2 with respect to an origin at the center of mass m_1.

1.5 Working Units and Constants

We are concerned with applying Equation 1.18 to orbital problems in which the motion of interest is about either the Sun or the Earth. As a matter of convenience, different frames of reference and separate systems of units are used for these two applications. Very brief descriptions of the two systems used in this text are given below. Additional information on this topic can be found in References 1 through 4, and a thorough discussion of the improved International Astronomical Union (IAU) 1976 System of Astronomical Constants is given in Section S of Reference 5.

1.5.1 The Heliocentric System

The motion of a body orbiting the Sun is referred to a rectangular coordinate system centered in the Sun. The fundamental defining constant of the heliocentric system of units is the *Gaussian gravitational constant* given exactly by

$$k \equiv 0.017\,202\,098\,95.$$

Length, mass, and time are expressed in *astronomical units* (au), *solar masses* (sm), and *days* (day), respectively. One day is defined to be 86,400 seconds, and the astronomical unit is that length for which the Gaussian gravitational constant takes the value defined above when the units of measurement are astronomical units, solar masses, and days. The resulting value is approximately equal to the Earth's average distance from the Sun. This practice of holding k fixed while allowing the astronomical unit to vary insures that whenever better data for

1.6. THE WORKING EQUATION OF MOTION

the masses of the Earth and Sun become available all calculations functionally dependent on k do not have to be repeated, but only scaled for the new value of the au.

The heliocentric working units are dimensionless quantities analogous to the radian (rad) unit for angular measure. Their IAU (1976) metric equivalents are given in Appendix C.

1.5.2 The Geocentric System

The geocentric system of units is used for computing the orbits of artificial Earth satellites, where the origin of the rectangular coordinate system is located at the center of the Earth. By analogy with the heliocentric system, we shall assume a fundamental *geocentric gravitational constant* given by

$$k_e = 0.07436680.$$

Length, mass, and time are expressed in *g-radii* (gr), *Earth masses* (em), and *minutes* (min), respectively. The g-radius is that distance for which k_e takes the value shown above when the units of measurement are g-radii, Earth masses, and minutes. Its value is very nearly equal to the Earth's equatorial radius (er). Therefore, for the purposes of this book, we will assume these two quantities are the same. When this system of units is employed to compute geocentric orbits, k_e is used instead of the Gaussian constant k.

The geocentric working units are also treated as dimensionless quantities. Their IAU (1976) metric equivalents are given in Appendix C.

1.6 The Working Equation of Motion

Consider again Equation 1.18 which describes the relative motion of m_2 with respect to m_1:

$$\mathbf{a}_2 = -\frac{k^2(m_1+m_2)\mathbf{r}_2}{r_2^3} + \sum_{q=3}^{N} k^2 m_q \left(\frac{\mathbf{p}_q}{p_q^3} - \frac{\mathbf{r}_q}{r_q^3}\right). \tag{1.19}$$

It is possible to write a simpler version of this equation if we consistently let m_1 represent the central mass and m_2 the mass of the orbiting body of primary interest. Then, in both heliocentric and geocentric problems, m_1 will be unity, so that we can define a *combined mass*

$$\mu \equiv 1 + m_2 \tag{1.20}$$

and drop the subscripts on $\mathbf{a}_2$ and $\mathbf{r}_2$. When this is accomplished, the summation indices can be adjusted to begin at $q = 1$ and end at a new value n which is equal only to the number of perturbing bodies. Thus, when all these changes are made, Equation 1.19 becomes

$$\mathbf{a} = -\frac{k^2 \mu \mathbf{r}}{r^3} + \sum_{q=1}^{n} k^2 m_q \left(\frac{\mathbf{p}_q}{p_q^3} - \frac{\mathbf{r}_q}{r_q^3} \right). \tag{1.21}$$

Simplification can be carried one step farther by defining *modified time* τ as follows [2]:

$$\tau \equiv k(t - t_0) \tag{1.22}$$

so that

$$d\tau = k\,dt\,, \tag{1.23}$$

where k is the appropriate gravitational constant, t is a given instant of time, and t_0 is an arbitrarily chosen initial time or *epoch*. If we use a dot to indicate differentiation with respect to modified time, then

$$\dot{\mathbf{r}} = \left(\frac{1}{k}\right) \mathbf{v} \tag{1.24}$$

$$\ddot{\mathbf{r}} = \left(\frac{1}{k^2}\right) \mathbf{a}. \tag{1.25}$$

It follows that if both sides of Equation 1.21 are multiplied by $1/k^2$, we are able to write

$$\ddot{\mathbf{r}} = -\frac{\mu \mathbf{r}}{r^3} + \sum_{q=1}^{n} m_q \left(\frac{\mathbf{p}_q}{p_q^3} - \frac{\mathbf{r}_q}{r_q^3} \right). \tag{1.26}$$

Equation 1.26 is our *working equation of motion*. If a celestial body's position and velocity are known for a given initial time, then, in principle, Equation 1.26 can be integrated to yield the body's position and velocity at some other time. We shall begin to describe various methods for accomplishing this in Chapter 3 after the fundamental celestial frame of reference is presented in more detail in Chapter 2.

Finally, as a matter of convenience, we redefine the symbols $\mathbf{v}$ and $\mathbf{a}$ as follows:

$$\mathbf{v} \equiv \dot{\mathbf{r}} \tag{1.27}$$

$$\mathbf{a} \equiv \ddot{\mathbf{r}}. \tag{1.28}$$

Throughout the remainder of the text all motion may be assumed to be with respect to modified time unless otherwise indicated.

1.7 Numerical Example

Problem

Compute the total relative acceleration of Mars for 1985 January 15 considering only the gravitational attractions of the Sun, Mars, and Jupiter. Assume an appropriate heliocentric rectangular coordinate system along with the following information:

- Mass of Sun: $m_1 = 1$
- Mass of Mars: $m_2 = 0.0000003$
- Mass of Jupiter: $m_q = 0.0009548$
- $\mathbf{r} = \{+1.3381043, +0.4271086, +0.1596756\}$
- $\mathbf{p}_q = \{+0.8335077, -4.7003356, -2.0442936\}$
- $\mathbf{r}_q = \{+2.1716120, -4.2732270, -1.8846180\}$

Solution

We use Equation 1.26, letting $n = 1$, and compute the x, y, and z components of $\ddot{\mathbf{r}}$ separately. First we obtain the combined mass according to Equation 1.20:

$$\mu = 1 + 0.0000003$$
$$\mu = 1.0000003.$$

Next, we calculate the cubes of r, p_1, and r_1:

$$r = \sqrt{(1.3381043)^2 + (0.4271086)^2 + (0.1596756)^2}$$
$$r = 1.4136623$$
$$r^3 = 2.8251208$$

$$p_1 = \sqrt{(0.8335077)^2 + (-4.7003356)^2 + (-2.0442936)^2}$$
$$p_1 = 5.1929785$$
$$p_1^3 = 140.00392$$

$$r_1 = \sqrt{(2.1716120)^2 + (-4.2732270)^2 + (-1.8846180)^2}$$
$$r_1 = 5.1505488$$
$$r_1^3 = 136.63455.$$

Finally, we substitute the given and computed quantities into Equation 1.26 and solve for $\ddot{x}$, $\ddot{y}$, and $\ddot{z}$ in turn:

$$\ddot{x} = -\frac{1.0000003(1.3381043)}{2.8251208} + 0.0009548 \left(\frac{0.8335077}{140.00392} - \frac{2.1716120}{136.63455} \right)$$
$$\ddot{x} = -0.4736451 - 0.0000094$$
$$\ddot{x} = -0.4736545$$

$$\ddot{y} = -\frac{1.0000003(0.4271086)}{2.8251208} + 0.0009548 \left(\frac{-4.7003356}{140.00392} - \frac{-4.2732270}{136.63455} \right)$$
$$\ddot{y} = -0.1511824 - 0.0000021$$
$$\ddot{y} = -0.1511845$$

$$\ddot{z} = -\frac{1.0000003(0.1596756)}{2.8251208} + 0.0009548 \left(\frac{-2.0442936}{140.00392} - \frac{-1.8846180}{136.63455} \right)$$
$$\ddot{z} = -0.0565199 - 0.0000007$$
$$\ddot{z} = -0.0565206.$$

Results

Under the stated conditions, the total acceleration of Mars relative to the center of the Sun is

$$\ddot{\mathbf{r}} = \{-0.4736545, -0.1511845, -0.0565206\}$$

with respect to modified time.

References

[1] *The Astronomical Almanac 1985*, U.S. Government Printing Office, 1984.

[2] Escobal, *Methods of Orbit Determination*, Krieger Publishing Co., 1976.

[3] Baker and Makemson, *An Introduction to Astrodynamics*, Academic Press, 1967.

[4] Fitzpatrick, *Principles of Celestial Mechanics*, Academic Press, 1970.

[5] *The Astronomical Almanac 1984*, U.S. Government Printing Office, 1983.

Chapter 2

Time and Position

2.1 Introduction

Nothing in the heavens is at rest, and not one of its movements is precisely uniform. Consequently, the specification of position is tied closely to a measure of time, and the convenient inertial reference frame we have thus far assumed does not exist. Of course, the situation is not really very serious. We know from experience that a carefully chosen origin along with a celestial coordinate system defined for a particular epoch can be used to accurately define a position in space at a given instant of time. This is achieved by determining initial values for certain fundamental parameters of the coordinate system and measuring the rate at which these quantities change with time [1; Section S].

2.2 The Fundamental References

Consider the *celestial sphere* depicted in Figure 2.1. The nighttime sky creates the strong impression that the stars are fixed to an enormous curved surface which appears to be equally distant in all directions regardless of our location on the Earth. Indeed, this imaginary sphere is so vast that any location within the solar system S can serve as the origin of its radius. The celestial sphere is banded by two fundamental reference circles which correspond to its intersections with the fundamental planes of the Earth's equator and orbit. These great circles are called the *celestial equator* and *ecliptic*, respectively. The two dimensional starry surface of the celestial sphere provides the background upon which the reference circles are traced. This is accomplished by meticulous observations of the motions of the Sun and planets relative to a network of fundamental reference stars [2]. The ecliptic crosses the celestial equator at an inclination of approximately 23.5 degrees, forming an angle ε known as the *obliquity of the*

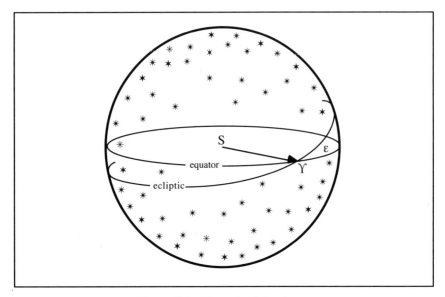

Figure 2.1: The celestial sphere.

ecliptic. The two points of intersection lie on a line passing through the center of the celestial sphere at S. One of these intersections has been defined as the fundamental direction. Originally named the *First Point of Aries* ♈, it is often symbolized by the horns of a ram. This designation was applied in the second century B.C. when ♈ was in the constellation of Aries.

The physical significance of ♈ is further illustrated in Figure 2.2, where the Earth is shown moving into spring (for its northern hemisphere) as it approaches the point E where day and night are of equal length (equinox). To an observer on the Earth, the Sun's projection S against the inside surface of the celestial sphere will travel along the path of the ecliptic toward the celestial equator, crossing it from south to north at the moment the Earth passes through E. Thus, the fundamental reference direction is also called the *vernal (spring) equinox*. The corresponding point on the opposite side of the celestial sphere is known as the *autumnal equinox*. When the term *equinox* is used in this text, it will always refer to the fundamental direction ♈.

2.3 The Empirical Frame of Reference

If the Earth were a perfect sphere traveling in a circular orbit as the only satellite of the Sun, the positions of the celestial equator, ecliptic, and vernal equinox would remain essentially fixed in inertial space. However, the Earth is an oblate

2.3. THE EMPIRICAL FRAME OF REFERENCE

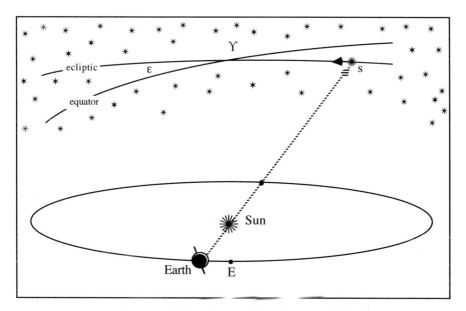

Figure 2.2: The physical significance of ♈.

spheroid (slightly flattened at the poles) moving in an elliptical orbit which is perturbed by the other bodies in the solar system.

The Sun and Moon exercise the greatest influence over the positions of the fundamental references. The Earth's equatorial bulge, coupled with the obliquity of the ecliptic, allow these two bodies to exert forces on the Earth which produce a torque about its center of mass. As illustrated in Figure 2.3, this torque causes the Earth's axis to precess with respect to the celestial sphere. The radius of this circle subtends an arc equal to ε, and the precessional period is approximately 26,000 years. The corresponding oscillation of the celestial equator causes ♈ to move slowly westward at roughly 50 arcseconds per year. When this is combined with much smaller (less than 1 arcsecond per year) contributions from the planets and general relativity, the total motion is known as *general precession*. Its magnitude for the year 2000 is approximately 50.3 arcseconds per year. Finally, there is another small oscillation of the equatorial plane which is superimposed upon the general precession. Known as *nutation*, this motion has a maximum amplitude of about 9 arcseconds and a period of approximately 18.6 years. This short-period effect is caused by the Moon and may be thought of as a rotation of the instantaneous axis of the Earth about its mean position on the circle of Figure 2.3 [3].

General precession and nutation are well understood, and the variations of the equatorial and ecliptic planes with respect to the celestial sphere are accurately

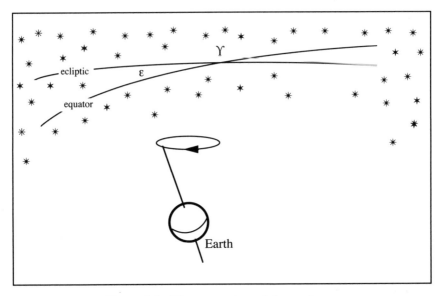

Figure 2.3: The precession of the equinox.

predictable. This makes it possible to freeze the motion of the fundamental references by arbitrarily choosing a particular orientation of the equator and equinox at a given instant of time as the basis for a preferred frame of reference. When this is done, the result is a very good approximation to an inertial frame of reference defined for the given epoch. The terminology *mean equator*, *mean ecliptic*, and *mean equinox* refer to the orientation which these references would have due to the effects of precession alone. In such a frame of reference, the location of a point on the celestial sphere is called a *mean place*. When the effect of nutation is added to that of general precession, the corresponding terms are *true equator*, *true ecliptic*, *true equinox*, and *true place*. The interested reader will find ample discussions of these subjects in References 2 and 3.

2.4 Time Scales

Before we can define an inertial frame of reference for some particular epoch, we must have at least one accurate method for measuring the passage of long intervals of time. Four different scales are now used in astronomy: *universal time*, *sidereal time*, *atomic time*, and *dynamical time*. For extended periods, each of these scales can be used with the continuous calendar of *Julian day numbers* which begins at noon on 1 January 4713 B.C. Starting in 1984 the standard epoch of the fundamental astronomical coordinate system is 2000 January 1.5,

2.4. TIME SCALES

dynamical time, which is defined as exactly Julian date 245 1545.0 and denoted J2000.0. The new Julian epoch supersedes the standard Besselian epochs of 1900 (B1900.0) and 1950 (B1950.0) [1].

The precise rationale for this assortment of time systems is not important to the computation of orbits. A working knowledge of the functional relationships between the various time scales is sufficient.

2.4.1 Universal Time

Universal time (UT) is a form of solar time which corresponds closely to the daily (diurnal) motion of the Sun across the sky as seen from a point on the Greenwich meridian of zero terrestrial longitude. It serves as the basis for worldwide civil timekeeping. Universal time is actually determined from observations of the diurnal motions of the stars and made to correspond to solar time by a formula which relates it to Greenwich mean sidereal time. Because of this, UT contains small nonuniformities due to slight changes in the rotational speed of the Earth with respect to inertial space. Furthermore, the UT scale derived directly from stellar observations, designated UT0, is dependent on the place of observation until corrected for an error caused by the variation of the observer's meridian due to the motion of the Earth's geographic pole. When this adjustment is made, the time scale is designated UT1, which is normally synonymous with the abbreviation UT [4].

Universal time can be easily computed for any given local standard civil time (CT) by using the following relationship:

$$\text{UT} = \text{CT} + \mathcal{Z}, \qquad (2.1)$$

where $\mathcal{Z}$ is the number of standard time zones which the locality is displaced to the *west* of the Greenwich meridian. For example, UT can be computed for points within the continental United States by using one of these four equations:

$$\begin{aligned} \text{UT} &= \text{EST} + 5 \\ \text{UT} &= \text{CST} + 6 \\ \text{UT} &= \text{MST} + 7 \\ \text{UT} &= \text{PST} + 8, \end{aligned} \qquad (2.2)$$

where time is measured in hours and EST, CST, MST, and PST are Eastern, Central, Mountain, and Pacific Standard Time, respectively.

2.4.2 Julian Date

The Julian date (JD) is an indispensable method of recording time because it avoids the troublesome discontinuity in the day count at the beginning of each month. A daily record of Julian dates at 0^h UT is included in a table published yearly in *The Astronomical Almanac*. Thus, at any given universal time,

$$\text{JD} = J_0 + \frac{\text{UT}}{24}, \qquad (2.3)$$

where UT is expressed in hours and J_0 is the tabular value of the Julian date at 0^h UT. As an alternative to using the table, Reference 5 gives a formula which can be used in a subroutine to automatically compute J_0 when the calendar date is given:

$$J_0 = 367Y - \left\langle \frac{7[Y + \langle (M+9)/12 \rangle]}{4} \right\rangle + \left\langle \frac{275M}{9} \right\rangle + D + 1721013.5, \quad (2.4)$$

where the symbolism is defined as follows:

- $\langle x \rangle$ represents a truncation function which extracts the integral part of x. For example, $\langle -7.32 \rangle = -7$ and $\langle 3.91 \rangle = 3$.
- Y is the year. It must be an integer in the range 1901 to 2099.
- M is the month. It must be an integer in the range 1 to 12.
- D is the day of the month. It must be an integer in the range 1 to 31.

The fact that the JD count was started at *noon* 4713 B.C., rather than midnight, opens the door for confusion if one is not careful. To get around this inconvenience and to reduce the number of digits in the Julian day number, the *modified Julian date* (MJD) is sometimes used. By definition

$$\text{MJD} \equiv \text{JD} - 2400000.5. \qquad (2.5)$$

In spite of the convenience of the MJD, we shall continue to use JD throughout the text and programs in order to be consistent with the practice of *The Astronomical Almanac*.

2.4.3 Sidereal Time

Sidereal time is determined by the Earth's diurnal rotation with respect to the vernal equinox of a given epoch. It is, therefore, a measure of the Earth's rotation with respect to the inertial reference frame of the celestial sphere rather than the Sun. This difference produces a significant disparity between the universal (solar) and sidereal rates.

Consider the situation illustrated in Figure 2.4. The point P is at the moment of sunrise on the surface of the Earth. This instant also corresponds to some particular value of the sidereal time. If sufficient time passes for the Earth

2.4. TIME SCALES

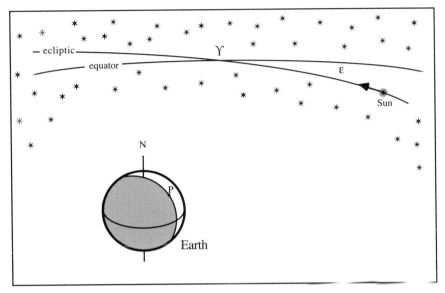

Figure 2.4. Initial relationship between sidereal and solar times at point P.

to make an integral number of revolutions with respect to the vernal equinox ♈, P will return to this same position with respect to the background of stars. However, during that interval, the Earth will have moved in its orbit around the Sun, causing the relative position of the Sun to move toward the left along the ecliptic to a new position as shown in Figure 2.5. Consequently, P is no longer at sunrise, and it is obvious that solar time does not proceed at the same rate as sidereal time. In fact, the Earth would have to rotate somewhat beyond the position shown in order to bring P back to dawn. It turns out that the motion of the Earth around the Sun (or the annual motion of the Sun relative to the Earth) causes the solar day to be about four minutes longer than the sidereal day.

Figure 2.6 shows that the *Greenwich mean sidereal time* (GMST) is equal to the angle measured eastward from the mean equinox ♈ along the celestial equator to the plane of the Greenwich meridian G. If θ_0 is the GMST measured in degrees at 0^h UT for the mean equinox of date, then, according to Reference 5,

$$\theta_0 = 100°.4606184 + 36000°.77004\, J + 0°.000387933\, J^2, \qquad (2.6)$$

where J is the number of *Julian centuries* from J2000.0. Thus,

$$J = \frac{J_0 - 2451545.0}{36525}. \qquad (2.7)$$

At an arbitrary UT on J_0, the GMST for the mean equinox of date is given in

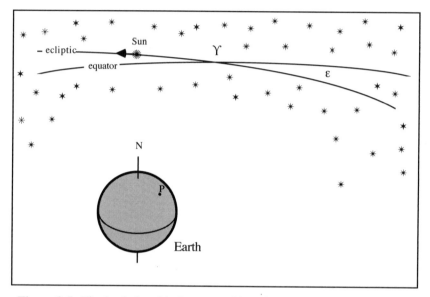

Figure 2.5: Final relationship between sidereal and solar times at point P.

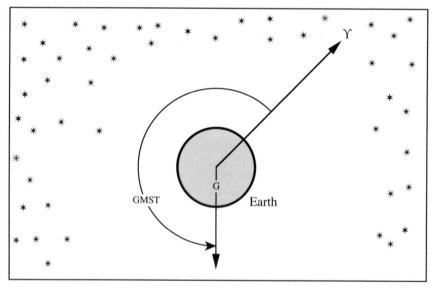

Figure 2.6: The angle of the Greenwich mean sidereal time.

2.4. TIME SCALES

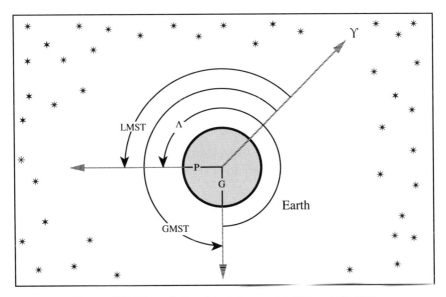

Figure 2.7: The relationship between LMST and GMST.

degrees by

$$\theta_g = \theta_0 + 360°.98564724 \frac{\text{UT}}{24}, \tag{2.8}$$

where UT is measured in hours and the numerical coefficient accounts for the difference between the solar and sidereal rates.

Figure 2.7 schematically depicts the relationship between GMST and the *local mean sidereal time* (LMST) at an arbitrary point P on the Earth's surface. If Λ is the longitude of P measured positively to the *east* from the Greenwich meridian, then the LMST for the mean equinox of date is given in degrees by

$$\theta = \theta_g + \Lambda. \tag{2.9}$$

Our interest in computing LMST stems from the fact that it can be used to link the position of a point on the rotating surface of the Earth to an inertial coordinate system anchored to the stars.

2.4.4 Atomic Time

The most precise time scale now available for astronomical use is *international atomic time* (TAI). Its fundamental unit is the SI second which is defined as the duration of 9,192,631,770 periods of the radiative transition between two hyperfine atomic energy levels of the ground state of cesium 133. Consequently,

TAI is independent of the Earth's diurnal rotation and, thus, not subject to the nonuniformities implicit in universal and sidereal times [6].

International atomic time is used to define the more familiar *coordinated universal time* (UTC), which is the basis for most radio broadcast time signals. UTC differs from TAI by an integral number of seconds and is kept within 0.9 seconds of UT by adding one-second steps whenever necessary. Precise values of the increment

$$\Delta AT \equiv TAI - UTC \qquad (2.10)$$

to be applied to UTC to obtain TAI are published in *The Astronomical Almanac*. Values from Reference 6 are shown in the table at the end of Section 2.4.5.

2.4.5 Dynamical Time

Dynamical time was established in 1984 along with the new Julian epoch J2000.0. Like its predecessor, *ephemeris time* (ET), dynamical time is a uniform measure of time determined empirically by comparing observations with the predictions of rigorously applied theories of celestial mechanics. In contrast to (ET), dynamical time exists in two forms: *terrestrial dynamical time* (TDT), which is used to calculate geocentric ephemerides and *barycentric dynamical time* (TDB), which is applied to those ephemerides referred to the barycenter (center of mass) of the solar system. TDB takes into account relativistic effects caused by variations in the gravitational field along the Earth's orbit around the Sun. The periodic differences between TDT and TDB are very small and not significant to our application. Accordingly, we will equate dynamical time with TDT [1,4,6].

In order to provide continuity with the pre-1984 use of ephemeris time, the difference between TDT and TAI was set equal to the 1 January 1984 estimate of the difference between ET and TAI. Since this difference was 32.184 seconds, then

$$TDT = TAI + 32^s.184, \qquad (2.11)$$

where, according to Equation 2.10,

$$TAI = UTC + \Delta AT. \qquad (2.12)$$

Therefore, procedures used with ephemerides tabulated as functions of ET are also applicable to those now based on TDT [1,4,6].

There is no functional connection between TDT and UT. Due to the small unpredictable changes in the Earth's rotation, their relationship must be determined by direct observation and calculation after the fact. It is possible, however, to extrapolate the difference

$$\Delta T \equiv TDT - UT \qquad (2.13)$$

2.5. COORDINATE SYSTEMS

for short intervals into the future. A table is published annually in *The Astronomical Almanac* listing the latest measured and predicted values for ΔT (ET−UT before 1984). The values below were taken from Reference 6:

Year	ΔAT sec	ΔT sec
1980.0	19.00	50.54
1981.0	20.00	51.38
1982.0	21.00	52.17
1983.0	22.00	52.96
1984.0		53.79
1985.0	23.00	54.34
1986.0		54.87
1987.0		55.32
1988.0	24.00	56.82
Extrapolated		
1989.0		56.3
1990.0		56.7
1991.0		57.2

Therefore, TDT may also be found from the equation

$$\text{TDT} = \text{UT} + \Delta T, \qquad (2.14)$$

where ΔT is taken from published tables. If UTC is known in lieu of UT, then

$$\text{TDT} \approx \text{UTC} + \Delta T. \qquad (2.15)$$

2.5 Coordinate Systems

We now have the pieces necessary to assemble the coordinate systems used for the computation of orbits. Based on what has been said up to this point, it should come as no surprise that the major astronomical coordinate systems are based on either the celestial equator or the ecliptic, and all share the vernal equinox as their principal coordinate direction. Furthermore, in the case of spherical coordinate systems, it is customary for the angle in the fundamental plane to be measured positively toward the east from Υ and to measure the perpendicular angle positively toward the north from the fundamental plane. Finally, the rectangular coordinate systems are all right-handed systems of three mutually perpendicular axes. The $+x$-axis is directed toward Υ, the xy-plane lies in the fundamental plane, and the $+z$-axis points northward.

It is convenient to group the primary coordinate systems used for orbit determination into three general categories: *celestial equatorial*, *terrestrial equatorial*, and *celestial ecliptic*. Within any one of these categories, the coordinate

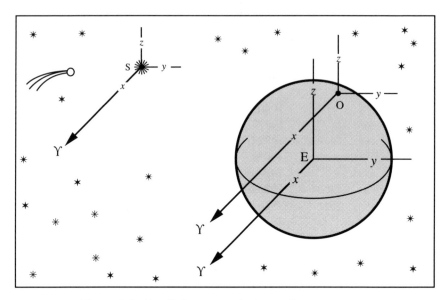

Figure 2.8: Parallel astronomical coordinate systems.

systems differ only by the location of their origins. As shown in Figure 2.8, the origin of a *geocentric* system is centered in the Earth at E, the origin of a *topocentric* system is at the observer O, and the origin of a *heliocentric* system is at S in the center of the Sun. It is important to realize that because ϒ is on the infinite celestial sphere, all axes pointing toward it are parallel. Therefore, although the coordinate systems in a particular category may be widely separated in space, their respective fundamental planes and axes are parallel.

2.5.1 Celestial Equatorial Systems

Figure 2.9 illustrates the characteristic features of celestial equatorial systems. The position of a given point P in space is specified by three spherical coordinates: the *radial distance* r from the origin C, the angle of *right ascension* α, and the angle of *declination* δ. Right ascension (RA) is measured eastward from ϒ around the celestial equator in units of time or degrees. Declination (DEC) is always measured in degrees and ranges from 0° at the equator to +90° at the north celestial pole (NCP) or −90° at the south celestial pole (SCP). Given the spherical coordinates of a point P, its rectangular coordinates can be found from the following:

$$x = r \cos \delta \cos \alpha \qquad (2.16)$$
$$y = r \cos \delta \sin \alpha \qquad (2.17)$$

2.5. COORDINATE SYSTEMS

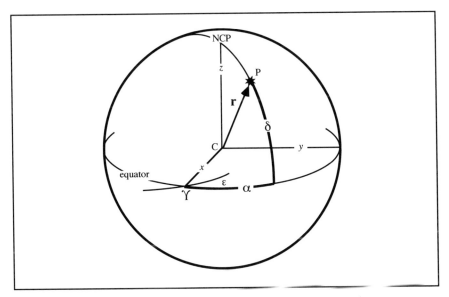

Figure 2.9: The celestial equatorial coordinate system.

$$z = r \sin \delta. \tag{2.18}$$

The process of orbit determination will also require converting equatorial rectangular coordinates to equatorial spherical coordinates. This can be accomplished by computing as follows:

$$r = \sqrt{x^2 + y^2 + z^2}, \tag{2.19}$$

so that the declination can be found from

$$\sin \delta = \frac{z}{r}. \tag{2.20}$$

Then, for the right ascension,

$$\cos \delta = \sqrt{1 - \sin^2 \delta} \tag{2.21}$$

$$\cos \alpha = \frac{x}{r \cos \delta} \tag{2.22}$$

$$\sin \alpha = \frac{y}{r \cos \delta}. \tag{2.23}$$

Equations 2.19 and 2.20 give δ unambiguously, because r is always positive and δ never exceeds 90°. Equations 2.21, 2.22, and 2.23 provide enough information to determine α in its proper quadrant.

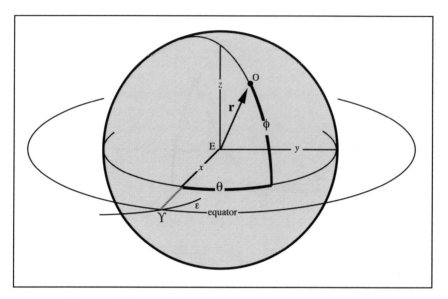

Figure 2.10: The terrestrial equatorial coordinate system.

2.5.2 Terrestrial Equatorial Systems

Consider the situation depicted in Figure 2.10, where O is the location of an observer on the surface of a spherical Earth which is centered at E. The diurnal motion of the Earth causes O to continually accelerate with respect to a geocentric terrestrial equatorial coordinate system anchored to the equinox ϒ. With respect to this inertial system, the instantaneous position of O is given by its *radial distance* r, *local mean sidereal time* θ, and *geodetic latitude* ϕ. When these quantities are known for a given instant of time, the rectangular coordinates of the observer's position can be computed as follows:

$$x = r \cos\phi \cos\theta \qquad (2.24)$$
$$y = r \cos\phi \sin\theta \qquad (2.25)$$
$$z = r \sin\phi. \qquad (2.26)$$

Equations 2.24 through 2.26 are quite satisfactory when applied to heliocentric orbits. However, in the case of geocentric orbits, the non-spherical shape of the Earth requires more complex relationships to be used. As indicated schematically in Figure 2.11, the actual shape of the Earth is similar to that of an ellipsoid of revolution about the z-axis (the illustration exaggerates the effect). This ellipsoidal shape can be characterized by a dimensionless parameter called the

2.5. COORDINATE SYSTEMS

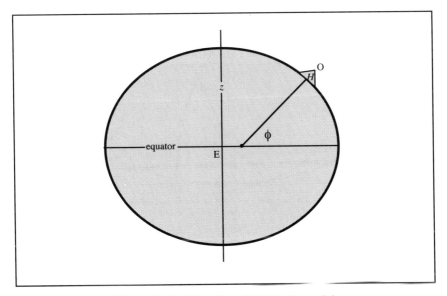

Figure 2.11: The ellipsoidal Earth model.

flattening f, where

$$f = \frac{1}{298.257}. \tag{2.27}$$

Flattening causes the vertex of the geodetic latitude angle to be displaced a small distance from the geocenter E. Furthermore, local terrain may elevate the observer to a significant *height* H above the surface of the reference ellipsoid. Taking all these factors into account, it can be shown that the geocentric position of the observer is more accurately described by equations of the form

$$x = G_c \cos\phi \cos\theta \tag{2.28}$$
$$y = G_c \cos\phi \sin\theta \tag{2.29}$$
$$z = G_s \sin\phi, \tag{2.30}$$

where

$$F = \sqrt{1 - (2f - f^2)\sin^2\phi} \tag{2.31}$$
$$G_c = \frac{1}{F} + H \tag{2.32}$$
$$G_s = \frac{(1-f)^2}{F} + H, \tag{2.33}$$

and all quantities are expressed in geocentric working units [2,3,7].

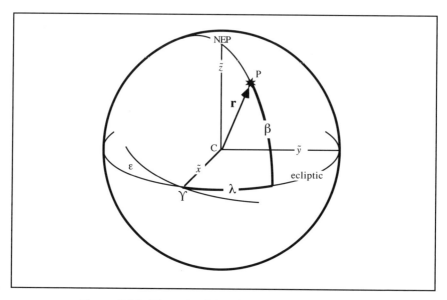

Figure 2.12: The celestial ecliptic coordinate system.

2.5.3 Celestial Ecliptic Systems

The spherical and rectangular celestial ecliptic systems are shown in Figure 2.12. The fundamental circle and plane are those defined by the ecliptic, and the origin C is usually centered in the Sun. The position of a point P in space is defined by its *radial distance* r, *ecliptic longitude* λ, and *ecliptic latitude* β. The angle λ is measured eastward from Υ around the ecliptic from 0° to 360°. The angle β is measured from the ecliptic plane to +90° at the north ecliptic pole (NEP) or −90° at the south ecliptic pole (SEP). Given these spherical coordinates, the rectangular coordinates are computed as follows:

$$\tilde{x} = r \cos\beta \cos\lambda \qquad (2.34)$$
$$\tilde{y} = r \cos\beta \sin\lambda \qquad (2.35)$$
$$\tilde{z} = r \sin\beta . \qquad (2.36)$$

2.6 Ecliptic-Equatorial Transformations

It is often convenient and sometimes necessary to transform rectangular coordinates between the ecliptic and equatorial frames of reference. As both frames share the same principal direction for their x-axes, the transformation is simply a rotation about the x-axis through an angle equal to the obliquity of the ecliptic ε. Because of the effects of general precession, the numerical value of ε is

2.7. THE FUNDAMENTAL VECTOR TRIANGLE

not constant. However, the obliquity of the ecliptic for date t, with respect to the equator of date t, can be computed from

$$\varepsilon = 23°.439291 - 0°.0130042\,T - 0°.00000016\,T^2, \tag{2.37}$$

where

$$T = \frac{t - 2000.0}{100} = \frac{JD - 2451545.0}{36525}. \tag{2.38}$$

Using the appropriate value of ε from Equation 2.37, the transformation of equatorial rectangular coordinates at epoch t to ecliptic rectangular coordinates at the same epoch is given by

$$\tilde{x} = x \tag{2.39}$$
$$\tilde{y} = z\sin\varepsilon + y\cos\varepsilon \tag{2.40}$$
$$\tilde{z} = z\cos\varepsilon - y\sin\varepsilon. \tag{2.41}$$

The reverse transformation from ecliptic rectangular coordinates to equatorial rectangular coordinates is given by

$$x = \tilde{x} \tag{2.42}$$
$$y = \tilde{y}\cos\varepsilon - \tilde{z}\sin\varepsilon \tag{2.43}$$
$$z = \tilde{y}\sin\varepsilon + \tilde{z}\cos\varepsilon. \tag{2.44}$$

A complete discussion of all ecliptic-equatorial transformations can be found in Reference 2.

2.7 The Fundamental Vector Triangle

Consider the general vector relationship illustrated in Figure 2.13, where $\mathbf{R}$, $\mathbf{r}$, and $\mathbf{p}$ define the relative positions of the observer O, center of force C, and object B. If it is assumed that all vectors are referred to the same inertial coordinate system, then

$$\mathbf{p} = \mathbf{r} + \mathbf{R}. \tag{2.45}$$

Equation 2.45 represents the *fundamental vector triangle* of orbit computation. If we let

$$\mathbf{r} = \{x, y, z\} \tag{2.46}$$
$$\mathbf{R} = \{X, Y, Z\}, \tag{2.47}$$

then the topocentric position vector to B can be expressed as

$$\mathbf{p} = \{x + X, y + Y, z + Z\}. \tag{2.48}$$

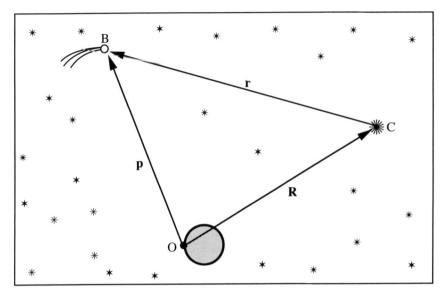

Figure 2.13: The fundamental vector triangle.

Let it now be assumed, as is almost always the case, that the object's topocentric position is measured in the celestial equatorial coordinate system. Then, by Equations 2.16 through 2.18, we can write

$$x + X = p \cos\delta \cos\alpha \qquad (2.49)$$
$$y + Y = p \cos\delta \sin\alpha \qquad (2.50)$$
$$z + Z = p \sin\delta, \qquad (2.51)$$

where $p = |\mathbf{p}|$ is the range of the object from the the observer. Dividing both sides of the above equations by the scalar p, the result is

$$\frac{x + X}{p} = \cos\delta \cos\alpha \qquad (2.52)$$
$$\frac{y + Y}{p} = \cos\delta \sin\alpha \qquad (2.53)$$
$$\frac{z + Z}{p} = \sin\delta. \qquad (2.54)$$

The ratios on the left sides of these equations are known as *direction cosines*. Since we have divided each component of $\mathbf{p}$ by p, we have created a unit vector

$$\mathbf{L} = \frac{\mathbf{p}}{p} \qquad (2.55)$$

pointing toward the position of the object on the celestial sphere. Thus,

$$\mathbf{L} = \{\cos\delta\cos\alpha, \cos\delta\sin\alpha, \sin\delta\}, \tag{2.56}$$

and

$$|\mathbf{L}| = 1. \tag{2.57}$$

Therefore, according to Equation 2.55,

$$\mathbf{p} = p\mathbf{L}, \tag{2.58}$$

and, substituting Equation 2.58 into Equation 2.45, we obtain

$$p\mathbf{L} = \mathbf{r} + \mathbf{R} \tag{2.59}$$

or, equivalently,

$$\mathbf{r} = p\mathbf{L} - \mathbf{R}. \tag{2.60}$$

Equations 2.59 and 2.60 are the most convenient forms of the equation of the fundamental vector triangle.

2.8 Reduction of Astronomical Coordinates

Before two or more sets of position data are compared, their coordinates should be based on a common inertial frame of reference. The choice of common reference frame is largely a matter of convenience according to the nature of the available data and the problem to be solved. From a physical standpoint, two general types of reductions are required: (1) those which correct for the motion of the fundamental reference planes and (2) those which correct for the motions of the celestial body and the observer with respect to the inertial reference frame. The former includes the reductions for general precession and nutation as briefly discussed in Section 2.3. The latter is the reduction for the aberration of light, which causes the observed direction toward the celestial body to depend on the motions of both the body and the observer during the time interval required for light to travel from the body to the observer.

2.8.1 Planetary Aberration

Because the velocity of light is finite, the *apparent direction* toward a moving celestial body as viewed by a moving observer is not the same as the *geometric direction* toward the object at the same instant of time. This displacement from the geometric position results from two separate effects. The first, caused by the motion of the celestial body independent of the motion of the observer, is known as the *correction for light-time*. The second, caused by the motion of

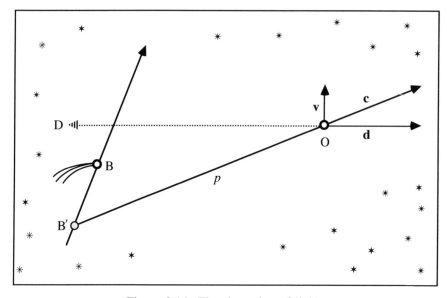

Figure 2.14: The aberration of light.

the observer independent of the motion of the celestial body, is called *stellar aberration* because it typically affects observations of the fixed stars. The total effect due to light-time and stellar aberration is called *planetary aberration* [8].

Consider the situation depicted in Figure 2.14, where a celestial body B and observer O are shown in the geometric positions they occupy at a time t when O observes B. Let B′ be the position of the celestial body at a previous instant t_c, when the light left the body to reach O at the observation time. If δt is the *light-time*, or interval required for light to travel from B′ to O, then

$$t_c = t - \delta t. \tag{2.61}$$

Furthermore, if c is the *speed of light* and p is the distance from B′ to O, we have

$$\delta t = \frac{p}{c} \tag{2.62}$$

and

$$t_c = t - \frac{p}{c}, \tag{2.63}$$

where, in heliocentric working units,

$$c = 173.1446 \text{ au/day}. \tag{2.64}$$

If it can be assumed that the observer is at rest, then the observed direction of the celestial body will be toward the point B′ which the body occupied previously

2.8. REDUCTION OF ASTRONOMICAL COORDINATES

at time t_c. Thus, an orbiting body's observed position may differ significantly from its geometric position due to its motion during the light-time interval.

In situations where the observer cannot be assumed to be at rest, stellar aberration causes the problem to be more complex. When the light which left the celestial body at time t_c arrives at the observer at time t, it will not appear to be coming from B' but from its direction *relative* to the moving observer. According to classical physics, the observed direction of the celestial body will be opposite that of the vector **d**, which is the difference between the velocity of light **c** and the velocity of the observer **v**. In other words,

$$\mathbf{d} = \mathbf{c} - \mathbf{v}. \tag{2.65}$$

Therefore, the apparent position of the celestial body will be some point D on the celestial sphere which is displaced from B' toward the direction of the observer's motion.

In practice, the correction for stellar aberration will be included with the corrections for precession and nutation during the initial reduction from apparent position. The correction for light-time will be handled separately when the distance to the celestial body is known. A comprehensive discussion of all aspects of aberration can be found in Reference 2.

2.8.2 The Instantaneous and Fixed Equator and Equinox

The *apparent place* of a moving celestial body is the point on the celestial sphere where the body would be seen from the center of the moving Earth and measured with respect to the *instantaneous* or *true equator and equinox of date* (the time of observation). The position of the true equator and equinox includes the effects of general precession and nutation. If planetary aberration were removed from the body's apparent position, the geometric direction to its actual position in space would be obtained. The *true place* of a celestial body is its geometric position on the celestial sphere measured with respect to the true equator and equinox of date. If the body's position were further reduced by removing the effects of nutation, the new location on the celestial sphere would be its *mean place* with respect to the mean equator and equinox of date. An additional reduction for general precession could be made to refer the celestial body's mean place to some *fixed equator and equinox* at a convenient epoch. The Julian epoch at the middle of the current year (Jxxxx.5) or the standard Julian epoch (J2000.0) is normally chosen.

2.8.3 Astrometric Positions

Consider the situation where the position of a moving celestial body is measured with respect to nearby reference stars, rather than the true equator and equinox of date. The mean places of stars are published in catalogs, so the positions of the reference stars are known. Since the mean place of a star does not include the effects of stellar aberration, the relative position of the celestial body will automatically have the stellar aberration removed, although the effects of light-time will remain. The position of a moving celestial body referred to the catalog mean places of comparison stars is sometimes called an *astrometric position* to distinguish it from a geometric position which is free of the light-time effect.

The geocentric ephemerides published in *The Astronomical Almanac* provide an excellent source of data with which to practice the skills of orbit determination. The ephemerides of the major planets, except Pluto, tabulate geocentric apparent positions for the true equinox of date. The ephemerides of Pluto and the minor planets Ceres, Pallas, Juno, and Vesta tabulate astrometric positions for the mean equinox J2000.0. The following three sections describe the formulas and procedures necessary to reduce published positions and observations to a form which can be used for orbit determination.

2.8.4 Reductions for Aberration and Nutation

Assume that a celestial body's apparent right ascension α' and declination δ' for the true equinox of date are known. Then, to the accuracy required for our application, its astrometric right ascension α and declination δ for the mean equinox Jxxxx.5 are given by the following equations:

$$\alpha = \alpha' - Ac_1 - Bc_2 - Cc_3 - Dc_4 - E \tag{2.66}$$
$$\delta = \delta' - Ac_5 - Bc_6 - Cc_7 - Dc_8, \tag{2.67}$$

where the *Besselian star constants* are

$$\begin{aligned} c_1 &= (m/n) + \sin\alpha' \tan\delta' \\ c_2 &= \cos\alpha' \tan\delta' \\ c_3 &= \cos\alpha' \sec\delta' \\ c_4 &= \sin\alpha' \sec\delta' \\ c_5 &= \cos\alpha' \\ c_6 &= -\sin\alpha' \\ c_7 &= \tan\varepsilon \cos\delta' - \sin\alpha' \sin\delta' \\ c_8 &= \cos\alpha' \sin\delta'. \end{aligned} \tag{2.68}$$

The *Besselian day numbers* A, B, C, D, E and the constants m/n and $\tan\varepsilon$ are taken from *The Astronomical Almanac*, Section B. The star constants and day numbers also take into account the small amount of precession required to adjust the mean equinox of date to the mean equinox at the middle of the current year.

2.8. REDUCTION OF ASTRONOMICAL COORDINATES

This reduction is reversible. In other words, given the astrometric position for the mean equinox Jxxxx.5, the apparent position for the true equinox of date is

$$\alpha' = \alpha + Ac_1 + Bc_2 + Cc_3 + Dc_4 + E \quad (2.69)$$
$$\delta' = \delta + Ac_5 + Bc_6 + Cc_7 + Dc_8, \quad (2.70)$$

where the Besselian star constants are computed using α and δ, rather than α' and δ'.

2.8.5 Reductions for Precession

Spherical Coordinates

Suppose we wish to reduce a set of spherical equatorial coordinates α and δ from an initial epoch t to their corresponding values α_0 and δ_0 at the epoch J2000.0. If we restrict our problem to situations where $|\delta| < 80°$ and $|t - 2000.0| < 25$, then, to a good approximation, the precession reduction can be accomplished by using the following equations:

$$\alpha_0 = \alpha - (\bar{m} + \bar{n} \sin \bar{\alpha} \tan \bar{\delta})T \quad (2.71)$$
$$\delta_0 = \delta - (\bar{n} \cos \bar{\alpha})T, \quad (2.72)$$

where

$$\bar{\alpha} = \frac{\alpha + \alpha_0}{2} \quad (2.73)$$

$$\bar{\delta} = \frac{\delta + \delta_0}{2} \quad (2.74)$$

are the average values of the right ascension and declination,

$$\bar{m} = 1°.28123227 + 0°.000775867 \frac{T}{2} - 0°.000000077 \frac{T^2}{4} \quad (2.75)$$

$$\bar{n} = 0°.55675303 - 0°.000237030 \frac{T}{2} - 0°.000000060 \frac{T^2}{4} \quad (2.76)$$

are the average rates of general precession for right ascension and declination, and

$$T = \frac{t - 2000.0}{100} = \frac{JD - 2451545.0}{36525} \quad (2.77)$$

is the number of Julian centuries between Jxxxx.x and J2000.0.

Finally, should the need arise to reduce α_0 and δ_0 from the standard epoch to their corresponding values α and δ at the arbitrary Julian epoch t, we simply use

$$\alpha = \alpha_0 + (\bar{m} + \bar{n} \sin \bar{\alpha} \tan \bar{\delta})T \quad (2.78)$$
$$\delta = \delta_0 + (\bar{n} \cos \bar{\alpha})T. \quad (2.79)$$

The precession reduction equations are solved by successive approximations because the average values of the right ascension and declination are not known in advance. This simple procedure is outlined in the algorithm for program ADCES in Section 2.9. The interested reader should consult Taff [3] for a thorough discussion of this and other methods for computing the general precession of spherical equatorial coordinates.

Rectangular Coordinates

Suppose we wish to reduce a set of rectangular equatorial coordinates $\{x, y, z\}$ from some given epoch t to their corresponding values $\{x_0, y_0, z_0\}$ at the standard epoch J2000.0. The solution can be expressed by the following three equations:

$$\begin{aligned} x_0 &= P_{11}x + P_{21}y + P_{31}z \\ y_0 &= P_{12}x + P_{22}y + P_{32}z \\ z_0 &= P_{13}x + P_{23}y + P_{33}z \,, \end{aligned} \quad (2.80)$$

where the coefficients

$$\begin{aligned} P_{11} &= +1.0 - 0.00029724T^2 - 0.00000013T^3 \\ P_{12} &= -0.02236172T - 0.00000677T^2 + 0.00000222T^3 \\ P_{13} &= -0.00971717T + 0.00000207T^2 + 0.00000096T^3 \\ P_{21} &= -P_{12} \\ P_{22} &= +1.0 - 0.00025002T^2 - 0.00000015T^3 \\ P_{23} &= -0.00010865T^2 \\ P_{31} &= -P_{13} \\ P_{32} &= +P_{23} \\ P_{33} &= +1.0 - 0.00004721T^2 \end{aligned} \quad (2.81)$$

are called the *elements of the precession matrix* **P**, and

$$T = \frac{t - 2000.0}{100} = \frac{JD - 2451545.0}{36525}.$$

The reverse transformation from J2000.0 to the arbitrary epoch of date t is given by

$$\begin{aligned} x &= P_{11}x_0 + P_{12}y_0 + P_{13}z_0 \\ y &= P_{21}x_0 + P_{22}y_0 + P_{23}z_0 \\ z &= P_{31}x_0 + P_{32}y_0 + P_{33}z_0 \,. \end{aligned} \quad (2.82)$$

Equations 2.81 were taken from Reference 9. Additionally, *The Astronomical Almanac* publishes numerical values for the elements of the precession matrix **P** for reductions between epochs J2000.0 and Jxxxx.5.

2.8. REDUCTION OF ASTRONOMICAL COORDINATES

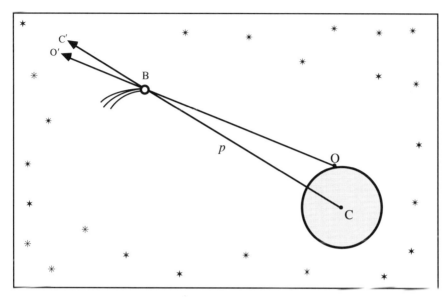

Figure 2.15: The geocentric parallax.

2.8.6 Reductions for Geocentric Parallax

In the case of objects within the solar system, there may be a small but significant difference between their observed positions on the celestial sphere and the positions they would occupy if seen from the center of the Earth. Figure 2.15 shows a celestial body B at a distance p from the geocenter C. An observer at O sees B projected against the celestial sphere at O'. However, if seen from C, body B would appear to be located at C'. The angular difference between O' and C' is the *geocentric parallax* of B at the time of observation. In order to reduce a given set of equatorial coordinates for geocentric parallax, we must be able to resolve the parallax into its angular components along the arcs of right ascension and declination.

Consider the situation where a celestial body's geocentric equatorial coordinates α_0 and δ_0 are known at a given instant of time and its distance p_0 is expressed in astronomical units. If the local mean sidereal time and the geodetic latitude of the observing station are θ and ϕ, respectively, then the topocentric right ascension α and declination δ are given by

$$\alpha = \alpha_0 - H_p(G_c \cos\phi \sin h_0 \sec\delta_0) \tag{2.83}$$
$$\delta = \delta_0 - H_p(G_s \sin\phi \cos\delta_0 - G_c \cos\phi \cos h_0 \sin\delta_0), \tag{2.84}$$

where h_0 is the *hour angle*

$$h_0 = \theta - \alpha_0, \tag{2.85}$$

H_p is the *horizontal parallax*

$$H_p = \frac{8''.794}{p_0}, \qquad (2.86)$$

and G_c and G_s are the same quantities given by Equations 2.31, 2.32, and 2.33 [8].

In the case where a celestial body's topocentric equatorial coordinates are known for a given local sidereal time and geodetic latitude, its geocentric right ascension and declination can be computed from

$$\alpha_0 = \alpha + H_p(G_c \cos\phi \sin h \sec\delta) \qquad (2.87)$$
$$\delta_0 = \delta + H_p(G_s \sin\phi \cos\delta - G_c \cos\phi \cos h \sin\delta), \qquad (2.88)$$

where

$$h = \theta - \alpha, \qquad (2.89)$$

and

$$H_p = \frac{8''.794}{p}. \qquad (2.90)$$

As before, G_c and G_s are given by Equations 2.31, 2.32, and 2.33.

2.9 Computer Programs

2.9.1 Program LMST

LMST is a utility program for computing local mean sidereal time. The results are expressed in hours. Most of the computation is accomplished in subroutine J0ST/SUB, which is listed separately on the page following the main program listing.

Program Algorithm

Given	Line Number
west longitude (°, ′, ″)	1080
latitude (°, ′, ″) for information only	1080
Y: year	1100
M: month	1100
D: day	1100
UT: universal time in hours	1100

Compute	Line Number
west longitude in degrees	1120
latitude in degrees for information only	1130
Y: year in four digits	14020
J_0: Equation 2.4	14030
J: Equation 2.7	14040
Λ: east longitude from west longitude	14050
θ_0: Equation 2.6	14060
θ_0: converted to a value $\leq 360°$	14070

θ_g: Equation 2.8	14080
θ: Equation 2.9	14090
θ: converted to a value $\leq 360°$	14100
θ_0: converted to hours	1240
θ: converted to hours	1260

End.

2.9. COMPUTER PROGRAMS

Program Listing

```
1000 CLS
1010 PRINT"# LMST # COMPUTES LOCAL MEAN SIDEREAL TIME"
1020 PRINT
1030 DEFDBL A-Z
1040 DEFINT I
1050 REM
1060 J$="########.######"
1070 REM
1080 READ W1,W2,W3, B1,B2,B3
1090 REM
1100 INPUT">>>YY,MM,DD,UT";IY,IM,ID,UT
1110 CLS
1120 WL=W1+W2/60+W3/3600
1130 BB=B1+B2/60+B3/3600
1140 REM
1150 GOSUB 14010 REM JOST/SUB
1160 REM
1170 PRINT"LOCAL MEAN SIDEREAL TIME"
1180 PRINT
1190 PRINT"DATE";TAB(15)IY;"/";IM;"/";ID;"/";UT
1200 PRINT"WL(DGS)";TAB(12);:PRINT USING J$;WL
1210 PRINT"BB(DGS)";TAB(12);:PRINT USING J$;BB
1220 PRINT
1230 PRINT"J0";TAB(12);:PRINT USING J$;J0
1240 PRINT"S0(HRS)";TAB(12);:PRINT USING J$;S0/15
1250 PRINT
1260 PRINT"LMST(HRS)";TAB(12);:PRINT USING J$;ST/15
1270 END
```

```
14000 STOP
14010 REM # JOST/SUB # CALCULATES J0 AND LMST
14020 IY=IY+1900
14030 J0=367*IY-INT((7*(IY+INT((IM+9)/12)))/4)+
      INT(275*IM/9)+ID+1721013.5#
14040 J=(J0-2451545#)/36525#
14050 EL=360-WL
14060 S0=100.4606184#+36000.77004#*J+.000387933#*J*J
14070 S0=S0-360*INT(S0/360)
14080 SG=S0+360.98564724#*(UT/24)
14090 ST=SG+EL
14100 ST=ST-360*INT(ST/360)
14110 RETURN
30000 DATA 80,22,55.79,  37,31,33.00
```

2.9. COMPUTER PROGRAMS

2.9.2 Program XYZ

XYZ is a utility program for converting equatorial spherical coordinates to equatorial rectangular coordinates.

Program Algorithm

Given	Line Number
radians per degree	1050
date for information only	1080
name of the object for information only	1100
equinox for information only	1120
r: distance to the object	1140
α: right ascension (^{h}, ^{m}, ^{s})	1160
δ: declination ($°$, $'$, $''$)	1180

Compute	Line Number
α: converted to radians	1200
δ: converted to radians	1210
x: Equation 2.16	1220
y: Equation 2.17	1230
z: Equation 2.18	1240

End.

Program Listing

```
1000 CLS
1010 PRINT"# XYZ # COMPUTES EQUATORIAL X Y Z FROM R A D"
1020 PRINT
1030 DEFDBL A-Z
1040 REM
1050 Q1=.0174532925#
1060 G$="####.#######"
1070 REM
1080 INPUT">>>DATE YY, MM, DD";IY,IM,ID
1090 PRINT
1100 INPUT">>>NAME OF OBJECT";N$
1110 PRINT
1120 INPUT">>>EQUINOX";E$
1130 PRINT
1140 INPUT">>>TRUE DISTANCE";R
1150 PRINT
1160 INPUT">>>RA : H,M,S";A1,A2,A3
1170 PRINT
1180 INPUT">>>DEC: D,M,S";D1,D2,D3
1190 CLS
1200 A=(A1+A2/60+A3/3600)*15*Q1
1210 D=(D1+D2/60+D3/3600)*Q1
1220 X=R*COS(D)*COS(A)
1230 Y=R*COS(D)*SIN(A)
1240 Z=R*SIN(D)
1250 REM
1260 PRINT"RECTANGULAR COORDINATES X Y Z FROM R A D"
1270 PRINT N$
1280 PRINT
1290 PRINT"DATE";TAB(13)IY;"/";IM;"/";ID
1300 PRINT
1310 PRINT E$
1320 PRINT"R";TAB(12);:PRINT USING G$;R
1330 PRINT"A(HRS)";TAB(12);:PRINT USING G$;A/(15*Q1)
1340 PRINT"D(DGS)";TAB(12);:PRINT USING G$;D/Q1
1350 PRINT
1360 PRINT"X Y Z";TAB(12);:PRINT USING G$;X,Y,Z
1370 END
```

2.9. COMPUTER PROGRAMS

2.9.3 Program RAD

RAD is a utility program for converting equatorial rectangular coordinates to equatorial spherical coordinates. This program contains a subroutine ARC/SUB which determines the correct quadrant for an angle given its sine and cosine.

Program Algorithm

Define	Line Number
FNASN(X): arcsine function	11020
FNACN(X): arccosine function	11030

Given	Line Number
radians per degree	1050
date for information only	1080
name of the object for information only	1100
equinox for information only	1120
x, y, z: rectangular coordinates	1140

Compute	Line Number
r: Equation 2.19	1160
$\sin \delta$: Equation 2.20	1170
$\cos \delta$: Equation 2.21	1180
$\cos \alpha$: Equation 2.22	1190
$\sin \alpha$: Equation 2.23	1200
α: using the smaller trig function for better accuracy	11040-11050
α: converted to the correct quadrant	11060-11090

δ: Equation 2.20	1250
α: converted to hours	1360
δ: converted to degrees	1370

End.

2.9. COMPUTER PROGRAMS

Program Listing

```
1000 CLS
1010 PRINT"# RAD # COMPUTES EQUATORIAL R A D FROM X Y Z"
1020 PRINT
1030 DEFDBL A-Z
1040 REM
1050 Q1=.0174532925#
1060 G$="####.#######"
1070 REM
1080 INPUT">>>DATE YY, MM, DD";IY,IM,ID
1090 PRINT
1100 INPUT">>>NAME OF OBJECT";N$
1110 PRINT
1120 INPUT">>>EQUINOX";E$
1130 PRINT
1140 INPUT">>>COORD X, Y, Z";RX,RY,RZ
1150 CLS
1160 R=SQR(RX*RX+RY*RY+RZ*RZ)
1170 SD=RZ/R
1180 CD=SQR(1-SD*SD)
1190 CX=RX/(R*CD)
1200 SX=RY/(R*CD)
1210 REM
1220 GOSUB 11010 REM ARC/SUB
1230 A=X
1240 REM
1250 D=FNASN(SD)
1260 REM
1270 PRINT"SPHERICAL COORDINATES R A D FROM X Y Z
1280 PRINT N$
1290 PRINT
1300 PRINT"DATE";TAB(13)IY;"/";IM;"/";ID
1310 PRINT
1320 PRINT E$
1330 PRINT"X Y Z";TAB(12);:PRINT USING G$;RX,RY,RZ
1340 PRINT
1350 PRINT"R";TAB(12);:PRINT USING G$;R
1360 PRINT"A(HRS)";TAB(12);:PRINT USING G$;A/(15*Q1)
1370 PRINT"D(DGS)";TAB(12);:PRINT USING G$;D/Q1
1380 END
```

```
11000 STOP
11010 REM # ARC/SUB # COMPUTES X FROM SIN(X) AND COS(X)
11020 DEF FNASN(X)=ATN(X/SQR(-X*X+1))
11030 DEF FNACN(X)=-ATN(X/SQR(-X*X+1))+1.5707963263#
11040 IF ABS(SX)<=.707107 THEN X=FNASN(ABS(SX))
11050 IF ABS(CX)<=.707107 THEN X=FNACN(ABS(CX))
11060 IF CX>=0 AND SX>=0 THEN X=X
11070 IF CX<0 AND SX>=0 THEN X=180*Q1-X
11080 IF CX<0 AND SX<0 THEN X=180*Q1+X
11090 IF CX>=0 AND SX<0 THEN X=360*Q1-X
11100 RETURN
```

2.9. COMPUTER PROGRAMS

2.9.4 Program CQTRAN

CQTRAN is a utility program for transforming equatorial rectangular coordinates to ecliptic rectangular coordinates and vice versa. Note that the magnitude of the vector remains unchanged by this transformation.

Program Algorithm

Given	Line Number
radians per degree	1050
direction of transformation	1080
date for information only	1100
name of the object for information only	1120
t: mean equinox of the initial coordinates	1140
x, y, z: initial coordinates	1160

Compute	Line Number
T: Equation 2.38	1180
ε: Equation 2.37	1190
$s = +1$ if transformation is *to* the ecliptic	1210
$s = -1$ if transformation is *from* the ecliptic	1210
x', y', z': final coordinates by Equations 2.39 through 2.44	1320-1340
r: for optional comparison with r'	1360
r': for optional comparison with r	1370
ε: converted to degrees for information only	1450

End.

Program Listing

```
1000 CLS
1010 PRINT"# CQTRAN # ECLIPTIC-EQUATORIAL TRANSFORMATION"
1020 PRINT
1030 DEFDBL A-Z
1040 REM
1050 Q1=.0174532925#
1060 G$="####.#######"
1070 REM
1080 INPUT">>>REDUCTION (T)O OR (F)ROM THE ECLIPTIC";A$
1090 PRINT
1100 INPUT">>>DATE YY, MM, DD";IY,IM,ID
1110 PRINT
1120 INPUT">>>NAME OF OBJECT";N$
1130 PRINT
1140 INPUT">>>EQUINOX(NUMBER)";EQ
1150 PRINT
1160 INPUT">>>INITIAL X, Y, Z";RI(1),RI(2),RI(3)
1170 CLS
1180 TT=(EQ-2000)/100
1190 EC=(23.439291#-.0130042*TT-.00000016#*TT*TT)*Q1
1200 REM
1210 IF A$="T" THEN 1230 ELSE 1280
1220 REM
1230 S=+1
1240 I$="EQUATORIAL"
1250 F$="ECLIPTIC"
1260 GOTO 1320
1270 REM
1280 S=-1
1290 I$="ECLIPTIC"
1300 F$="EQUATORIAL"
1310 REM
1320 RF(1)=+RI(1)
1330 RF(2)=+RI(2)*COS(EC)+S*RI(3)*SIN(EC)
1340 RF(3)=-S*RI(2)*SIN(EC)+RI(3)*COS(EC)
1350 REM
1360 RI=SQR(RI(1)*RI(1)+RI(2)*RI(2)+RI(3)*RI(3))
1370 RF=SQR(RF(1)*RF(1)+RF(2)*RF(2)+RF(3)*RF(3))
1380 REM
1390 PRINT"ECLIPTIC-EQUATORIAL TRANSFORMATION"
```

2.9. COMPUTER PROGRAMS

```
1400 PRINT N$
1410 PRINT
1420 PRINT"DATE";TAB(14)IY;"/";IM;"/";ID
1430 PRINT
1440 PRINT"J"+STR$(EQ)
1450 PRINT"EC(DGS)";TAB(12);:PRINT USING G$;EC/Q1
1460 PRINT
1470 PRINT I$
1480 PRINT"VECTOR";TAB(12);:PRINT USING G$;RI(1),RI(2),RI(3)
1490 PRINT"MAGNITUDE";TAB(12);:PRINT USING G$;RI
1500 PRINT
1510 PRINT F$
1520 PRINT"VECTOR";TAB(12);:PRINT USING G$;RF(1),RF(2),RF(3)
1530 PRINT"MAGNITUDE";TAB(12);:PRINT USING G$;RF
1540 END
```

2.9.5 Program ADAPP

ADAPP is a utility program for reducing right ascension and declination coordinates from an apparent position for the true equinox of date to an astrometric position for the mean equinox at the middle of the current year and vice versa.

Program Algorithm

Given	Line Number
radians per degree	1060
equinox Jxxxx.5 for information only	1090
m/n	1090
$\tan \varepsilon$	1090
date for information only	1100
A, B, C, D, E: Besselian day numbers for the date	1110
direction of the reduction	1130
name of the object for information only	1150
initial RA (^{h}, ^{m}, ^{s})	1170
initial DEC (°, ′, ″)	1190

Compute	Line Number
$s = +1$ if reduction is *to* apparent place	1210
$s = -1$ if reduction is *from* apparent place	1210
initial RA in hours	1320
initial DEC in degrees	1330
initial RA in radians	1340

2.9. COMPUTER PROGRAMS

initial DEC in radians	1350
$c_1, c_2, c_3, c_4, c_5, c_6, c_7, c_8$: Equations 2.68	1370-1440
final RA: Equation 2.66 or 2.69	1460
final DEC: Equation 2.67 or 2.70	1470

End.

Program Listing

```
1000 CLS
1010 PRINT"# ADAPP # REDUCTION OF RA AND DEC"
1020 PRINT"# BETWEEN APPARENT PLACE AND JXXXX.5"
1030 PRINT
1040 DEFDBL A-Z
1050 REM
1060 Q1=.0174532925#
1070 G$="####.#######"
1080 REM
1090 READ E$,MN,TE
1100 READ IY,IM,ID
1110 READ NA,NB,NC,ND,NE
1120 REM
1130 INPUT">>>REDUCTION (T)O OR (F)ROM APPARENT PLACE";A$
1140 PRINT
1150 INPUT">>>NAME OF OBJECT";N$
1160 PRINT
1170 INPUT">>>INITIAL RA : H,M,S";A1,A2,A3
1180 PRINT
1190 INPUT">>>INITIAL DEC: D,M,S";D1,D2,D3
1200 CLS
1210 IF A$="T" THEN 1230 ELSE 1280
1220 REM
1230 S=+1
1240 EI$=E$
1250 EF$="APPARENT"
1260 GOTO 1320
1270 REM
1280 S=-1
1290 EI$="APPARENT"
1300 EF$=E$
1310 REM
1320 AI=A1+A2/60+A3/3600
1330 DI=D1+D2/60+D3/3600
1340 A=AI*15*Q1
1350 D=DI*Q1
1360 REM
1370 C1=(MN+SIN(A)*TAN(D))/15
1380 C2=(COS(A)*TAN(D))/15
1390 C3=(COS(A)/COS(D))/15
```

2.9. COMPUTER PROGRAMS

```
1400 C4=(SIN(A)/COS(D))/15
1410 C5=COS(A)
1420 C6=-SIN(A)
1430 C7=TE*COS(D)-SIN(A)*SIN(D)
1440 C8=COS(A)*SIN(D)
1450 REM
1460 AF=AI+S*(NA*C1+NB*C2+NC*C3+ND*C4+NE)/3600
1470 DF=DI+S*(NA*C5+NB*C6+NC*C7+ND*C8)/3600
1480 REM
1490 PRINT"REDUCTION OF RA AND DEC"
1500 PRINT"BETWEEN APPARENT PLACE AND JXXXX.5"
1510 PRINT N$
1520 PRINT
1530 PRINT"DATE";TAB(13)IY;"/";IM;"/";ID
1540 PRINT
1550 PRINT EI$
1560 PRINT"A(HRS)";TAB(12);:PRINT USING G$;AI
1570 PRINT"D(DGS)";TAB(12);:PRINT USING G$;DI
1580 PRINT
1590 PRINT EF$
1600 PRINT"A(HRS)";TAB(12);:PRINT USING G$;AF
1610 PRINT"D(DGS)";TAB(12);:PRINT USING G$;DF
1620 END
30000 DATA "J1987.5", 2.30006, 0.43359
30010 DATA 87, 07, 24
30020 DATA +0.970, -8.698, +9.495, -17.315, -0.0001
```

2.9.6 Program ADCES

ADCES is a utility program for reducing right ascension and declination coordinates from the mean equinox Jxxxx.x to the standard mean equinox J2000.0 and vice versa.

Program Algorithm

Given	Line Number
radians per degree	1060
date for information only	1090
name of the object for information only	1110
initial equinox	1130
initial RA (^{h}, ^{m}, ^{s})	1150
initial DEC ($^\circ$, $'$, $''$)	1170
final equinox	1190

Compute	Line Number
T: Equation 2.77	1230
initial RA in radians	1250
initial DEC in radians	1260
$\bar{m}$: Equation 2.75	1280
$\bar{n}$: Equation 2.76	1290
$s = +1$ if reduction is *from* J2000.0	1310
$s = -1$ if reduction is *to* J2000.0	1310
first approximation for the final RA	1330

2.9. COMPUTER PROGRAMS 59

first approximation for the final DEC	1340
By successive approximation:	1350
$\bar{\alpha}$: Equation 2.73	1360
$\bar{\delta}$: Equation 2.74	1370
final RA: Equation 2.71 or 2.78	1380
final DEC: Equation 2.72 or 2.79	1390
test for solution	1400
final RA in hours	1450
final DEC in degrees	1460

End.

Program Listing

```
1000 CLS
1010 PRINT"# ADCES # PRECESSION OF RA AND DEC"
1020 PRINT"# BETWEEN JXXXX.X AND J2000.0"
1030 PRINT
1040 DEFDBL A-Z
1050 REM
1060 Q1=.0174532925#
1070 G$="####.#######"
1080 REM
1090 INPUT">>>DATE YY, MM, DD";IY,IM,ID
1100 PRINT
1110 INPUT">>>NAME OF OBJECT"; N$
1120 PRINT
1130 INPUT">>>INITIAL EQUINOX (NUMBER)";EI
1140 PRINT
1150 INPUT">>>INITIAL RA : H,M,S";A1,A2,A3
1160 PRINT
1170 INPUT">>>INITIAL DEC: D,M,S";D1,D2,D3
1180 PRINT
1190 INPUT">>>FINAL EQUINOX (NUMBER)";EF
1200 CLS
1210 IF EI=2000 THEN EQ=EF ELSE EQ=EI
1220 REM
1230 TT=(EQ-2000)/100
1240 REM
1250 AI=(A1+A2/60+A3/3600)*15*Q1
1260 DI=(D1+D2/60+D3/3600)*Q1
1270 REM
1280 MB=(1.28123227#+.000775867#*(TT/2)-
     .0000000772#*(TT*TT/4))*Q1
1290 NB=(.55675303#-.00023703#*(TT/2)-
     .0000000603#*(TT*TT/4))*Q1
1300 REM
1310 IF EI=2000 THEN S=+1 ELSE S=-1
1320 REM
1330 AF=AI
1340 DF=DI
1350 REM
1360    AB=(AI+AF)/2
1370    DB=(DI+DF)/2
```

2.9. COMPUTER PROGRAMS

```
1380    AX=AI+S*(MB+NB*SIN(AB)*TAN(DB))*TT
1390    DX=DI+S*(NB*COS(AB))*TT
1400    IF ABS(AX-AF)<.000001# AND ABS(DX-DF)<.000001#
        THEN 1450
1410    AF=AX
1420    DF=DX
1430    GOTO 1360
1440 REM
1450 AF=AX/(Q1*15)
1460 DF=DX/Q1
1470 REM
1480 PRINT"PRECESSION OF RA AND DEC"
1490 PRINT"BETWEEN JXXXX.X AND J2000.0"
1500 PRINT N$
1510 PRINT
1520 PRINT"DATE";TAB(13);IY;"/";IM;"/";ID
1530 PRINT
1540 PRINT"J"+STR$(EI)
1550 PRINT"A(HRS)";TAB(12);:PRINT USING G$;AI/(Q1*15)
1560 PRINT"D(DGS)";TAB(12);:PRINT USING G$;DI/Q1
1570 PRINT
1580 PRINT"J"+STR$(EF)
1590 PRINT"A(HRS)";TAB(12);:PRINT USING G$;AF
1600 PRINT"D(DGS)";TAB(12);:PRINT USING G$;DF
1610 END
```

2.9.7 Program XYZCES

XYZCES is a utility program for reducing rectangular equatorial coordinates from the mean equinox Jxxxx.x to the standard equinox J2000.0 and vice versa.

Program Algorithm

Given	Line Number
date for information only	1090
name of the object for information only	1110
initial equinox	1130
initial rectangular coordinates	1150
final equinox	1170

Compute	Line Number
T: Equation 2.77	12020
P_{11} through P_{33}: Equations 2.81	12030-12110
final rectangular coordinates: Equations 2.82 or 2.80	1260 or 1320

End.

2.9. COMPUTER PROGRAMS

Program Listing

```
1000 CLS
1010 PRINT"# XYZCES # PRECESSION OF X Y Z"
1020 PRINT"# BETWEEN JXXXX.X AND J2000.0"
1030 PRINT
1040 DEFDBL A-Z
1050 DEFINT K
1060 REM
1070 G$="####.#######"
1080 REM
1090 INPUT">>>DATE YY, MM, DD";IY,IM,ID
1100 PRINT
1110 INPUT">>>NAME OF OBJECT";N$
1120 PRINT
1130 INPUT">>>INITIAL EQUINOX (NUMBER)";EI
1140 PRINT
1150 INPUT">>>INITIAL X, Y, Z";RI(1),RI(2),RI(3)
1160 PRINT
1170 INPUT">>>FINAL EQUINOX (NUMBER)";EF
1180 CLS
1190 IF EI=2000 THEN EQ=EF ELSE EQ=EI
1200 REM
1210 GOSUB 12010 REM PMATRX/SUB
1220 REM
1230 IF EI=2000 THEN 1250 ELSE 1310
1240 REM
1250 FOR K=1 TO 3
1260    RF(K)=PP(K,1)*RI(1)+PP(K,2)*RI(2)+PP(K,3)*RI(3)
1270 NEXT K
1280 REM
1290 GOTO 1350
1300 REM
1310 FOR K=1 TO 3
1320    RF(K)=PP(1,K)*RI(1)+PP(2,K)*RI(2)+PP(3,K)*RI(3)
1330 NEXT K
1340 REM
1350 PRINT"PRECESSION OF X Y Z"
1360 PRINT"BETWEEN JXXXX.X AND J2000.0"
1370 PRINT N$
1380 PRINT
1390 PRINT"DATE";TAB(13);IY;"/";IM;"/";ID
```

```
1400 PRINT
1410 PRINT"J"+STR$(EI)
1420 PRINT"X  Y  Z";TAB(12);:PRINT USING G$;RI(1),RI(2),RI(3)
1430 PRINT
1440 PRINT"J"+STR$(EF)
1450 PRINT"X  Y  Z";TAB(12);:PRINT USING G$;RF(1),RF(2),RF(3)
1460 END
12000 STOP
12010 REM# PMATRX/SUB # PRECESSION ROTATION MATRIX
12020 TT=(EQ-2000)/100
12030 PP(1,1)=1+(-29724*TT-13*TT*TT)*TT*.00000001#
12040 PP(1,2)=(-2236172-677*TT+222*TT*TT)*TT*.00000001#
12050 PP(1,3)=(-971717+207*TT+96*TT*TT)*TT*.00000001#
12060 PP(2,1)=-PP(1,2)
12070 PP(2,2)=1+(-25002*TT-15*TT*TT)*TT*.00000001#
12080 PP(2,3)=(-10865*TT)*TT*.00000001#
12090 PP(3,1)=-PP(1,3)
12100 PP(3,2)=PP(2,3)
12110 PP(3,3)=1+(-4721*TT)*TT*.00000001#
12120 RETURN
```

2.9. COMPUTER PROGRAMS

2.9.8 Program ADLAX

ADLAX is a utility program for reducing right ascension and declination coordinates for geocentric parallax.

Program Algorithm

Given	Line Number
radians per degree	1070
f: Equation 2.27	1080
west longitude (°, ′, ″)	1110
geodetic latitude (°, ′, ″)	1110
direction of reduction	1130
date	1150
UT in hours	1150
name of the object for information only	1170
equinox for information only	1190
initial RA (^{h}, ^{m}, ^{s})	1210
initial DEC (°, ′, ″)	1230
p: geocentric distance measured in astronomical units	1250

Compute	Line Number
$s = +1$ if reduction is *to* the geocenter	1270
$s = -1$ if reduction is *from* the geocenter	1270
west longitude in degrees	1380
ϕ: geodetic latitude in radians	1390

θ: sidereal time	14020-14100
initial RA in radians	1430
initial DEC in radians	1440
hour angle in radians	1450
horizontal parallax in radians	1460
F: Equation 2.31	1470
G_c: Equation 2.32	1480
G_s: Equation 2.33	1490
final RA: Equation 2.83 or 2.87	1510
final DEC: Equation 2.84 or 2.88	1520
final RA converted to hours	1670
final DEC converted to degrees	1680

End.

2.9. COMPUTER PROGRAMS

Program Listing

```
1000 CLS
1010 PRINT"# ADLAX # REDUCTION OF RA AND DEC"
1020 PRINT"# FOR GEOCENTRIC PARALLAX"
1030 PRINT
1040 DEFDBL A-Z
1050 DEFINT I
1060 REM
1070 Q1=.0174532925#
1080 F=1/298.257#
1090 G$="####.#######"
1100 REM
1110 READ W1,W2,W3, B1,B2,B3
1120 REM
1130 INPUT">>>REDUCTION (T)O OR (F)ROM GEOCENTER";A$
1140 PRINT
1150 INPUT">>>DATE YY, MM, DD, UT";IY,IM,ID,UT
1160 PRINT
1170 INPUT">>>NAME OF OBJECT";N$
1180 PRINT
1190 INPUT">>>EQUINOX";E$
1200 PRINT
1210 INPUT">>>INITIAL RA : H,M,S";A1,A2,A3
1220 PRINT
1230 INPUT">>>INITIAL DEC: D,M,S";D1,D2,D3
1240 PRINT
1250 INPUT">>>DISTANCE";P
1260 CLS
1270 IF A$="T" THEN 1290 ELSE 1340
1280 REM
1290 S=+1
1300 I$="TOPOCENTRIC"
1310 F$="GEOCENTRIC"
1320 GOTO 1380
1330 REM
1340 S=-1
1350 I$="GEOCENTRIC"
1360 F$="TOPOCENTRIC"
1370 REM
1380 WL=W1+W2/60+W3/3600
1390 BB=(B1+B2/60+B3/3600)*Q1
```

```
1400 REM
1410 GOSUB 14010 REM JOST/SUB
1420 REM
1430 AI=(A1+A2/60+A3/3600)*15*Q1
1440 DI=(D1+D2/60+D3/3600)*Q1
1450 H=ST*Q1-AI
1460 HP=(8.794/P)*Q1/3600
1470 FF=SQR(1-(2*F-F*F)*SIN(BB)*SIN(BB))
1480 GC=1/FF
1490 GS=(1-F)*(1-F)/FF
1500 REM
1510 AF=AI+S*HP*(GC*COS(BB)*SIN(H)/COS(DI))
1520 DF=DI+S*HP*(GS*SIN(BB)*COS(DI)-
     GC*COS(BB)*COS(H)*SIN(DI))
1530 REM
1540 PRINT"REDUCTION OF RA AND DEC"
1550 PRINT"FOR GEOCENTRIC PARALLAX"
1560 PRINT N$
1570 PRINT"DATE(UT)";TAB(11);IY;"/";IM;"/";ID;"/";UT
1580 PRINT"WL(DGS)";TAB(12);:PRINT USING G$;WL
1590 PRINT"BB(DGS)";TAB(12);:PRINT USING G$;BB/Q1
1600 PRINT
1610 PRINT I$
1620 PRINT"A(HRS)";TAB(12);:PRINT USING G$;AI/(Q1*15)
1630 PRINT"D(DGS)";TAB(12);:PRINT USING G$;DI/Q1
1640 PRINT"DISTANCE";TAB(12);:PRINT USING G$;P
1650 PRINT
1660 PRINT F$
1670 PRINT"A(HRS)";TAB(12);:PRINT USING G$;AF/(Q1*15)
1680 PRINT"D(DGS)";TAB(12);:PRINT USING G$;DF/Q1
1690 END
```

2.9. COMPUTER PROGRAMS

```
14000 STOP
14010 REM # JOST/SUB # CALCULATES JO AND LMST
14020 IY=IY+1900
14030 JO=367*IY-INT((7*(IY+INT((IM+9)/12)))/4)+
      INT(275*IM/9)+ID+1721013.5#
14040 J=(JO-2451545#)/36525#
14050 EL=360-WL
14060 S0=100.4606184#+36000.77004#*J+.000387933#*J*J
14070 S0=S0-360*INT(S0/360)
14080 SG=S0+360.98564724#*(UT/24)
14090 ST=SG+EL
14100 ST=ST-360*INT(ST/360)
14110 RETURN
30000 DATA 100,0,0, 50,0,0
```

2.10 Numerical Examples

2.10.1 Computing Local Mean Sidereal Time

Problem

Find the local mean sidereal time (LMST) which corresponds to the solar time $9^h.741667$ UT on 1985 July 8 for a location at west longitude $80°22'55''.79$ and latitude $37°31'33''.00$ [4; Page B7].

Solution

Use the given longitude and latitude to write data line 30000 as shown at the end of program LMST. Run the program and answer the prompt as follows:

```
>>>YY,MM,DD,UT? 85,7,8,9.741667#
```

Results

```
LOCAL MEAN SIDEREAL TIME

DATE            1985 / 7 / 8 / 9.741667
WL(DGS)            80.382164
BB(DGS)            37.525833

J0              2446254.500000
S0(HRS)            19.059549

LMST(HRS)          23.469077
```

2.10. NUMERICAL EXAMPLES

2.10.2 Converting Spherical to Rectangular Coordinates

Problem

On 1985 January 1, a celestial body's geocentric mean place for equinox J2000.0 was RA $14^h.6600241$ and DEC $-60°.8353166$. Its distance is unknown. Find the direction cosines of the body's position [4; Page B40].

Solution

Let the distance to the object equal 1. Run program XYZ and use the given data to answer the prompts as follows:

\>\>\>DATE YY, MM, DD? 85,1,1

\>\>\>NAME OF OBJECT? OBJECT

\>\>\>EQUINOX? J2000.0

\>\>\>TRUE DISTANCE? 1

\>\>\>RA : H,M,S? 14.6600241#,0,0

\>\>\>DEC: D,M,S? -60.8353166#,0,0

Results

RECTANGULAR COORDINATES X Y Z FROM R A D
OBJECT

DATE   85 / 1 / 1

J2000.0
R   1.0000000
A(HRS)   14.6600241
D(DGS)   -60.8353166

X Y Z   0.3738541   -0.3125946   -0.8732226

2.10.3 Converting Rectangular to Spherical Coordinates

Problem

On 1985 January 1, a celestial body's geocentric direction cosines for the mean equinox J2000.0 were as follows:

$$\mathbf{L} = \{-0.3738541, -0.3125946, -0.8732226\}.$$

Find the body's right ascension and declination [4; Page B40].

Solution

Run program RAD and use the given data to answer the prompts as follows:

```
>>>DATE YY, MM, DD? 85,1,1

>>>OBJECT? OBJECT

>>>EQUINOX? J2000.0

>>>COORD X, Y, Z? -0.3738541#,-0.3125946#,-0.8732226#
```

Results

```
SPHERICAL COORDINATES R A D FROM X Y Z
OBJECT

DATE          85 / 1 / 1

J2000.0
X Y Z         -0.3738541  -0.3125946  -0.8732226

R              1.0000000
A(HRS)        14.6600244
D(DGS)       -60.8353148
```

2.10. NUMERICAL EXAMPLES

2.10.4 Converting Equatorial to Ecliptic Coordinates

Problem

On 1985 January 15, the heliocentric radius vector of the planet Mercury was

$$\mathbf{r} = \{-0.3200095, -0.2890390, -0.1211825\}$$

with respect to the mean equinox and equator of J2000.0. Find the ecliptic rectangular coordinates of Mercury on this date.

Solution

Run program CQTRAN and use the given data to answer the prompts as follows:

```
>>>REDUCTION (T)O OR (F)ROM THE ECLIPTIC? T

>>>DATE YY, MM, DD? 85,1,15

>>>NAME OF OBJECT? MERCURY

>>>EQUINOX(NUMBER)? 2000

>>>INITIAL X, Y, Z? -0.3200095#,-0.2890390#,-0.1211825#
```

Results

```
ECLIPTIC-EQUATORIAL TRANSFORMATION
MERCURY

DATE            85 / 1 / 15

J 2000
EC(DGS)         23.4392910

EQUATORIAL
VECTOR          -0.3200095   -0.2890390   -0.1211825
MAGNITUDE       0.4479228

ECLIPTIC
VECTOR          -0.3200095   -0.3133917    0.0037903
MAGNITUDE       0.4479228
```

2.10.5 Reducing Apparent Place to Astrometric Place

Problem

At 0^h TDT on 1987 Jul 24, the apparent place of the planet Mars for the true equinox of date was right ascension $8^h 55^m 16^s.620$ and declination $18°34'42''.21$ [10]. The following data is also published in the reference:

- $m/n = 2.30096$
- $\tan \varepsilon = 0.43359$
- $A = +0''.970$
- $B = -8''.698$
- $C = +9''.495$
- $D = -17''.315$
- $E = -0^s.0001$

Find the astrometric position of Mars for the mean equinox J1987.5.

Solution

Use the given data to write data lines 30000 to 30020 as shown at the end of program ADAPP. Run the program and answer the prompts as follows:

```
>>>REDUCTION (T)O OR (F)ROM APPARENT PLACE? F

>>>NAME OF OBJECT? MARS

>>>INITIAL RA : H,M,S? 8,55,16.620#

>>>INITIAL DEC: D,M,S? 18,34,42.21#
```

2.10. NUMERICAL EXAMPLES

Results

```
REDUCTION OF RA AND DEC
BETWEEN APPARENT PLACE AND JXXXX.5
MARS

DATE           87 / 7 / 24

APPARENT
A(HRS)         8.9212833
D(DGS)        18.5783917

J1987.5
A(HRS)         8.9215727
D(DGS)        18.5752962
```

2.10.6 Reducing RA and DEC from Jxxxx.x to J2000.0

Problem

In the previous numerical example, we found that the astrometric position of Mars at 0^h TDT on 1987 Jul 24 was right ascension $8^h.9215727$ and declination $18°.5752962$ with respect to the mean equinox J1987.5. Find the astrometric position of Mars with respect to the standard equinox J2000.0.

Solution

Run program ADCES and use the given data to answer the prompts as follows:

>>>DATE YY, MM, DD? 87,7,24

>>>NAME OF OBJECT? MARS

>>>INITIAL EQUINOX (NUMBER)? 1987.5

>>>INITIAL RA : H,M,S? 8.9215727#,0,0

>>>INITIAL DEC: D,M,S? 18.5752962#,0,0

>>>FINAL EQUINOX (NUMBER)? 2000.0

Results

```
PRECESSION OF RA AND DEC
BETWEEN JXXXX.X AND J2000.0
MARS

DATE          87 / 7 / 24

J 1987.5
A(HRS)        8.9215727
D(DGS)        18.5752962

J 2000
A(HRS)        8.9333709
D(DGS)        18.5270277
```

2.10. NUMERICAL EXAMPLES

2.10.7 Reducing Rectangular Coordinates from J2000.0 to Jxxxx.x

Problem

In the previous numerical examples, the right ascension and declination of Mars for the date 1987 Jul 24 were reduced to the following astrometric values with respect to the standard equinox J2000.0:

- $\alpha = 8.9333709$
- $\delta = 18.5270277$

When these angular coordinates are used in program XYZ to compute a geocentric unit position vector for J2000.0, the result is

$$\mathbf{L} = \{-0.6586636, +0.6820527, +0.3177520\}.$$

Reduce the rectangular coordinates of the vector $\mathbf{L}$ to the mean equinox J1987.5.

Solution

Run program XYZCLS and use the given data to answer the prompts as follows:

```
>>>DATE YY, MM, DD? 87,7,24

>>>NAME OF OBJECT? MARS

>>>INITIAL EQUINOX (NUMBER)? 2000.0

>>>INITIAL X, Y, Z? -0.6586636#,+0.6820527#,+0.3177520#

>>>FINAL EQUINOX (NUMBER)? 1987.5
```

Results

```
PRECESSION OF X Y Z
BETWEEN JXXXX.X AND J2000.0
MARS

DATE           87 / 7 / 24

J 2000
X Y Z          -0.6586636    0.6820527    0.3177520

J 1987.5
X Y Z          -0.6563682    0.6838905    0.3185507
```

78   CHAPTER 2. TIME AND POSITION

2.10.8   Reducing Geocentric Place to Topocentric Place

Problem

At $16^h.75$ UT on 1979 Feb 26, a celestial body's geocentric position was right ascension $22^h36^m44^s$ and declination $-8°44'24''$ with respect to equinox J2000.0. Its geocentric distance was 0.9901 au. An observer was located on the Earth's surface at west longitude 100° and geodetic latitude 50°. Find the body's topocentric position with respect to the observer's location [11; pages 66 to 69].

Solution

Use the given longitude and latitude to write data line 30000 as shown at the end of program ADLAX. Run the program and answer the prompts as follows:

>>>REDUCTION (T)O OR (F)ROM GEOCENTER? F

>>>DATE YY, MM, DD, UT? 79,2,26,16.75

>>>NAME OF OBJECT? OBJECT

>>>EQUINOX? J2000.0

>>>INITIAL RA : H,M,S? 22,36,44

>>>INITIAL DEC: D,M,S? -8,-44,-24

>>>DISTANCE? 0.9901

2.10. NUMERICAL EXAMPLES

Results

```
REDUCTION OF RA AND DEC
FOR GEOCENTRIC PARALLAX
OBJECT
DATE(UT)    1979 / 2 / 26 / 16.75
WL(DGS)      100.0000000
BB(DGS)       50.0000000

GEOCENTRIC
A(HRS)        22.6122222
D(DGS)        -8.7400000
DISTANCE       0.9901000

TOPOCENTRIC
A(HRS)        22.6122790
D(DGS)        -8.7420640
```

References

[1] *The Astronomical Almanac 1984*, U.S. Government Printing Office, 1983.

[2] Woolard and Clemence, *Spherical Astronomy*, Academic Press, 1966.

[3] Taff, *Computational Spherical Astronomy*, John Wiley and Sons, 1981.

[4] *The Astronomical Almanac 1985*, U.S. Government Printing Office, 1984.

[5] *Almanac for Computers 1985*, U.S. Government Printing Office, 1984.

[6] *The Astronomical Almanac 1990*, U.S. Government Printing Office, 1989.

[7] Escobal, *Methods of Orbit Determination*, Krieger Publishing Co., 1976.

[8] *Explanatory Supplement to the Astronomical Ephemeris and the American Ephemeris and Nautical Almanac*, Her Majesty's Stationary Office, 1961.

[9] *Planetary and Lunar Coordinates for the Years 1984-2000*, U.S. Government Printing Office, 1983.

[10] *The Astronomical Almanac 1987*, U.S. Government Printing Office, 1986.

[11] Duffett-Smith, *Practical Astronomy with Your Calculator*, Cambridge University Press, 1981.

Chapter 3

The Two-Body Problem

3.1 Introduction

We have at our disposal a celestial frame of reference and a general expression for orbital motion which takes into account the gravitational perturbations of a given number of other bodies. We shall now apply that equation to a specific type of orbital problem which is of great practical significance. The opportunity presents itself because our interest in heliocentric and geocentric orbits restricts the application of the general equation to orbits dominated by the mutual gravitational attraction of two celestial bodies. In this context, theory and experience have shown that a *two-body orbit* can be computed using a simple mathematical model which ignores all perturbations and considers only the attraction between the orbiting and central masses. Such two-body motion provides a *preliminary orbit* which can later be improved, if required.

3.2 The Two-Body Equation of Motion

Two-body motion is nothing more than a special case of the many-body motion modeled by Equation 1.26, namely,

$$\ddot{\mathbf{r}} = -\frac{\mu \mathbf{r}}{r^3} + \sum_{q=1}^{n} m_q \left(\frac{\mathbf{p}_q}{p_q^3} - \frac{\mathbf{r}_q}{r_q^3} \right). \tag{3.1}$$

The two-body equation follows immediately from the above when we make the simplifying assumption that all terms involving masses m_q can be neglected when computing a first approximation of the orbit. Thus,

$$\ddot{\mathbf{r}} = -\frac{\mu \mathbf{r}}{r^3} \tag{3.2}$$

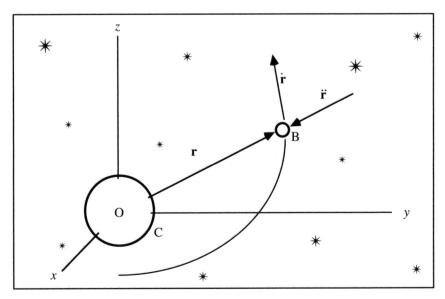

Figure 3.1: Two-body orbital motion.

is the *two-body equation of motion*. As shown in Figure 3.1, the acceleration $\ddot{\mathbf{r}}$ will always point directly toward the origin O at the center of the central body C because that is the direction of the net force acting on body B. As a consequence of this, there is no tendency for B to move out of the plane formed by $\mathbf{r}$, $\dot{\mathbf{r}}$, and O. Therefore, *a two-body orbit is always confined to a plane which passes through the center of the central body.*

The justification for Equation 3.2 may not be obvious. It is definitely not because the perturbing bodies are so far away that their attractions are negligible! For in the case of the Earth-Moon system, the Sun exerts a greater gravitational force on the Moon than does the Earth. The reason we can take this dramatic step is revealed by considering the algebraic form of the perturbation terms in light of Figure 3.2. By Equation 3.1, the net perturbative acceleration $\mathbf{a}_q$ produced by an arbitrary mass m_q is given by

$$\mathbf{a}_q = m_q \left(\frac{\mathbf{p}_q}{p_q^3} - \frac{\mathbf{r}_q}{r_q^3} \right),$$

which can also be written

$$\mathbf{a}_q = m_q \frac{\mathbf{p}_q}{p_q^3} - m_q \frac{\mathbf{r}_q}{r_q^3}. \tag{3.3}$$

As shown in Figure 3.2, $\mathbf{p}_q$ and $\mathbf{r}_q$ are the positions of m_q in relation to the orbiting body B and the central body C, respectively. The first term in Equa-

3.3. THE ORBITAL AND RADIAL RATES

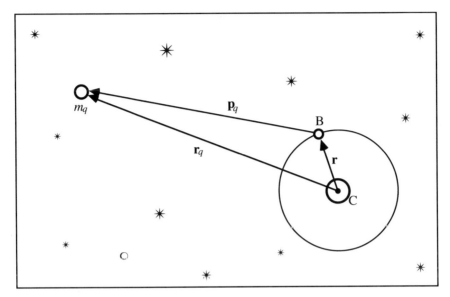

Figure 3.2: The perturbation geometry.

tion 3.3 is the acceleration which m_q produces on B, while the second term is the acceleration which m_q gives to C. Therefore, even though these two accelerations may individually be very great, it is only their *difference* which enters into the equation of motion. In other words, continuing the Earth-Moon example, the Moon is orbiting the Earth as both fall freely about the Sun. Thus, the perturbation caused by m_q is only a differential or secondary effect which can be neglected for a first approximation of the orbit of B.

3.3 The Orbital and Radial Rates

Before taking up the solution of the two-body equation of motion, we must digress to discuss the potentially confusing relationship between the rate of motion along the orbital path and the rate of change of the magnitude of the radius vector **r**. As illustrated in Figure 3.3, a body B has velocity $\dot{\mathbf{r}}$ which is tangent to the orbit at **r**. According to Equations 1.4 and 1.27, the magnitude of $\dot{\mathbf{r}}$ is the orbital speed v. Thus,

$$v = |\dot{\mathbf{r}}|. \tag{3.4}$$

Letting $\dot{r}$ represent the rate of change of the scalar r with respect to modified time, then

$$\dot{r} = \frac{dr}{d\tau}, \tag{3.5}$$

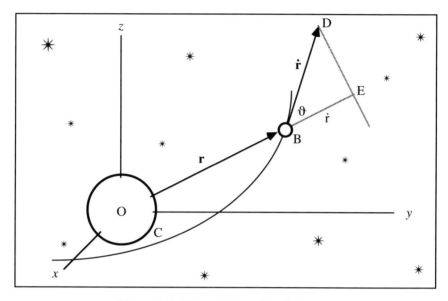

Figure 3.3: The orbital and radial rates.

where
$$r = |\mathbf{r}| \tag{3.6}$$
and dr is the incremental change in r during an infinitesimal interval of modified time $d\tau$. Equation 3.5 is a scalar relationship which expresses only the rate at which the *distance* between bodies B and C changes with time.

In the right triangle BDE of Figure 3.3, ϑ represents the angle between two sides whose lengths are in the ratio $\dot{r}/v$. Thus,
$$\dot{r} = v\cos\vartheta. \tag{3.7}$$
By the definition of the dot product given in Appendix A, we may write
$$\mathbf{r}\cdot\dot{\mathbf{r}} = |\mathbf{r}||\dot{\mathbf{r}}|\cos\vartheta. \tag{3.8}$$
Being careful to note that
$$\dot{r} \neq |\dot{\mathbf{r}}|,$$
but, rather,
$$v = |\dot{\mathbf{r}}|,$$
substituting Equations 3.4 and 3.6 into Equation 3.8 produces
$$\mathbf{r}\cdot\dot{\mathbf{r}} = rv\cos\vartheta. \tag{3.9}$$

3.4. THE LAWS OF TWO-BODY MOTION

Therefore, by Equation 3.7 we have the interesting and useful expression

$$\mathbf{r} \cdot \dot{\mathbf{r}} = r\dot{r}. \tag{3.10}$$

One can also make use of the fact that

$$\mathbf{r} \cdot \dot{\mathbf{r}} = x\dot{x} + y\dot{y} + z\dot{z} \tag{3.11}$$

to obtain the alternate expression,

$$r\dot{r} = x\dot{x} + y\dot{y} + z\dot{z}. \tag{3.12}$$

Additionally, the following are useful identities which can be applied to any vector:

$$\mathbf{r} \cdot \mathbf{r} = xx + yy + zz \tag{3.13}$$
$$r^2 = \mathbf{r} \cdot \mathbf{r} \tag{3.14}$$
$$r = \sqrt{\mathbf{r} \cdot \mathbf{r}} \tag{3.15}$$

3.4 The Laws of Two-Body Motion

When the techniques of integral calculus are used to solve the Newtonian two-body differential equation of motion, the results confirm the laws derived empirically by Johannes Kepler:

1. *The orbits of the planets are ellipses, with the Sun at one focus.*

2. *The line joining a planet to the Sun sweeps out equal areas in equal times.*

3. *The square of a planet's period is proportional to the cube of its mean distance from the Sun.*

Moreover, the Newtonian formulation expresses these original principles in a general fashion, and an important relationship is shown to exist between the speed and position of a celestial body in any two-body orbit.

3.4.1 The Conic Section Law

Let the fundamental equation of two-body motion be written as follows [1]:

$$\ddot{\mathbf{r}} = -\frac{\mu}{r^2}\mathbf{U}, \tag{3.16}$$

where

$$\mathbf{U} = \frac{\mathbf{r}}{r} \tag{3.17}$$

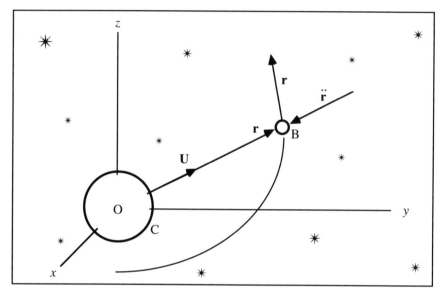

Figure 3.4: The two-body problem.

is a unit vector pointing in the direction of the radius vector as shown in Figure 3.4. Taking the cross product of Equation 3.16 with **r**, we obtain

$$\mathbf{r} \times \ddot{\mathbf{r}} = -\frac{\mu}{r^2}(\mathbf{r} \times \mathbf{U}),$$

so that

$$\mathbf{r} \times \ddot{\mathbf{r}} = \mathbf{0} \qquad (3.18)$$

because **r** and **U** are parallel, causing their cross product to be the *null vector* **0**. Now consider the following differentiation with respect to modified time:

$$\frac{d}{d\tau}(\mathbf{r} \times \dot{\mathbf{r}}) = (\mathbf{r} \times \ddot{\mathbf{r}}) + (\dot{\mathbf{r}} \times \dot{\mathbf{r}}),$$

which simplifies to

$$\frac{d}{d\tau}(\mathbf{r} \times \dot{\mathbf{r}}) = \mathbf{0} \qquad (3.19)$$

because $\mathbf{r} \times \ddot{\mathbf{r}} = \mathbf{0}$ by Equation 3.18, and $\dot{\mathbf{r}} \times \dot{\mathbf{r}} = \mathbf{0}$ for the same reason. If we integrate Equation 3.19 to reverse the differentiation process, we obtain

$$\mathbf{r} \times \dot{\mathbf{r}} = \mathbf{h}, \qquad (3.20)$$

where **h** is a vector constant of integration which is equal to the *angular momentum per unit mass* of the two-body system.

3.4. THE LAWS OF TWO-BODY MOTION

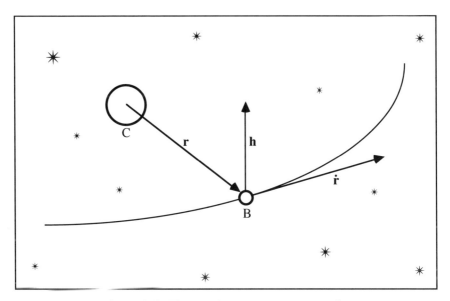

Figure 3.5: The angular momentum vector **h**.

The physical significance of this important result is illustrated in Figure 3.5. The fact that **h** is constant implies that its magnitude and direction in inertial space never change. Therefore, as noted in Section 3.2, vectors **r** and **ṙ** always lie in a fixed plane which passes through the central body C. The direction of **h** establishes the orientation of the orbital plane with respect to the inertial rectangular coordinate system. When the z component of **h** is positive, the celestial body B is moving in a counter-clockwise direction as viewed from the positive z-axis, and the motion is called *direct*. When the z component of **h** is negative, the orbital motion is clockwise as seen from the positive z-axis, and the motion is called *retrograde*.

The angular momentum vector can be used to transform the fundamental equation of motion into an expression which can be easily integrated. Returning to Equation 3.16 and taking the cross product of that expression with **h**, we have

$$\ddot{\mathbf{r}} \times \mathbf{h} = -\frac{\mu}{r^2}(\mathbf{U} \times \mathbf{h}), \tag{3.21}$$

so that

$$\ddot{\mathbf{r}} \times \mathbf{h} = -\frac{\mu}{r^2}[\mathbf{U} \times (\mathbf{r} \times \dot{\mathbf{r}})] \tag{3.22}$$

because $\mathbf{h} = \mathbf{r} \times \dot{\mathbf{r}}$ by definition. Now, there is a vector identity which states that for three given vectors **A**, **B**, and **C**

$$\mathbf{A} \times (\mathbf{B} \times \mathbf{C}) = (\mathbf{A} \cdot \mathbf{C})\mathbf{B} - (\mathbf{A} \cdot \mathbf{B})\mathbf{C}. \tag{3.23}$$

Applying this identity to Equation 3.22, the result is

$$\ddot{\mathbf{r}} \times \mathbf{h} = -\frac{\mu}{r^2}[(\mathbf{U} \cdot \dot{\mathbf{r}})\mathbf{r} - (\mathbf{U} \cdot \mathbf{r})\dot{\mathbf{r}}].$$

Replacing **U** by its definition and using Equations 3.10 and 3.14 to simplify the result yields

$$\ddot{\mathbf{r}} \times \mathbf{h} = \frac{\mu}{r^2}[r\dot{\mathbf{r}} - \dot{r}\mathbf{r}]. \tag{3.24}$$

Before we can integrate Equation 3.24, we must pause to show that the expression on each side of the equal sign represents a perfect differential. Consider the following:

$$\frac{d\mathbf{U}}{d\tau} = \frac{d}{d\tau}\left(\frac{\mathbf{r}}{r}\right). \tag{3.25}$$

Carrying out the differentiation on the right side of Equation 3.25 produces

$$\frac{d}{d\tau}\left(\frac{\mathbf{r}}{r}\right) = \frac{r\dot{\mathbf{r}} - \dot{r}\mathbf{r}}{r^2}. \tag{3.26}$$

Consider also the following:

$$\frac{d}{d\tau}(\dot{\mathbf{r}} \times \mathbf{h}) = (\ddot{\mathbf{r}} \times \mathbf{h}) + (\dot{\mathbf{r}} \times \dot{\mathbf{h}}). \tag{3.27}$$

However, by Equations 3.19 and 3.20, **h** is a constant vector so that

$$\dot{\mathbf{h}} = \mathbf{0}. \tag{3.28}$$

Therefore, Equation 3.27 becomes simply

$$\frac{d}{d\tau}(\dot{\mathbf{r}} \times \mathbf{h}) = (\ddot{\mathbf{r}} \times \mathbf{h}). \tag{3.29}$$

Returning to Equation 3.24, we see that its right and left sides can be replaced by Equations 3.26 and 3.29, respectively. Upon making those substitutions, we have

$$\frac{d}{d\tau}(\dot{\mathbf{r}} \times \mathbf{h}) = \mu \frac{d}{d\tau}\left(\frac{\mathbf{r}}{r}\right). \tag{3.30}$$

Equation 3.30 can be immediately integrated to eliminate the differentiation. The result can be written as follows:

$$\dot{\mathbf{r}} \times \mathbf{h} = \mu\left(\frac{\mathbf{r}}{r} + \mathbf{e}\right), \tag{3.31}$$

where **e** is an arbitrary vector constant of integration.

3.4. THE LAWS OF TWO-BODY MOTION

If we take the dot product of Equation 3.31 with **r**, the result can be manipulated to obtain a key relationship. First,

$$(\dot{\mathbf{r}} \times \mathbf{h}) \cdot \mathbf{r} = \mu \left(\frac{\mathbf{r} \cdot \mathbf{r}}{r} + \mathbf{e} \cdot \mathbf{r} \right). \tag{3.32}$$

Using the vector identity

$$(\mathbf{A} \times \mathbf{B}) \cdot \mathbf{C} = (\mathbf{C} \times \mathbf{A}) \cdot \mathbf{B} \tag{3.33}$$

and Equation 3.14, Equation 3.32 can be rewritten to obtain

$$(\mathbf{r} \times \dot{\mathbf{r}}) \cdot \mathbf{h} = \mu (r + \mathbf{e} \cdot \mathbf{r}), \tag{3.34}$$

which, according to Equation 3.20, becomes

$$h^2 = \mu (r + \mathbf{e} \cdot \mathbf{r}), \tag{3.35}$$

Now, using the definition of the dot product produces

$$\mathbf{e} \cdot \mathbf{r} = er \cos \nu, \tag{3.36}$$

where $e = |\mathbf{e}|$ and ν is the angle between **e** and **r**. Therefore, substituting Equation 3.36 into Equation 3.35, one obtains the scalar relationship

$$h^2 = \mu r (1 + e \cos \nu), \tag{3.37}$$

which can also be written

$$r = \frac{h^2/\mu}{1 + e \cos \nu}. \tag{3.38}$$

The geometric significance of Equation 3.38 can be seen by comparing it to the general equation of a conic section written in polar coordinates

$$r = \frac{\wp}{1 + e \cos \nu}, \tag{3.39}$$

where the origin is at a focus, the polar angle ν is the angle between the radius vector and the point on the conic nearest the focus, and

$$\wp = \frac{h^2}{\mu}. \tag{3.40}$$

The conclusion is that a two-body orbit is always a conic section which lies in a fixed plane which passes through the central body at the focus. This is a generalized statement of Kepler's first law.

The elements commonly used to describe conic sections are illustrated for the case of an ellipse in Figure 3.6, where **r** is the radius vector and **e** defines the

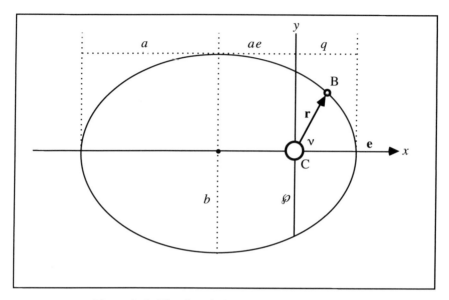

Figure 3.6: The descriptive elements of a conic.

direction of the perifocus. The angle ν is called the *true anomaly*, the quantity $\wp$ is the *semiparameter*, the constant e is the *eccentricity*, the length q is the *perifocal distance*, and the length a is the *semimajor axis*. These elements are related as follows:

$$\wp = q(1+e) \qquad (3.41)$$
$$q = a(1-e) \qquad (3.42)$$
$$\wp = a(1-e^2). \qquad (3.43)$$

Additionally, the *semiminor axis* b is related to the above by

$$b = a\sqrt{1-e^2} \qquad (3.44)$$
$$b = \sqrt{\wp a}. \qquad (3.45)$$

The value of the eccentricity determines the specific type of conic section represented by Equation 3.38. Four possibilities are depicted in Figure 3.7, and Equation 3.31 provides the useful vector relationship,

$$\mathbf{e} = \frac{\dot{\mathbf{r}} \times \mathbf{h}}{\mu} - \frac{\mathbf{r}}{r}, \qquad (3.46)$$

which permits us to find the magnitude $e = |\mathbf{e}|$ and the direction of the perifocus in inertial space when $\mathbf{r}$ and $\dot{\mathbf{r}}$ are known at any point on the orbit.

3.4. THE LAWS OF TWO-BODY MOTION

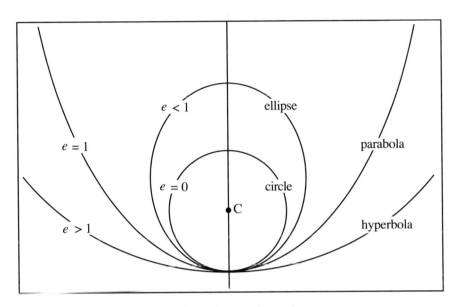

Figure 3.7: Four conic sections.

3.4.2 The Law of Areas

Consider the vector cross product shown below in light of the geometric relationships depicted in Figure 3.8:

$$\mathbf{W} dA = \frac{1}{2}(\mathbf{r} \times d\mathbf{r}), \tag{3.47}$$

where dA is a very narrow triangular area swept out by the radius vector $\mathbf{r}$ during an infinitesimal modified time interval $d\tau$, $d\mathbf{r}$ is the incremental change in $\mathbf{r}$ during the interval, and $\mathbf{W}$ is a unit vector normal to the orbital plane. The numerical constant $1/2$ accounts for the fact that the cross product is equivalent to the vector area of the entire parallelogram ABCD. The unit normal $\mathbf{W}$ defines the direction associated with the vector area.

Dividing both sides of Equation 3.47 by the interval $d\tau$, during which the triangular area is swept out, we obtain

$$\mathbf{W}\frac{dA}{d\tau} = \frac{1}{2}\left(\mathbf{r} \times \frac{d\mathbf{r}}{d\tau}\right),$$

so that

$$\mathbf{W}\frac{dA}{d\tau} = \frac{1}{2}(\mathbf{r} \times \dot{\mathbf{r}}),$$

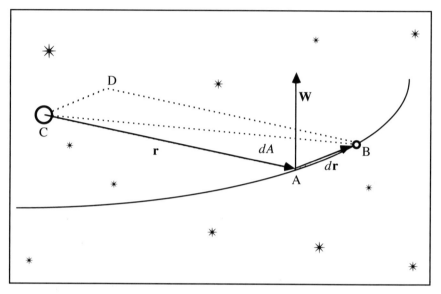

Figure 3.8: The vector area $\mathbf{W}dA$.

which, by Equation 3.20, is equivalent to

$$\mathbf{W}\frac{dA}{d\tau} = \frac{1}{2}\mathbf{h}. \tag{3.48}$$

Taking the magnitude of Equation 3.48, the result is

$$\frac{dA}{d\tau} = \frac{h}{2}. \tag{3.49}$$

Therefore, the rate at which the radius vector sweeps out area is constant. In other words, the radius vector generates equal areas in equal times, which is Kepler's second law.

3.4.3 The Harmonic Law

Let Equation 3.49 be rearranged and the modified time interval replaced by the corresponding normal time interval kdt as follows:

$$2(dA) = hk(dt). \tag{3.50}$$

If we assume the orbit to be an ellipse, integrating the area swept out by the radius vector over a time interval of exactly one orbital period will yield the following expression:

$$2(\pi ab) = hkP, \tag{3.51}$$

3.4. THE LAWS OF TWO-BODY MOTION

where the term in parentheses is the area of an ellipse, and P is the orbital period. Recalling that, by Equations 3.40 and 3.45,

$$h = \sqrt{\mu \wp}$$
$$b = \sqrt{\wp a},$$

we can write

$$P^2 = \left[\frac{1}{\mu}\left(\frac{2\pi}{k}\right)^2\right] a^3, \tag{3.52}$$

which is the generalized statement of Kepler's third law.

3.4.4 The Vis-viva Law

An extremely important relationship between orbital speed and position can be derived through another integration of the two-body equation of motion

$$\ddot{\mathbf{r}} = -\frac{\mu \mathbf{r}}{r^3}.$$

If we take the dot product of this equation with $2\dot{\mathbf{r}}$ and recall Equation 3.10, then we can obtain

$$2(\ddot{\mathbf{r}} \cdot \dot{\mathbf{r}}) = 2\left(-\frac{\mu \dot{r}}{r^2}\right). \tag{3.53}$$

Now consider the following two derivatives:

$$\frac{d}{d\tau}(\dot{\mathbf{r}} \cdot \dot{\mathbf{r}}) = (\dot{\mathbf{r}} \cdot \ddot{\mathbf{r}}) + (\ddot{\mathbf{r}} \cdot \dot{\mathbf{r}}) = 2(\ddot{\mathbf{r}} \cdot \dot{\mathbf{r}}) \tag{3.54}$$

and

$$\frac{d}{d\tau}\left(\frac{\mu}{r}\right) = -\frac{\mu \dot{r}}{r^2}. \tag{3.55}$$

Therefore, substituting Equations 3.54 and 3.55 into 3.53 yields

$$\frac{d}{d\tau}(\dot{\mathbf{r}} \cdot \dot{\mathbf{r}}) = 2\frac{d}{d\tau}\left(\frac{\mu}{r}\right). \tag{3.56}$$

Integrating Equation 3.56 will reverse the differentiation and produce an arbitrary constant $\mathcal{E}$. Thus,

$$\dot{\mathbf{r}} \cdot \dot{\mathbf{r}} = \frac{2\mu}{r} + \mathcal{E},$$

which is equivalent to

$$v^2 = \frac{2\mu}{r} + \mathcal{E}. \tag{3.57}$$

The constant $\mathcal{E}$ can be evaluated by imposing the conditions which exist when the orbiting body is at perifocus, as shown in Figure 3.9. In this situation,

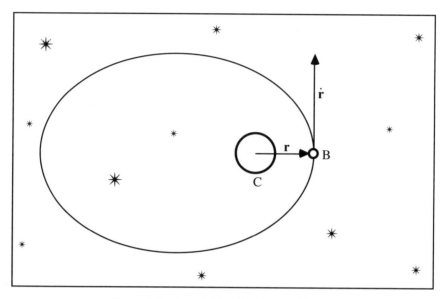

Figure 3.9: An orbiting body at perifocus.

$r = q$, so that

$$\mathcal{E} = v^2 - \frac{2\mu}{q}. \tag{3.58}$$

Furthermore, because $\mathbf{r}$ and $\dot{\mathbf{r}}$ are perpendicular at the perifocus, we may use the definition of the vector cross product to obtain the simple relationship

$$h = |\mathbf{r} \times \dot{\mathbf{r}}| = rv \sin 90° = qv \quad \text{(at perifocus)}. \tag{3.59}$$

Squaring Equation 3.59 and substituting Equations 3.40 and 3.41, we can write

$$v^2 = \frac{\mu(1+e)}{q}. \tag{3.60}$$

Using the above expression for v^2 and the fact that $q = a(1-e)$, Equation 3.58 yields the following expression for $\mathcal{E}$:

$$\mathcal{E} = -\frac{\mu}{a}. \tag{3.61}$$

Finally, employing Equation 3.61 to replace $\mathcal{E}$ in Equation 3.57, the result is

$$v^2 = \mu \left(\frac{2}{r} - \frac{1}{a} \right). \tag{3.62}$$

3.5. TWO-BODY MOTION BY NUMERICAL INTEGRATION

Equation 3.62 is called the *vis-viva equation*. It reflects the fact that the sum of the kinetic and potential energy of the two-body system remains constant. The vis-viva equation is particularly useful as a means of determining the semimajor axis when the position and velocity vectors are known for any point on the orbit. Thus,

$$\frac{1}{a} = \frac{2}{r} - \frac{v^2}{\mu}. \tag{3.63}$$

3.5 Two-Body Motion by Numerical Integration

When a celestial body's position $\mathbf{r}_0$ and velocity $\dot{\mathbf{r}}_0$ are given for some epoch time t_0, then its acceleration $\ddot{\mathbf{r}}_0$ is also known since

$$\ddot{\mathbf{r}}_0 = -\frac{\mu \mathbf{r}_0}{r_0^3}.$$

Knowing the position, velocity, and acceleration at some epoch makes it possible to extrapolate the orbiting body's trajectory over a short interval of time to yield a new position and velocity. Furthermore, the new position permits the calculation of a new acceleration, and the whole process can be repeated. Therefore, in principle, it is possible to calculate the body's motion over an extended period of time by means of a sequence of relatively short steps if its initial position and velocity are known for a given epoch. This process is called *numerical integration*, which is a broad designation for a variety of clever procedures that are among the most powerful tools of celestial mechanics. Although some examples of numerical integration are introduced here and applied to two-body motion in order to illustrate a practical application of calculus, the true utility of the general concept is realized in the computation of perturbed orbital motion [2,3,4].

3.5.1 The f and g Series

Consider the free-fall motion illustrated in Figure 3.10, where body B has position $\mathbf{r}_0$ and velocity $\dot{\mathbf{r}}_0$ at an arbitrary epoch t_0. If we want to compute the position of B at some other time t, we must have an expression for $\mathbf{r}$ which will satisfy the initial conditions at t_0 and the equation of motion

$$\ddot{\mathbf{r}} = -\frac{\mu \mathbf{r}}{r^3} \tag{3.64}$$

for any given time before or after t_0. It can be shown that the solution of Equation 3.64 may be written as an *infinite power series* expansion in modified time about the position $\mathbf{r}_0$ [2, 5]. Therefore,

$$\mathbf{r} = \mathbf{C}_0 + \mathbf{C}_1 \tau + \mathbf{C}_2 \tau^2 + \mathbf{C}_3 \tau^3 + \mathbf{C}_4 \tau^4 + \cdots,$$

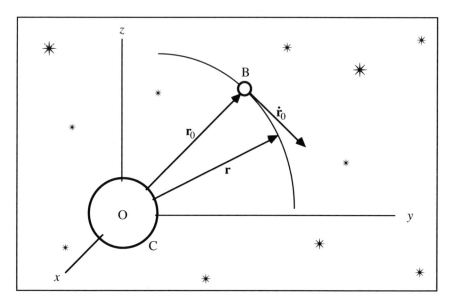

Figure 3.10: The two-body free-fall trajectory.

where $\tau = k(t - t_0)$ is the modified time interval, and the coefficients are vector constants which must be determined by applying the constraints imposed by the initial conditions. If this infinite series is repeatedly differentiated with respect to modified time, we obtain

$$\begin{aligned}
\mathbf{r} &= \mathbf{C}_0 + \mathbf{C}_1 \tau + \mathbf{C}_2 \tau^2 + \mathbf{C}_3 \tau^3 + \mathbf{C}_4 \tau^4 + \cdots \\
\dot{\mathbf{r}} &= \mathbf{C}_1 + 2\mathbf{C}_2 \tau + 3\mathbf{C}_3 \tau^2 + 4\mathbf{C}_4 \tau^3 + \cdots \\
\ddot{\mathbf{r}} &= 2\mathbf{C}_2 + 6\mathbf{C}_3 \tau + 12\mathbf{C}_4 \tau^2 + \cdots \\
\dddot{\mathbf{r}} &= 6\mathbf{C}_3 + 24\mathbf{C}_4 \tau + \cdots \\
&\vdots
\end{aligned} \qquad (3.65)$$

Evaluating these equations at t_0 when $\tau = 0$ yields the following values for the vector constants:

$$\begin{aligned}
\mathbf{C}_0 &= \mathbf{r}_0 \\
\mathbf{C}_1 &= \dot{\mathbf{r}}_0 \\
\mathbf{C}_2 &= \frac{\ddot{\mathbf{r}}_0}{2!} \\
\mathbf{C}_3 &= \frac{\dddot{\mathbf{r}}_0}{3!} \\
&\vdots
\end{aligned}$$

3.5. TWO-BODY MOTION BY NUMERICAL INTEGRATION

where, for any integer $n > 0$, $n! = 1 \cdot 2 \cdot 3 \cdots (n-1) \cdot n$. Replacing the vector coefficients in Equation 3.65 with the values given above, the result is

$$\mathbf{r} = \mathbf{r}_0 + \dot{\mathbf{r}}_0 \tau + \ddot{\mathbf{r}}_0 \frac{\tau^2}{2!} + \dddot{\mathbf{r}}_0 \frac{\tau^3}{3!} + \ddddot{\mathbf{r}}_0 \frac{\tau^4}{4!} + \cdots \qquad (3.66)$$

Notice that the first three terms are similar to the more familiar scalar equation for free fall near the surface of the Earth. Equation 3.66 is a very useful model for two-body orbital motion because the higher derivatives can be found by successively differentiating Equation 3.64 and evaluating the results at the epoch t_0.

Rather than attacking the problem head-on, it is more convenient to differentiate Equation 3.64 by means of auxiliary quantities. Let u, z, and q (not to be confused with the perifocal distance) be defined as follows [1,5]:

$$u = \frac{\mu}{r^3} \qquad (3.67)$$

$$z = \frac{\mathbf{r} \cdot \dot{\mathbf{r}}}{r^2} \qquad (3.68)$$

$$q = \frac{\dot{\mathbf{r}} \cdot \dot{\mathbf{r}}}{r^2} - u, \qquad (3.69)$$

and let Equation 3.64 be rewritten in the simpler form

$$\ddot{\mathbf{r}} = -u\mathbf{r}. \qquad (3.70)$$

Now, Equation 3.70 is the expression we want to differentiate; however, for reasons which will be apparent shortly, we shall delay its differentiation until after we have developed expressions for the first derivatives of u, z, and q. Commencing with u, we have

$$\begin{aligned} u &= \mu r^{-3} \\ \dot{u} &= \mu(-3r^{-4}\dot{r}) \\ \dot{u} &= -3\mu r^{-3}(r\dot{r})r^{-2} \\ \dot{u} &= -3u(\mathbf{r} \cdot \dot{\mathbf{r}})r^{-2}, \end{aligned}$$

which becomes

$$\dot{u} = -3uz. \qquad (3.71)$$

Next, for the quantity z,

$$\begin{aligned} z &= (\mathbf{r} \cdot \dot{\mathbf{r}})r^{-2} \\ \dot{z} &= [(\dot{\mathbf{r}} \cdot \dot{\mathbf{r}}) + (\mathbf{r} \cdot \ddot{\mathbf{r}})]r^{-2} + (\mathbf{r} \cdot \dot{\mathbf{r}})(-2r^{-3}\dot{r}). \end{aligned}$$

Using Equation 3.70 to replace $\ddot{\mathbf{r}}$, and substituting r^2 for the resulting $\mathbf{r} \cdot \mathbf{r}$, the outcome is

$$\dot{z} = [(\dot{\mathbf{r}} \cdot \dot{\mathbf{r}})r^{-2} - u] - 2(\mathbf{r} \cdot \dot{\mathbf{r}})(r\dot{r})r^{-4},$$

which can be written
$$\dot{z} = q - 2z^2 \,. \tag{3.72}$$
Finally, by means of a similar process for q, we have
$$\begin{aligned} q &= (\dot{\mathbf{r}} \cdot \dot{\mathbf{r}})r^{-2} - u \\ \dot{q} &= 2(\ddot{\mathbf{r}} \cdot \dot{\mathbf{r}})r^{-2} - 2(\dot{\mathbf{r}} \cdot \dot{\mathbf{r}})(r\dot{r})r^{-4} - \dot{u} \,, \end{aligned}$$
and Equations 3.10, 3.70, and 3.71 can be used to replace their corresponding identities to obtain
$$\dot{q} = -2u(\dot{\mathbf{r}} \cdot \mathbf{r})r^{-2} - 2(\dot{\mathbf{r}} \cdot \dot{\mathbf{r}})(\mathbf{r} \cdot \dot{\mathbf{r}})r^{-4} + 3uz \,.$$
Now, according to Equations 3.68 and 3.69, the above expression is equivalent to
$$\dot{q} = -2uz - 2z(q+u) + 3uz \,,$$
and the result is
$$\dot{q} = -(uz + 2zq) \,. \tag{3.73}$$
The derivatives $\dot{u}$, $\dot{z}$, and $\dot{q}$ are now given in terms of the original quantities u, z, and q.

At last we are ready to complete the series solution by finding expressions for the derivatives required by Equation 3.66. To do this, we differentiate
$$\ddot{\mathbf{r}} = -u\mathbf{r} \tag{3.74}$$
repeatedly, replacing $\dot{u}$, $\dot{z}$, $\dot{q}$, and $\ddot{\mathbf{r}}$ by their corresponding values in terms of u, z, q, and $-u\mathbf{r}$ whenever they appear during the differentiation process. Thus,
$$\begin{aligned} \dddot{\mathbf{r}} &= 3uz\mathbf{r} - u\dot{\mathbf{r}} \\ \ddddot{\mathbf{r}} &= (3uq - 15uz^2 + u^2)\mathbf{r} + 6uz\dot{\mathbf{r}} \\ &\vdots \end{aligned}$$

Evaluating all the derivatives at t_0 and substituting the resulting expressions into the power series of Equation 3.66 yields
$$\begin{aligned} \mathbf{r} = {} & \mathbf{r}_0 + \dot{\mathbf{r}}_0 \tau + (-u)\mathbf{r}_0 \frac{\tau^2}{2} + (3uz\mathbf{r}_0 - u\dot{\mathbf{r}}_0)\frac{\tau^3}{6} \\ & + [(3uq - 15uz^2 + u^2)\mathbf{r}_0 + 6uz\dot{\mathbf{r}}_0]\frac{\tau^4}{24} + \cdots \,, \end{aligned} \tag{3.75}$$
where, for $t = t_0$,
$$u = \frac{\mu}{r_0^3} \tag{3.76}$$
$$z = \frac{\mathbf{r}_0 \cdot \dot{\mathbf{r}}_0}{r_0^2} \tag{3.77}$$
$$q = \frac{\dot{\mathbf{r}}_0 \cdot \dot{\mathbf{r}}_0}{r_0^2} - u \,. \tag{3.78}$$

3.5. TWO-BODY MOTION BY NUMERICAL INTEGRATION

The solution of the two-body equation of motion can be now written in a very convenient form by rearranging and collecting the terms of Equation 3.75 into coefficients of $\mathbf{r}_0$ and $\dot{\mathbf{r}}_0$. The result is

$$\mathbf{r} = f\mathbf{r}_0 + g\dot{\mathbf{r}}_0, \qquad (3.79)$$

where the *the f and g series* are given by

$$f = 1 - \frac{u}{2}\tau^2 + \frac{uz}{2}\tau^3 + \frac{3uq - 15uz^2 + u^2}{24}\tau^4 + \cdots \qquad (3.80)$$

$$g = \tau - \frac{u}{6}\tau^3 + \frac{uz}{4}\tau^4 + \cdots \qquad (3.81)$$

For numerical calculations, it is helpful to reduce the above expressions by defining the quantities:

$$F_0 = 1 \qquad\qquad G_0 = 0$$

$$F_1 = 0 \qquad\qquad G_1 = 1$$

$$F_2 = -u/2 \qquad\qquad G_2 = 0$$

$$F_3 = (uz)/2 \qquad\qquad G_3 = -u/6$$

$$F_4 = (u/24)(3q - 15z^2 + u) \quad G_4 = (uz)/4$$

and so on. The series for f and g then become

$$f = 1 + F_2\tau^2 + F_3\tau^3 + F_4\tau^4 + \cdots \qquad (3.82)$$
$$g = \tau + G_3\tau^3 + G_4\tau^4 + \cdots \qquad (3.83)$$

Finally, an expression for velocity is obtained by differentiating the equation for position. Thus,

$$\dot{\mathbf{r}} = \dot{f}\mathbf{r}_0 + \dot{g}\dot{\mathbf{r}}_0, \qquad (3.84)$$

where, by differentiating Equations 3.82 and 3.83,

$$\dot{f} = 2F_2\tau + 3F_3\tau^2 + 4F_4\tau^3 + \cdots \qquad (3.85)$$
$$\dot{g} = 1 + 3G_3\tau^2 + 4G_4\tau^3 + \cdots \qquad (3.86)$$

The BASIC program FANDG listed in Section 3.6.1 computes two-body orbital motion using an eighth-order f and g series.

3.5.2 Taylor Series

We have shown that a power series solution for two-body orbital motion can be written in the form

$$\mathbf{r} = \mathbf{r}_0 + \dot{\mathbf{r}}_0 \tau + \ddot{\mathbf{r}}_0 \frac{\tau^2}{2!} + \dddot{\mathbf{r}}_0 \frac{\tau^3}{3!} + \ddddot{\mathbf{r}}_0 \frac{\tau^4}{4!} + \cdots \qquad (3.87)$$

where τ represents the modified time interval. An expression of this type is known as a *Taylor series* expansion in time. In the case where

$$\ddot{\mathbf{r}} = -\mu \frac{\mathbf{r}}{r^3}, \qquad (3.88)$$

the Taylor series can be cast into the form of recurrence relations which permit the expansion to be extended to any order [3]. Let new auxiliary quantities now be defined as follows:

$$\begin{aligned} u &= \frac{\mu}{r^3} \\ w &= \frac{1}{r^2} \\ s &= \mathbf{r} \cdot \dot{\mathbf{r}} \\ z &= ws. \end{aligned} \qquad (3.89)$$

Substituting u into Equation 3.88, differentiating the expressions for u and w, and retaining the expressions for s and z, we can obtain the set

$$\begin{aligned} \ddot{\mathbf{r}} &= -u\mathbf{r} \\ \dot{u} &= -3uz \\ \dot{w} &= -2wz \\ s &= \mathbf{r} \cdot \dot{\mathbf{r}} \\ z &= ws. \end{aligned} \qquad (3.90)$$

Notice that the right side of each equation contains the product of two variables.

Now, let $\mathbf{r}$, u, w, s, and z be expressed in the form of infinite series as follows:

$$\begin{aligned} \mathbf{r} &= \mathbf{c}_0 + \mathbf{c}_1 \tau + \mathbf{c}_2 \tau^2 + \mathbf{c}_3 \tau^3 + \cdots \\ u &= u_0 + u_1 \tau + u_2 \tau^2 + u_3 \tau^3 + \cdots \\ w &= w_0 + w_1 \tau + w_2 \tau^2 + w_3 \tau^3 + \cdots \\ s &= s_0 + s_1 \tau + s_2 \tau^2 + s_3 \tau^3 + \cdots \\ z &= z_0 + z_1 \tau + z_2 \tau^2 + z_3 \tau^3 + \cdots \end{aligned} \qquad (3.91)$$

The recurrence relations are developed by substituting Equations 3.91 into Equations 3.90, carrying out the multiplication of the two series, and then equating

3.5. TWO-BODY MOTION BY NUMERICAL INTEGRATION

the constant coefficients of like powers of τ. For example, to find an expression for c_j we first differentiate the series for $\mathbf{r}$ twice with respect to τ to produce

$$\dot{\mathbf{r}} = \mathbf{c}_1 + 2\mathbf{c}_2\tau + 3\mathbf{c}_3\tau^2 + \cdots \qquad (3.92)$$
$$\ddot{\mathbf{r}} = 2\mathbf{c}_2 + 6\mathbf{c}_3\tau + 12\mathbf{c}_4\tau^2 + \cdots . \qquad (3.93)$$

Next we substitute the series for $\mathbf{r}$ and u from Equations 3.91 into the first of Equations 3.90 to obtain

$$\ddot{\mathbf{r}} = -(u_0 + u_1\tau + \cdots)(\mathbf{c}_0 + \mathbf{c}_1\tau + \cdots).$$

Then, carrying out the multiplication and equating coefficients of like powers of τ in the resulting products with those in Equation 3.93, we find that

$$2\mathbf{c}_2 = -\mathbf{c}_0 u_0$$
$$6\mathbf{c}_3 = -(\mathbf{c}_0 u_1 + \mathbf{c}_1 u_0)$$
$$12\mathbf{c}_4 = -(\mathbf{c}_0 u_2 + \mathbf{c}_1 u_1 + \mathbf{c}_2 u_0)$$
$$\vdots$$

From this process, it can be deduced that, in general,

$$\mathbf{c}_{j+2} = -\frac{1}{(j+1)(j+2)} \sum_{n=0}^{j} \mathbf{c}_n u_{j-n} . \qquad (3.94)$$

In a similar manner, we can derive

$$u_j = -\frac{3}{j} \sum_{n=0}^{j-1} u_n z_{j-n-1}$$
$$w_j = -\frac{2}{j} \sum_{n=0}^{j-1} w_n z_{j-n-1}$$
$$s_j = \sum_{n=0}^{j} (n+1)\mathbf{c}_{n+1} \cdot \mathbf{c}_{j-n}$$
$$z_j = \sum_{n=0}^{j} w_n s_{j-n} . \qquad (3.95)$$

Thus, given the position $\mathbf{r}_0$ and velocity $\dot{\mathbf{r}}_0$ at $\tau = 0$, Equations 3.89 through 3.93 yield

$$\mathbf{c}_0 = \mathbf{r}_0$$
$$\mathbf{c}_1 = \dot{\mathbf{r}}_0$$
$$\mathbf{c}_2 = -\frac{1}{2} u_0 \mathbf{r}_0 , \qquad (3.96)$$

as well as u_0, w_0, s_0, and z_0. Therefore, using Equations 3.94 and 3.95, the higher coefficients can be built up by successive computation loops, and we can calculate **r** at the end of the time interval τ from

$$\mathbf{r} = \mathbf{c}_0 + \mathbf{c}_1 \tau + \mathbf{c}_2 \tau^2 + \cdots + \mathbf{c}_m \tau^m, \tag{3.97}$$

where m is the order of the Taylor series expansion and the number of the highest derivative used. The corresponding series for the velocity is found by differentiating Equation 3.97 with respect to modified time:

$$\dot{\mathbf{r}} = \mathbf{c}_1 + 2\mathbf{c}_2 \tau + 3\mathbf{c}_3 \tau^2 + \cdots + m\mathbf{c}_m \tau^{m-1}. \tag{3.98}$$

The recurrence process is ideally suited to implementation by a computer routine. Program TAYLOR in Section 3.6.2 contains a subroutine which will generate the $\mathbf{c}_j$ coefficients.

3.5.3 Runge-Kutta Five

Runge-Kutta (RK) methods can achieve the accuracy of a medium-order Taylor series without requiring the calculation of derivatives beyond the first. Although many variations of the Runge-Kutta method exist, we shall only consider *Butcher's fifth-order (RK5) method* [6]. The basic RK5 procedure will first be presented without derivation and then adapted to the special case of two-body orbital motion.

Given a first-order ordinary differential equation having the general form

$$\frac{dx}{dt} = f(t, x), \tag{3.99}$$

the RK5 numerical integration with step-size h can be represented as follows:

$$x = x_0 + \delta x, \tag{3.100}$$

where the zero subscript indicates the value of the variable at the beginning of the step and δx is the *increment function*

$$\delta x = \frac{1}{90}(7F_1 + 32F_3 + 12F_4 + 32F_5 + 7F_6), \tag{3.101}$$

in which

$$F_1 = hf(t_0, x_0)$$

$$F_2 = hf(t_0 + \frac{1}{4}h, x_0 + \frac{1}{4}F_1)$$

3.5. TWO-BODY MOTION BY NUMERICAL INTEGRATION

$$F_3 = hf(t_0 + \frac{1}{4}h, x_0 + \frac{1}{8}F_1 + \frac{1}{8}F_2)$$

$$F_4 = hf(t_0 + \frac{1}{2}h, x_0 - \frac{1}{2}F_2 + F_3)$$

$$F_5 = hf(t_0 + \frac{3}{4}h, x_0 + \frac{3}{16}F_1 + \frac{9}{16}F_4)$$

$$F_6 = hf(t_0 + h, x_0 - \frac{3}{7}F_1 + \frac{2}{7}F_2 + \frac{12}{7}F_3 - \frac{12}{7}F_4 + \frac{8}{7}F_5). \quad (3.102)$$

Consider now the special case of two-body orbital motion where the position r_0, and velocity v_0 are known for a given epoch t_0. If we let

$$f(\mathbf{v}) = \mathbf{v} \quad (3.103)$$

$$g(\mathbf{r}) = -\frac{\mu \mathbf{r}}{r^3}, \quad (3.104)$$

then the given information can be used to write two simultaneous first-order differential equations as follows:

$$\frac{d\mathbf{r}}{d\tau} = f(\mathbf{v})$$

$$\frac{d\mathbf{v}}{d\tau} = g(\mathbf{r})$$

Since these expressions are similar in form to Equation 3.99, they can be integrated by a Runge-Kutta method. Applying Equations 3.102 to each differential equation yields the following vectors:

$$\mathbf{F}_1 = hf(\mathbf{v}_0)$$

$$\mathbf{G}_1 = hg(\mathbf{r}_0)$$

$$\mathbf{F}_2 = hf(\mathbf{v}_0 + \frac{1}{4}\mathbf{G}_1)$$

$$\mathbf{G}_2 = hg(\mathbf{r}_0 + \frac{1}{4}\mathbf{F}_1)$$

$$\mathbf{F}_3 = hf(\mathbf{v}_0 + \frac{1}{8}\mathbf{G}_1 + \frac{1}{8}\mathbf{G}_2)$$

$$\mathbf{G}_3 = hg(\mathbf{r}_0 + \frac{1}{8}\mathbf{F}_1 + \frac{1}{8}\mathbf{F}_2)$$

$$\mathbf{F}_4 = hf(\mathbf{v}_0 - \frac{1}{2}\mathbf{G}_2 + \mathbf{G}_3)$$

$$\mathbf{G}_4 = hg(\mathbf{r}_0 - \frac{1}{2}\mathbf{F}_2 + \mathbf{F}_3)$$

$$\mathbf{F}_5 = hf(\mathbf{v}_0 + \frac{3}{16}\mathbf{G}_1 + \frac{9}{16}\mathbf{G}_4)$$

$$\mathbf{G}_5 = hg(\mathbf{r}_0 + \frac{3}{16}\mathbf{F}_1 + \frac{9}{16}\mathbf{F}_4)$$

$$\mathbf{F}_6 = hf(\mathbf{v}_0 - \frac{3}{7}\mathbf{G}_1 + \frac{2}{7}\mathbf{G}_2 + \frac{12}{7}\mathbf{G}_3 - \frac{12}{7}\mathbf{G}_4 + \frac{8}{7}\mathbf{G}_5)$$

$$\mathbf{G}_6 = hg(\mathbf{r}_0 - \frac{3}{7}\mathbf{F}_1 + \frac{2}{7}\mathbf{F}_2 + \frac{12}{7}\mathbf{F}_3 - \frac{12}{7}\mathbf{F}_4 + \frac{8}{7}\mathbf{F}_5), \qquad (3.105)$$

where $h = k(t - t_0)$ is the modified time interval. Therefore, according to Equation 3.101, the vector increments are

$$\delta\mathbf{r} = \frac{1}{90}(7\mathbf{F}_1 + 32\mathbf{F}_3 + 12\mathbf{F}_4 + 32\mathbf{F}_5 + 7\mathbf{F}_6) \qquad (3.106)$$

$$\delta\mathbf{v} = \frac{1}{90}(7\mathbf{G}_1 + 32\mathbf{G}_3 + 12\mathbf{G}_4 + 32\mathbf{G}_5 + 7\mathbf{G}_6). \qquad (3.107)$$

Finally, applying the general relationship of Equation 3.100, the position and velocity vectors at the end of the integration step are given by

$$\mathbf{r} = \mathbf{r}_0 + \delta\mathbf{r} \qquad (3.108)$$
$$\mathbf{v} = \mathbf{v}_0 + \delta\mathbf{v}, \qquad (3.109)$$

which become the starting point for the next step.

Program RUNGE in Section 3.6.3 is an example of an RK5 numerical integration procedure for two-body motion. A more extensive RK5 algorithm is used to integrate perturbed orbital motion in Chapter 6.

3.5.4 Numerical Error

Numerical integration is an attempt to approximate the process of analytical integration. Although numerical methods are powerful, they may only be satisfactory for one or two revolutions about the central body. After this, the accumulation of truncation and rounding errors can begin to seriously degrade the accuracy of the results. The first error comes from the fact that an infinite

3.5. TWO-BODY MOTION BY NUMERICAL INTEGRATION

power series has been cut short (truncated), so the solution is not exact. The second arises from the small numerical rounding errors which occur at each step in the computation. Therefore, it is desirable to make the interval τ as large as possible in order to minimize the number of computation steps. Unfortunately, the allowable length of the computation interval is limited by the order of the numerical method being employed. In practice, it is often possible to determine a satisfactory step size through experience and some trial-and-error on the microcomputer. If required, more sophisticated procedures for automatic step-size control can be included in the computer routine [4,6,8].

3.6 Computer Programs

3.6.1 Program FANDG

Program FANDG computes two-body orbital motion by numerical integration using an eighth-order f and g series. This procedure is described in Section 3.5.1.

Program Algorithm

Define	Line Number
FNVS(X,Y,Z): vector squaring function	1070
FNMG(X,Y,Z): vector magnitude function	1080

Given	Line Number
name of the object for information only	1160
equinox for information only	1160
k: gravitational constant	1160
μ: combined mass	1170
t_0: epoch	1180
x_0, y_0, z_0: position vector	1190
$\dot{x}_0, \dot{y}_0, \dot{z}_0$: velocity vector	1200
$dt = t - t_0$: interval between positions	1220
n: number of positions	1240

Compute	Line Number
$h = kdt$: modified time interval equivalent to $\tau = k(t - t_0)$	1330
For $i = 0$ to n:	1370
time at the end of the integration step	1410

3.6. COMPUTER PROGRAMS

r_{i-1}: magnitude of **r** at the beginning of the step	16020
u: Equation 3.76	16030
z: Equation 3.77	16040
q: Equation 3.78	16050
$F_2 = -u/2$	16060
$F_3 = (uz)/2$	16070
$F_4 = (u/24)(3q - 15z^2 + u)$	16080

⋮

$G_3 = -u/6$	16130
$G_4 = (uz)/4$	16140

⋮

f: Equation 3.82	16190
g: Equation 3.83	16200
$\dot{f}$: Equation 3.85	16210
$\dot{g}$: Equation 3.86	16220
$\mathbf{r}_i$: **r** at the end of the step by Equation 3.79	1460
$\dot{\mathbf{r}}_i$: $\dot{\mathbf{r}}$ at the end of the step by Equation 3.84	1470
r_i: magnitude of **r** at the end of the step	1500

End.

Program Listing

```
1000 CLS
1010 PRINT"# FANDG # TWO-BODY MOTION"
1020 PRINT"# BY F&G SERIES"
1030 PRINT
1040 DEFDBL A-Z
1050 DEFINT I,K,N
1060 REM
1070 DEF FNVS(X,Y,Z)=X*X+Y*Y+Z*Z
1080 DEF FNMG(X,Y,Z)=SQR(X*X+Y*Y+Z*Z)
1090 REM
1100 DIM T(100),R(100,3),V(100,3)
1110 REM
1120 REM
1130 G$="###.#######"
1140 J$="####.#####"
1150 REM
1160 READ N$, E$, K#
1170 READ M
1180 READ T(0)
1190 READ R(0,1),R(0,2),R(0,3)
1200 READ V(0,1),V(0,2),V(0,3)
1210 REM
1220 INPUT">>>INTERVAL BETWEEN POSITIONS";DT
1230 PRINT
1240 INPUT">>>NUMBER OF POSITIONS";NP
1250 CLS
1260 PRINT"TWO-BODY MOTION BY F&G SERIES"
1270 PRINT N$
1280 PRINT E$
1290 PRINT
1300 PRINT TAB(4)"T(I)";
1310 PRINT TAB(17)"X";TAB(28)"Y";TAB(39)"Z";TAB(50)"R"
1320 REM
1330 H=K#*DT
1340 REM
1350 REM
1360 REM
1370 FOR I=0 TO NP
1380     REM
1390     IF I=0 THEN 1500
```

3.6. COMPUTER PROGRAMS

```
1400    REM
1410    T(I)=T(I-1)+DT
1420    REM
1430    GOSUB 16010 REM FG/SUB
1440    REM
1450    FOR K=1 TO 3
1460       R(I,K)=F*R(I-1,K)+G*V(I-1,K)
1470       V(I,K)=FP*R(I-1,K)+GP*V(I-1,K)
1480    NEXT K
1490    REM
1500    R(I,0)=FNMG(R(I,1),R(I,2),R(I,3))
1510    REM
1520    PRINT USING J$;T(I);
1530    PRINT USING G$;R(I,1),R(I,2),R(I,3),R(I,0);
1540    REM
1550    REM *** PAUSE FOR SCREEN SPACE ***
1560    REM
1570    LINE INPUT"";A$
1580    REM
1590    REM
1600 NEXT I
1610 REM
1620 I=I-1
1630 REM
1640 PRINT TAB(11);
1650 PRINT USING G$;V(I,1),V(I,2),V(I,3)
1660 END
16000 STOP
16010 REM # FG/SUB # COMPUTES 8TH ORDER F&G SERIES
16020 R=FNMG(R(I-1,1),R(I-1,2),R(I-1,3))
16030 U=M/(R*R*R)
16040 Z=(R(I-1,1)*V(I-1,1)+R(I-1,2)*V(I-1,2)+
     R(I-1,3)*V(I-1,3))/(R*R)
16050 Q=FNVS(V(I-1,1),V(I-1,2),V(I-1,3))/(R*R)-U
16060 F2=-U/2
16070 F3=U*Z/2
16080 F4=(U/24)*(3*Q-15*Z*Z+U)
16090 F5=(U*Z/8)*(7*Z*Z-3*Q-U)
16100 F6=(U/720)*(Z*Z*(630*Q+210*U-945*Z*Z)-U*(24*Q+U)-
     45*Q*Q)
16110 F7=(U*Z/5040)*(Z*Z*(10395*Z*Z-3150*U-9450*Q)+
     U*(882*Q+63*U)+1575*Q*Q)
```

```
16120 F8=(U/40320#)*(Z*Z*(Z*Z*(51975#*U+155925#*Q-
      135135#*Z*Z)-U*(24570#*Q+2205*U)-42525#*Q*Q)+
      Q*Q*(1107*U+1575*Q)+U*U*(117*Q+U))
16130 G3=-U/6
16140 G4=U*Z/4
16150 G5=(U/120)*(9*Q-45*Z*Z+U)
16160 G6=(U*Z/360)*(210*Z*Z-90*Q-15*U)
16170 G7=(U/5040)*(Z*Z*(3150*Q+630*U-4725*Z*Z)-
      Q*(54*U+225*Q)-U*U)
16180 G8=(U*Z/40320#)*(Z*Z*(62370#*Z*Z-56700#*Q-12600#*U)+
      Q*(3024*U+9450*Q)+126*U*U)
16190 F=1+H*H*(F2+H*(F3+H*(F4+H*(F5+H*(F6+H*(F7+H*(F8)))))))
16200 G=H*(1+H*H*(G3+H*(G4+H*(G5+H*(G6+H*(G7+H*(G8)))))))
16210 FP=H*(2*F2+H*(3*F3+H*(4*F4+H*(5*F5+H*(6*F6+H*(7*F7+
      H*(8*F8)))))))
16220 GP=1+H*H*(3*G3+H*(4*G4+H*(5*G5+H*(6*G6+H*(7*G7+
      H*(8*G8))))))
16230 RETURN
30000 DATA "MERCURY", "J2000.0", 0.01720209895#
30010 DATA 1.000000166#
30020 DATA 6280.5#
30030 DATA +0.1693419#,-0.3559908#,-0.2077172#
30040 DATA +1.1837591#,+0.6697770#,+0.2349312#
```

3.6. COMPUTER PROGRAMS

3.6.2 Program TAYLOR

Program TAYLOR computes two-body orbital motion by numerical integration of a Taylor series using subroutine RECUR/SUB as part of a program similar to FANDG. The recurrence algorithm of RECUR/SUB is discussed in Section 3.5.2.

Program Algorithm

Define	Line Number
FNMG(X,Y,Z): vector magnitude function	1070

Given	Line Number
m: order of the Taylor series	1090
$dt = t - t_0$: interval between positions	1110
n: number of positions	1130
name of the object for information only	1230
equinox for information only	1230
k: gravitational constant	1230
μ: combined mass	1240
t_0: epoch	1250
x_0, y_0, z_0: position vector	1260
$\dot{x}_0, \dot{y}_0, \dot{z}_0$: velocity vector	1260

Compute	Line Number
$h = kdt$: modified time interval equivalent to $\tau = k(t - t_0)$	1360
$h^0, h^1, h^2, h^3, \ldots, h^m$: powers of h	1370-1400
For $i = 0$ to n:	1440

time at the end of the integration step	1480
r_{i-1}: magnitude of **r** at the beginning of the step	16020
u_0: Equations 3.89	16030
w_0: Equations 3.89	16040
s_0: Equations 3.89	16050
z_0: Equations 3.89	16060
$\mathbf{c}_0, \mathbf{c}_1, \mathbf{c}_2$: Equations 3.96	16070-16110
u_j, w_j, s_j, z_j: for $j = 1, 2, 3, \ldots$ by Equations 3.95	16130-16280
$\mathbf{c}_j$: for $j = 3, 4, 5, \ldots$ by Equation 3.94	16290-16340
$\mathbf{r}_i$: **r** at the end of the step by Equation 3.97	1560
$\dot{\mathbf{r}}_i$: $\dot{\mathbf{r}}$ at the end of the step by Equation 3.98	1570
r_i: magnitude of **r** at the end of the step	1610

End.

3.6. COMPUTER PROGRAMS

Program Listing

```
1000 CLS
1010 PRINT"# TAYLOR # TWO-BODY MOTION"
1020 PRINT"# BY TAYLOR SERIES"
1030 PRINT
1040 DEFDBL A-Z
1050 DEFINT I,J,K,N
1060 REM
1070 DEF FNMG(X,Y,Z)=SQR(X*X+Y*Y+Z*Z)
1080 REM
1090 INPUT">>>ORDER OF TAYLOR SERIES";NM
1100 PRINT
1110 INPUT">>>INTERVAL BETWEEN POSITIONS";DT
1120 PRINT
1130 INPUT">>>NUMBER OF POSITIONS";NP
1140 REM
1150 DIM T(NP),R(NP,3),V(NP,3)
1160 DIM H(NM),U(NM-2),W(NM-2),S(NM-2),Z(NM-2)
1170 DIM CJ(NM,3)
1180 REM
1190 REM
1200 G$="###.#######"
1210 J$="####.#####"
1220 REM
1230 READ N$, E$, K#
1240 READ M
1250 READ T(0)
1260 READ R(0,1),R(0,2),R(0,3)
1270 READ V(0,1),V(0,2),V(0,3)
1280 CLS
1290 PRINT"TWO-BODY MOTION BY TAYLOR SERIES"
1300 PRINT N$
1310 PRINT E$
1320 PRINT
1330 PRINT TAB(4)"T(I)";
1340 PRINT TAB(17)"X";TAB(28)"Y";TAB(39)"Z";TAB(50)"R"
1350 REM
1360 H=K#*DT
1370 H(0)=1
1380 FOR N=1 TO NM
1390    H(N)=H(N-1)*H
```

```
1400 NEXT N
1410 REM
1420 REM
1430 REM
1440 FOR I=0 TO NP
1450   REM
1460   IF I=0 THEN 1610
1470   REM
1480   T(I)=T(I-1)+DT
1490   REM
1500   GOSUB 16010 REM RECUR/SUB
1510   REM
1520   FOR K=1 TO 3
1530     R(I,K)=0
1540     V(I,K)=0
1550     FOR N=0 TO NM
1560       R(I,K)=R(I,K)+CJ(N,K)*H(N)
1570       V(I,K)=V(I,K)+N*CJ(N,K)*H(N)/H
1580     NEXT N
1590   NEXT K
1600   REM
1610   R(I,0)=FNMG(R(I,1),R(I,2),R(I,3))
1620   REM
1630   PRINT USING J$;T(I);
1640   PRINT USING G$;R(I,1),R(I,2),R(I,3),R(I,0);
1650   REM
1660   REM *** PAUSE FOR SCREEN SPACE ***
1670   REM
1680   LINE INPUT"";A$
1690   REM
1700   REM
1710 NEXT I
1720 REM
1730 I=I-1
1740 REM
1750 PRINT TAB(11);
1760 PRINT USING G$;V(I,1),V(I,2),V(I,3)
1770 REM
1780 PRINT"ORDER";NM
1790 END
16000 STOP
16010 REM # RECUR/SUB # TAYLOR RECURRENCE RELATIONS
```

3.6. COMPUTER PROGRAMS

```
16020 R=FNMG(R(I-1,1),R(I-1,2),R(I-1,3))
16030 U(0)=M/(R*R*R)
16040 W(0)=1/(R*R)
16050 S(0)=V(I-1,1)*R(I-1,1)+V(I-1,2)*R(I-1,2)+
      V(I-1,3)*R(I-1,3)
16060 Z(0)=W(0)*S(0)
16070 FOR K=1 TO 3
16080   CJ(0,K)=R(I-1,K)
16090   CJ(1,K)=V(I-1,K)
16100   CJ(2,K)=(-1/2)*U(0)*R(I-1,K)
16110 NEXT K
16120 FOR J=1 TO (NM-2)
16130   U(J)=0
16140   W(J)=0
16150   S(J)=0
16160   Z(J)=0
16170   FOR N=0 TO (J-1)
16180     U(J)=U(J)+(-3/J)*U(N)*Z(J-N-1)
16190   NEXT N
16200   FOR N=0 TO (J-1)
16210     W(J)=W(J)+(-2/J)*W(N)*Z(J-N-1)
16220   NEXT N
16230   FOR N=0 TO J
16240     S(J)=S(J)+(N+1)*(CJ(N+1,1)*CJ(J-N,1)+
          CJ(N+1,2)*CJ(J-N,2)+CJ(N+1,3)*CJ(J-N,3))
16250   NEXT N
16260   FOR N=0 TO J
16270     Z(J)=Z(J)+W(N)*S(J-N)
16280   NEXT N
16290   FOR K=1 TO 3
16300     CJ(J+2,K)=0
16310     FOR N=0 TO J
16320       CJ(J+2,K)=CJ(J+2,K)+
            (-1/((J+1)*(J+2)))*CJ(N,K)*U(J-N)
16330     NEXT N
16340   NEXT K
16350 NEXT J
16360 RETURN
```

```
30000 DATA "MERCURY", "J2000.0", 0.01720209895#
30010 DATA 1.000000166#
30020 DATA 6280.5#
30030 DATA +0.1693419#,-0.3559908#,-0.2077172#
30040 DATA +1.1837591#,+0.6697770#,+0.2349312#
```

3.6.3 Program RUNGE

Program RUNGE computes two-body orbital motion by a fifth-order Runge-Kutta (RK5) numerical integration. RUNGE uses subroutine RK5/SUB as part of a program similar to FANDG and TAYLOR. The RK5/SUB algorithm is discussed in Section 3.5.3.

Program Algorithm

Define	Line Number
FNMG(X,Y,Z): vector magnitude function	1070
FNF(VK): function for Equation 3.103	1080
FNG(M,RK,R): function for Equation 3.104	1090

Given	Line Number
name of the object for information only	1170
equinox for information only	1170
k: gravitational constant	1170
μ: combined mass	1180
t_0: epoch	1180
x_0, y_0, z_0: position vector	1200
$\dot{x}_0, \dot{y}_0, \dot{z}_0$: velocity vector	1210
$dt = t - t_0$: interval between positions	1230
n: number of positions	1250

Compute	Line Number
$h = kdt$: modified time interval equivalent to $\tau = k(t - t_0)$	1340
For $i = 0$ to n:	1360

time at the end of the integration step	1400
F_1, G_1: Equations 3.105	16050-16150
F_2, G_2: Equations 3.105	16190-16290
F_3, G_3: Equations 3.105	16330-16430
F_4, G_4: Equations 3.105	16470-16570
F_5, G_5: Equations 3.105	16610-16710
F_6, G_6: Equations 3.105	16750-16850
position by Equations 3.106 and 3.108	16900
velocity by Equations 3.107 and 3.109	16910
r_i: r at the end of the step	1450
$\dot{r}_i$: $\dot{r}$ at the end of the step	1460
r_i: magnitude of r at the end of the step	1490

End.

3.6. COMPUTER PROGRAMS

Program Listing

```
1000 CLS
1010 PRINT"# RUNGE # TWO-BODY MOTION"
1020 PRINT"# BY RK5 INTEGRATION"
1030 PRINT
1040 DEFDBL A-Z
1050 DEFINT I,K,N
1060 REM
1070 DEF FNMG(X,Y,Z)=SQR(X*X+Y*Y+Z*Z)
1080 DEF FNF(VK)=VK
1090 DEF FNG(M,RK,R)=-M*RK/(R*R*R)
1100 REM
1110 DIM T(200),R(200,3),V(200,3)
1120 REM
1130 REM
1140 C$="###.#######"
1150 J$="####.#####"
1160 REM
1170 READ N$, E$, K#
1180 READ M
1190 READ T(0)
1200 READ R(0,1),R(0,2),R(0,3)
1210 READ V(0,1),V(0,2),V(0,3)
1220 REM
1230 INPUT">>>INTERVAL BETWEEN POSITIONS";DT
1240 PRINT
1250 INPUT">>>NUMBER OF POSITIONS";NP
1260 CLS
1270 PRINT"TWO-BODY MOTION BY RK5 INTEGRATION"
1280 PRINT N$
1290 PRINT E$
1300 PRINT
1310 PRINT TAB(4)"T(I)";
1320 PRINT TAB(17)"X";TAB(28)"Y";TAB(39)"Z";TAB(50)"R"
1330 REM
1340 H=K#*DT
1350 REM
1360 FOR I=0 TO NP
1370    REM
1380    IF I=0 THEN 1490
1390    REM
```

```
1400    T(I)=T(I-1)+DT
1410    REM
1420    GOSUB 16010 REM RK5/SUB
1430    REM
1440    FOR K=1 TO 3
1450       R(I,K)=RK(K)
1460       V(I,K)=VK(K)
1470    NEXT K
1480    REM
1490    R(I,0)=FNMG(R(I,1),R(I,2),R(I,3))
1500    REM
1510    PRINT USING J$;T(I);
1520    PRINT USING G$;R(I,1),R(I,2),R(I,3),R(I,0);
1530    REM
1540    REM *** PAUSE FOR SCREEN SPACE ***
1550    REM
1560    LINE INPUT"";A$
1570    REM
1580    REM
1590 NEXT I
1600 REM
1610 I=I-1
1620 REM
1630 PRINT TAB(11);
1640 PRINT USING G$;V(I,1),V(I,2),V(I,3)
1650 END
16000 STOP
16010 REM # RK5/SUB # RUNGE-KUTTA 5 INTEGRATION
16020 REM
16030 REM ********** STEP 1 **********
16040 REM
16050 FOR K=1 TO 3
16060    RK(K)=R(I-1,K)
16070    VK(K)=V(I-1,K)
16080 NEXT K
16090 REM
16100 GOSUB 17010 REM RK5/SUBSUB
16110 REM
16120 FOR K=1 TO 3
16130    F1(K)=FX(K)
16140    G1(K)=GX(K)
16150 NEXT K
```

3.6. COMPUTER PROGRAMS

```
16160 REM
16170 REM ********** STEP 2 **********
16180 REM
16190 FOR K=1 TO 3
16200   RK(K)=R(I-1,K)+F1(K)/4
16210   VK(K)=V(I-1,K)+G1(K)/4
16220 NEXT K
16230 REM
16240 GOSUB 17010 REM RK5/SUBSUB
16250 REM
16260 FOR K=1 TO 3
16270   F2(K)=FX(K)
16280   G2(K)=GX(K)
16290 NEXT K
16300 REM
16310 REM ********** STEP 3 **********
16320 REM
16330 FOR K=1 TO 3
16340   RK(K)=R(I-1,K)+(F1(K)+F2(K))/8
16350   VK(K)=V(I-1,K)+(G1(K)+G2(K))/8
16360 NEXT K
16370 REM
16380 GOSUB 17010 REM RK5/SUBSUB
16390 REM
16400 FOR K=1 TO 3
16410   F3(K)=FX(K)
16420   G3(K)=GX(K)
16430 NEXT K
16440 REM
16450 REM ********** STEP 4 **********
16460 REM
16470 FOR K=1 TO 3
16480   RK(K)=R(I-1,K)-(F2(K)-2*F3(K))/2
16490   VK(K)=V(I-1,K)-(G2(K)-2*G3(K))/2
16500 NEXT K
16510 REM
16520 GOSUB 17010 REM RK5/SUBSUB
16530 REM
16540 FOR K=1 TO 3
16550   F4(K)=FX(K)
16560   G4(K)=GX(K)
16570 NEXT K
```

```
16580 REM
16590 REM ********** STEP 5 **********
16600 REM
16610 FOR K=1 TO 3
16620   RK(K)=R(I-1,K)+(3*F1(K)+9*F4(K))/16
16630   VK(K)=V(I-1,K)+(3*G1(K)+9*G4(K))/16
16640 NEXT K
16650 REM
16660 GOSUB 17010 REM RK5/SUBSUB
16670 REM
16680 FOR K=1 TO 3
16690   F5(K)=FX(K)
16700   G5(K)=GX(K)
16710 NEXT K
16720 REM
16730 REM ********** STEP 6 **********
16740 REM
16750 FOR K=1 TO 3
16760   RK(K)=R(I-1,K)-(3*F1(K)-2*F2(K)-12*F3(K)+12*F4(K)-
        8*F5(K))/7
16770   VK(K)=V(I-1,K)-(3*G1(K)-2*G2(K)-12*G3(K)+12*G4(K)-
        8*G5(K))/7
16780 NEXT K
16790 REM
16800 GOSUB 17010 REM RK5/SUBSUB
16810 REM
16820 FOR K=1 TO 3
16830   F6(K)=FX(K)
16840   G6(K)=GX(K)
16850 NEXT K
16860 REM
16870 REM ********** RESULT **********
16880 REM
16890 FOR K=1 TO 3
16900   RK(K)=R(I-1,K)+(7*F1(K)+32*F3(K)+12*F4(K)+32*F5(K)+
        7*F6(K))/90
16910   VK(K)=V(I-1,K)+(7*G1(K)+32*G3(K)+12*G4(K)+32*G5(K)+
        7*G6(K))/90
16920 NEXT K
16930 RETURN
17000 STOP
17010 REM # RK5/SUBSUB #
```

3.6. COMPUTER PROGRAMS

```
17020 REM
17030 RK=FNMG(RK(1),RK(2),RK(3))
17040 REM
17050 FOR K=1 TO 3
17060   FX(K)=H*FNF(VK(K))
17070   GX(K)=H*FNG(M,RK(K),RK)
17080 NEXT K
17090 RETURN
30000 DATA  "MARS", "J2000.0", 0.017202099#
30010 DATA  1.000000323#
30020 DATA  6280.5#
30030 DATA -0.8888462#,+1.2418782#,+0.5936583#
30040 DATA -0.6523424#,-0.3449870#,-0.1405741#
```

3.7 Numerical Examples

3.7.1 Two-Body Motion by f and g Series

Problem

Compute the heliocentric motion of Mercury for 100 days using an interval of two days between positions. Assume the following data for the arbitrary epoch time 1985 August 3:

- equinox of coordinates: J2000.0
- $k = 0.01720209895$
- mass of Sun: 1
- mass of Mercury: 0.000000166
- $t_0 = 2446280.5$ (this JD corresponds to 1985 August 3)
- $\mathbf{r}_0 = \{+0.1693419, -0.3559908, -0.2077172\}$
- $\dot{\mathbf{r}}_0 = \{+1.1837591, +0.6697770, +0.2349312\}$

Solution

Use the given information to write data lines 30000 to 30040 as shown at the end of program FANDG. Run the program and answer the two prompts as follows:

```
>>>INTERVAL BETWEEN POSITIONS? 2

>>>NUMBER OF POSITIONS? 50
```

Results

```
TWO-BODY MOTION BY F&G SERIES
MERCURY
J2000.0
```

T(I)	X	Y	Z	R
6280.50000	0.1693419	-0.3559908	-0.2077172	0.4455924
6282.50000	0.2088258	-0.3305839	-0.1982431	0.4383997
6284.50000	0.2453725	-0.3005380	-0.1859864	0.4302577
6286.50000	0.2782687	-0.2660321	-0.1709687	0.4212326

3.7. NUMERICAL EXAMPLES

6288.50000	0.3067543	−0.2273212	−0.1532474	0.4114096
6290.50000	0.3300221	−0.1847563	−0.1329261	0.4008976
6292.50000	0.3472251	−0.1388099	−0.1101694	0.3898343
6294.50000	0.3574907	−0.0901055	−0.0852198	0.3783927
6296.50000	0.3599502	−0.0394507	−0.0584184	0.3667877
6298.50000	0.3537854	0.0121300	−0.0302278	0.3552815
6300.50000	0.3383010	0.0633693	−0.0012525	0.3441872
6302.50000	0.3130249	0.1127499	0.0277460	0.3338667
6304.50000	0.2778334	0.1585318	0.0558510	0.3247200
6306.50000	0.2330881	0.1988329	0.0820197	0.3171621
6308.50000	0.1797503	0.2317740	0.1051488	0.3115857
6310.50000	0.1194295	0.2556827	0.1241778	0.3083134
6312.50000	0.0543192	0.2693176	0.1382161	0.3075488
6314.50000	−0.0129915	0.2720540	0.1466615	0.3093409
6316.50000	−0.0797985	0.2639679	0.1492740	0.3135755
6318.50000	−0.1435777	0.2457912	0.1461826	0.3199957
6320.50000	−0.2022161	0.2187609	0.1378287	0.3282445
6322.50000	−0.2541331	0.1844178	0.1248715	0.3379148
6324.50000	−0.2982951	0.1444136	0.1080859	0.3485941
6326.50000	−0.3341546	0.1003628	0.0882773	0.3598956
6328.50000	−0.3615530	0.0537484	0.0662217	0.3714765
6330.50000	−0.3806193	0.0058771	0.0426301	0.3830443
6332.50000	−0.3916798	−0.0421331	0.0181338	0.3943565
6334.50000	−0.3951868	−0.0893435	−0.0067191	0.4052160
6336.50000	−0.3916666	−0.1349760	−0.0314583	0.4154647
6338.50000	−0.3816830	−0.1783904	−0.0556833	0.4249772
6340.50000	−0.3658130	−0.2190615	−0.0790537	0.4336549
6342.50000	−0.3446328	−0.2565596	−0.1012803	0.4414208
6344.50000	−0.3187087	−0.2905323	−0.1221161	0.4482149
6346.50000	−0.2885940	−0.3206901	−0.1413490	0.4539914
6348.50000	−0.2548276	−0.3467948	−0.1587958	0.4587154
6350.50000	−0.2179354	−0.3686496	−0.1742969	0.4623611
6352.50000	−0.1784327	−0.3860921	−0.1877122	0.4649099
6354.50000	−0.1368280	−0.3989888	−0.1989175	0.4663498
6356.50000	−0.0936266	−0.4072311	−0.2078023	0.4666743
6358.50000	−0.0493354	−0.4107324	−0.2142678	0.4658817
6360.50000	−0.0044676	−0.4094275	−0.2182261	0.4639757
6362.50000	0.0404517	−0.4032726	−0.2195992	0.4609652
6364.50000	0.0848803	−0.3922467	−0.2183195	0.4568649
6366.50000	0.1282528	−0.3763545	−0.2143310	0.4516960
6368.50000	0.1699753	−0.3556317	−0.2075911	0.4454880
6370.50000	0.2094188	−0.3301516	−0.1980737	0.4382801

```
6372.50000  0.2459142  -0.3000347  -0.1857738  0.4301238
6374.50000  0.2787478  -0.2654612  -0.1707135  0.4210856
6376.50000  0.3071584  -0.2266874  -0.1529508  0.4112510
6378.50000  0.3303385  -0.1840659  -0.1325902  0.4007293
6380.50000  0.3474402  -0.1380714  -0.1097973  0.3896586
            0.4003164   1.3802560   0.6957087
```

Discussion of Results

The FANDG numerical integration yields 100 days of two-body motion which carries the planet more than once around the Sun. Its final position and velocity are very close to that computed by an analytical procedure which will be developed in Chapter 5:

- $t = 2446380.5$ (this JD corresponds to 1985 November 11)
- $\mathbf{r} = \{+0.3474402, -0.1380714, -0.1097973\}$
- $\dot{\mathbf{r}} = \{+0.4003166, +1.3802559, +0.6957087\}$
- $r = 0.3896586$

The table below compares the FANDG computation with the corresponding values of Mercury's radius vector given in Reference 9. The column on the right is the residual found by subtracting the computed values from those of the reference:

Date	Reference	Computed	Residual
6280.5	0.4455924	0.4455924	0.000000
6290.5	0.4008978	0.4008976	0.000002
6300.5	0.3441876	0.3441872	0.000004
6310.5	0.3083137	0.3083134	0.000003
6320.5	0.3282443	0.3282445	−0.000002
6330.5	0.3830440	0.3830443	−0.000003
6340.5	0.4336545	0.4336549	−0.000004
6350.5	0.4623603	0.4623611	−0.000008
6360.5	0.4639746	0.4639757	−0.000011
6370.5	0.4382787	0.4382801	−0.000014
6380.5	0.3896576	0.3896586	−0.000010

It can be seen from the residuals that, at least for Mercury, the two-body motion agrees fairly well with the actual perturbed orbit.

3.7. NUMERICAL EXAMPLES

3.7.2 Two-Body Motion by Taylor Series

Problem

Compute the heliocentric motion of Mercury for 100 days using an interval of four days between positions. As in the previous numerical example, assume the following data for 1985 August 3:

- equinox of coordinates: J2000.0
- $k = 0.01720209895$
- mass of Sun: 1
- mass of Mercury: 0.000000166
- $t_0 = 2446280.5$
- $\mathbf{r}_0 = \{+0.1693419, -0.3559908, -0.2077172\}$
- $\dot{\mathbf{r}}_0 = \{+1.1837591, +0.6697770, +0.2349312\}$

Solution

Use the given information to write data lines 30000 to 30040 as shown at the end of program TAYLOR. Run the program and answer the three prompts as follows:

```
>>>ORDER OF TAYLOR SERIES? 10

>>>INTERVAL BETWEEN POSITIONS? 4

>>>NUMBER OF POSITIONS? 25
```

Results

```
TWO-BODY MOTION BY TAYLOR SERIES
MERCURY
J2000.0
```

T(I)	X	Y	Z	R
6280.50000	0.1693419	-0.3559908	-0.2077172	0.4455924
6284.50000	0.2453725	-0.3005380	-0.1859864	0.4302577
6288.50000	0.3067543	-0.2273212	-0.1532474	0.4114096

6292.50000	0.3472251	-0.1388099	-0.1101694	0.3898343
6296.50000	0.3599502	-0.0394507	-0.0584184	0.3667877
6300.50000	0.3383010	0.0633693	-0.0012525	0.3441872
6304.50000	0.2778334	0.1585318	0.0558510	0.3247200
6308.50000	0.1797503	0.2317740	0.1051488	0.3115857
6312.50000	0.0543192	0.2693176	0.1382161	0.3075487
6316.50000	-0.0797985	0.2639679	0.1492740	0.3135755
6320.50000	-0.2022161	0.2187609	0.1378287	0.3282445
6324.50000	-0.2982951	0.1444136	0.1080858	0.3485941
6328.50000	-0.3615530	0.0537484	0.0662216	0.3714765
6332.50000	-0.3916797	-0.0421331	0.0181337	0.3943565
6336.50000	-0.3916666	-0.1349761	-0.0314583	0.4154647
6340.50000	-0.3658130	-0.2190616	-0.0790538	0.4336549
6344.50000	-0.3187087	-0.2905323	-0.1221161	0.4482149
6348.50000	-0.2548275	-0.3467948	-0.1587958	0.4587154
6352.50000	-0.1784327	-0.3860921	-0.1877122	0.4649099
6356.50000	-0.0936266	-0.4072311	-0.2078023	0.4666743
6360.50000	-0.0044675	-0.4094275	-0.2182261	0.4639757
6364.50000	0.0848803	-0.3922466	-0.2183195	0.4568648
6368.50000	0.1699754	-0.3556316	-0.2075911	0.4454880
6372.50000	0.2459142	-0.3000346	-0.1857738	0.4301238
6376.50000	0.3071585	-0.2266873	-0.1529507	0.4112510
6380.50000	0.3474402	-0.1380713	-0.1097972	0.3896586
	0.4003160	1.3802561	0.6957088	

ORDER 10

3.7. NUMERICAL EXAMPLES

3.7.3 Two-Body Motion by Runge-Kutta Five

Problem

Compute the heliocentric motion of Mars for 720 days using an interval of five days between positions. Assume the following data for the epoch time 1985 August 3:

- equinox of coordinates: J2000.0
- $k = 0.01720209895$
- mass of Sun: 1
- mass of Mars: 0.000000323
- $t_0 = 2446280.5$
- $\mathbf{r}_0 = \{-0.8888462, +1.2418782, +0.5936583\}$
- $\dot{\mathbf{r}}_0 = \{-0.6523424, -0.3449870, -0.1405741\}$

Solution

Use the given information to write data lines 30000 to 30040 as shown at the end of program RUNGE. Run the program and answer the prompts as follows:

>>>INTERVAL BETWEEN POSITIONS? 5

>>>NUMBER OF POSITIONS? 144

Results

```
TWO-BODY MOTION BY RK5 INTEGRATION
MARS
J2000.0
```

T(I)	X	Y	Z	R
6280.50000	-0.8888462	1.2418782	0.5936583	1.6385174
6285.50000	-0.9441931	1.2111721	0.5810728	1.6419757
6290.50000	-0.9979625	1.1784422	0.5675163	1.6452143
6295.50000	-1.0500742	1.1437548	0.5530171	1.6482290
6300.50000	-1.1004513	1.1071778	0.5376044	1.6510161
6305.50000	-1.1490197	1.0687810	0.5213079	1.6535722

6310.50000	-1.1957083	1.0286355	0.5041586	1.6558941
6315.50000	-1.2404490	0.9868143	0.4861879	1.6579792
6320.50000	-1.2831765	0.9433913	0.4674281	1.6598247
6325.50000	-1.3238283	0.8984423	0.4479122	1.6614287
6330.50000	-1.3623449	0.8520441	0.4276738	1.6627892
6335.50000	-1.3986696	0.8042750	0.4067472	1.6639045
6340.50000	-1.4327484	0.7552144	0.3851676	1.6647735
6345.50000	-1.4645301	0.7049430	0.3629704	1.6653950
6350.50000	-1.4939665	0.6535427	0.3401919	1.6657684
6355.50000	-1.5210120	0.6010964	0.3168690	1.6658933
6360.50000	-1.5456240	0.5476884	0.2930391	1.6657695
6365.50000	-1.5677623	0.4934039	0.2687402	1.6653971
6370.50000	-1.5873900	0.4383293	0.2440110	1.6647766
6375.50000	-1.6044728	0.3825519	0.2188906	1.6639086
6380.50000	-1.6189791	0.3261604	0.1934187	1.6627943
6385.50000	-1.6308806	0.2692440	0.1676356	1.6614348
6390.50000	-1.6401517	0.2118934	0.1415822	1.6598319
6395.50000	-1.6467696	0.1542002	0.1152998	1.6579873
6400.50000	-1.6507148	0.0962567	0.0888302	1.6559033
6405.50000	-1.6519708	0.0381564	0.0622160	1.6535823
6410.50000	-1.6505242	-0.0200061	0.0355000	1.6510272
6415.50000	-1.6463648	-0.0781358	0.0087257	1.6482410
6420.50000	-1.6394857	-0.1361362	-0.0180629	1.6452272
6425.50000	-1.6298833	-0.1939105	-0.0448216	1.6419896
6430.50000	-1.6175576	-0.2513606	-0.0715053	1.6385322
6435.50000	-1.6025120	-0.3083880	-0.0980687	1.6348594
6440.50000	-1.5847536	-0.3648933	-0.1244660	1.6309760
6445.50000	-1.5642933	-0.4207762	-0.1506511	1.6268872
6450.50000	-1.5411459	-0.4759362	-0.1765773	1.6225984
6455.50000	-1.5153300	-0.5302721	-0.2021977	1.6181154
6460.50000	-1.4868685	-0.5836819	-0.2274650	1.6134444
6465.50000	-1.4557885	-0.6360637	-0.2523317	1.6085921
6470.50000	-1.4221215	-0.6873150	-0.2767499	1.6035654
6475.50000	-1.3859035	-0.7373331	-0.3006715	1.5983717
6480.50000	-1.3471751	-0.7860155	-0.3240484	1.5930187
6485.50000	-1.3059819	-0.8332596	-0.3468324	1.5875147
6490.50000	-1.2623743	-0.8789629	-0.3689750	1.5818682
6495.50000	-1.2164078	-0.9230236	-0.3904281	1.5760883
6500.50000	-1.1681433	-0.9653402	-0.4111434	1.5701845
6505.50000	-1.1176470	-1.0058122	-0.4310731	1.5641665
6510.50000	-1.0649906	-1.0443399	-0.4501695	1.5580447
6515.50000	-1.0102516	-1.0808251	-0.4683854	1.5518299

3.7. NUMERICAL EXAMPLES

```
6520.50000  -0.9535133  -1.1151708  -0.4856742  1.5455332
6525.50000  -0.8948647  -1.1472821  -0.5019898  1.5391663
6530.50000  -0.8344012  -1.1770659  -0.5172871  1.5327411
6535.50000  -0.7722239  -1.2044318  -0.5315217  1.5262702
6540.50000  -0.7084403  -1.2292918  -0.5446504  1.5197665
6545.50000  -0.6431642  -1.2515614  -0.5566315  1.5132431
6550.50000  -0.5765156  -1.2711593  -0.5674243  1.5067138
6555.50000  -0.5086206  -1.2880084  -0.5769900  1.5001926
6560.50000  -0.4396117  -1.3020357  -0.5852917  1.4936940
6565.50000  -0.3696274  -1.3131731  -0.5922943  1.4872326
6570.50000  -0.2988124  -1.3213579  -0.5979652  1.4808234
6575.50000  -0.2273170  -1.3265329  -0.6022739  1.4744817
6580.50000  -0.1552976  -1.3286471  -0.6051930  1.4682231
6585.50000  -0.0829156  -1.3276564  -0.6066979  1.4620632
6590.50000  -0.0103377  -1.3235238  -0.6067669  1.4560179
6595.50000   0.0622646  -1.3162198  -0.6053820  1.4501030
6600.50000   0.1347155  -1.3057232  -0.6025288  1.4443340
6605.50000   0.2068354  -1.2920216  -0.5981965  1.4387285
6610.50000   0.2784412  -1.2751112  -0.5923786  1.4333005
6615.50000   0.3493474  -1.2549980  -0.5850728  1.4280664
6620.50000   0.4193660  -1.2316978  -0.5762812  1.4230415
6625.50000   0.4883080  -1.2052365  -0.5660106  1.4182410
6630.50000   0.5559836  -1.1756506  -0.5542725  1.4136796
6635.50000   0.6222036  -1.1429871  -0.5410836  1.4093716
6640.50000   0.6867795  -1.1073042  -0.5264652  1.4053306
6645.50000   0.7495251  -1.0686707  -0.5104439  1.4015698
6650.50000   0.8102570  -1.0271667  -0.4930516  1.3981015
6655.50000   0.8687961  -0.9828831  -0.4743250  1.3949372
6660.50000   0.9249677  -0.9359215  -0.4543061  1.3920877
6665.50000   0.9786035  -0.8863943  -0.4330417  1.3895628
6670.50000   1.0295417  -0.8344239  -0.4105838  1.3873710
6675.50000   1.0776285  -0.7801427  -0.3869888  1.3855201
6680.50000   1.1227187  -0.7236923  -0.3623178  1.3840167
6685.50000   1.1646769  -0.6652232  -0.3366361  1.3828659
6690.50000   1.2033780  -0.6048937  -0.3100130  1.3820720
6695.50000   1.2387078  -0.5428697  -0.2825214  1.3816377
6700.50000   1.2705641  -0.4793235  -0.2542376  1.3815647
6705.50000   1.2988571  -0.4144330  -0.2252408  1.3818531
6710.50000   1.3235098  -0.3483809  -0.1956127  1.3825020
6715.50000   1.3444582  -0.2813537  -0.1654370  1.3835090
6720.50000   1.3616518  -0.2135407  -0.1347993  1.3848705
6725.50000   1.3750540  -0.1451329  -0.1037862  1.3865817
```

6730.50000	1.3846416	-0.0763224	-0.0724852	1.3886366
6735.50000	1.3904048	-0.0073009	-0.0409838	1.3910279
6740.50000	1.3923477	0.0617408	-0.0093698	1.3937474
6745.50000	1.3904871	0.1306140	0.0222699	1.3967857
6750.50000	1.3848528	0.1991330	0.0538493	1.4001325
6755.50000	1.3754870	0.2671156	0.0852837	1.4037765
6760.50000	1.3624437	0.3343843	0.1164902	1.4077059
6765.50000	1.3457882	0.4007668	0.1473880	1.4119076
6770.50000	1.3255963	0.4660964	0.1778986	1.4163683
6775.50000	1.3019538	0.5302130	0.2079462	1.4210739
6780.50000	1.2749558	0.5929635	0.2374581	1.4260099
6785.50000	1.2447058	0.6542019	0.2663645	1.4311614
6790.50000	1.2113152	0.7137899	0.2945989	1.4365129
6795.50000	1.1749022	0.7715973	0.3220984	1.4420489
6800.50000	1.1355911	0.8275019	0.3488037	1.4477536
6805.50000	1.0935120	0.8813898	0.3746588	1.4536113
6810.50000	1.0487994	0.9331555	0.3996119	1.4596058
6815.50000	1.0015920	0.9827019	0.4236147	1.4657213
6820.50000	0.9520317	1.0299403	0.4466225	1.4719419
6825.50000	0.9002632	1.0747902	0.4685946	1.4782519
6830.50000	0.8464330	1.1171794	0.4894939	1.4846356
6835.50000	0.7906894	1.1570437	0.5092868	1.4910778
6840.50000	0.7331814	1.1943266	0.5279436	1.4975632
6845.50000	0.6740585	1.2289794	0.5454378	1.5040770
6850.50000	0.6134700	1.2609608	0.5617464	1.5106047
6855.50000	0.5515648	1.2902367	0.5768496	1.5171321
6860.50000	0.4884911	1.3167797	0.5907311	1.5236455
6865.50000	0.4243955	1.3405692	0.6033773	1.5301312
6870.50000	0.3594234	1.3615910	0.6147778	1.5365764
6875.50000	0.2937183	1.3798368	0.6249249	1.5429683
6880.50000	0.2274214	1.3953043	0.6338137	1.5492948
6885.50000	0.1606720	1.4079965	0.6414419	1.5555441
6890.50000	0.0936065	1.4179218	0.6478096	1.5617048
6895.50000	0.0263590	1.4250936	0.6529192	1.5677659
6900.50000	-0.0409394	1.4295298	0.6567755	1.5737171
6905.50000	-0.1081603	1.4312530	0.6593854	1.5795483
6910.50000	-0.1751784	1.4302897	0.6607576	1.5852498
6915.50000	-0.2418715	1.4266708	0.6609030	1.5908124
6920.50000	-0.3081204	1.4204305	0.6598341	1.5962274
6925.50000	-0.3738091	1.4116068	0.6575650	1.6014864
6930.50000	-0.4388248	1.4002410	0.6541118	1.6065815
6935.50000	-0.5030579	1.3863774	0.6494918	1.6115052

3.7. NUMERICAL EXAMPLES

```
6940.50000  -0.5664020   1.3700633   0.6437237   1.6162502
6945.50000  -0.6287540   1.3513488   0.6368279   1.6208100
6950.50000  -0.6900138   1.3302867   0.6288257   1.6251780
6955.50000  -0.7500846   1.3069321   0.6197399   1.6293483
6960.50000  -0.8088729   1.2813424   0.6095942   1.6333153
6965.50000  -0.8662881   1.2535775   0.5984137   1.6370738
6970.50000  -0.9222429   1.2236989   0.5862242   1.6406187
6975.50000  -0.9766529   1.1917706   0.5730527   1.6439457
6980.50000  -1.0294370   1.1578579   0.5589271   1.6470504
6985.50000  -1.0805169   1.1220284   0.5438761   1.6499290
6990.50000  -1.1298173   1.0843509   0.5279295   1.6525779
6995.50000  -1.1772661   1.0448961   0.5111175   1.6549938
7000.50000  -1.2227938   1.0037363   0.4934715   1.6571739
            -0.5179000  -0.4881732  -0.2098869
```

Discussion of Results

Since the orbital period of Mars is approximately 687 days, the RUNGE integration takes the planet a little more than once around its orbit. This numerical solution agrees well with the final position and velocity computed by an analytical procedure developed in Chapter 5:

- $t = 2447000.5$ (this JD corresponds to 1987 July 24)
- $\mathbf{r} = \{-1.2227938, +1.0037362, +0.4934715\}$
- $\dot{\mathbf{r}} = \{-0.5179000, -0.4881732, -0.2098869\}$
- $r = 1.6571739$

References

[1] Escobal, *Methods of Orbit Determination*, Krieger Publishing Co., 1976.

[2] Brouwer and Clemence, *Methods of Celestial Mechanics*, Academic Press, 1961.

[3] Roy, *Orbital Motion*, Adam Hilger Ltd., 1978.

[4] Danby, *Fundamentals of Celestial Mechanics*, Willmann-Bell, Inc., 1988.

[5] Taff, *Celestial Mechanics*, John Wiley and Sons, 1985.

[6] Chapra and Canale, *Numerical Methods for Engineers with Personal Computer Applications*, McGraw-Hill Book Co., 1985.

[7] Carnahan, Luther, and Wilkes, *Applied Numerical Methods*, John Wiley and Sons, 1969.

[8] Danby, *Computing Applications to Differential Equations*, Reston Publishing Co., Inc., 1985.

[9] *The Astronomical Almanac 1985*, U.S. Government Printing Office, 1984.

Chapter 4

Orbit Geometry

4.1 Introduction

The solution of the two-body problem can be characterized by six numerical quantities which are related to the arbitrary constants resulting from the integration of the differential equation

$$\ddot{\mathbf{r}} = -\frac{\mu \mathbf{r}}{r^3}.$$

These fundamental parameters are called the *orbital elements,* and, when they are known, the orbiting body's motion can be computed. The set of elements we shall use most often consists of the six scalar components of position and velocity evaluated at a given instant of time. However, although this set uniquely defines the size and orientation of the orbit in space, there are a number of other convenient sets from which to choose.

This chapter continues the discussion of the two-body problem by introducing an orbit-plane coordinate system which facilitates the derivation of elliptic, hyperbolic, and parabolic equations of motion, which can then be used to change the form of the orbital elements. The mathematical manipulations are somewhat tedious; however, the result is two very practical computer programs which convert the position and velocity elements into a set of descriptive geometric parameters or vice versa.

4.2 General Relationships

Figure 4.1 depicts the two-body orbit of a celestial body B about a dynamical center C located at the origin of an inertial rectangular coordinate system. The $\bar{x}\bar{y}$-plane coincides with the orbit plane, and the $\bar{x}$-axis is aligned with the orbit's semimajor axis. The vector $\mathbf{v}$ is the velocity of the celestial body at a given

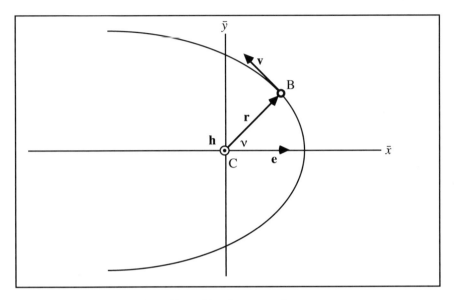

Figure 4.1: The orbit-plane coordinate system.

instant when its radius vector **r** is displaced from the $\bar{x}$-axis by the angle of the true anomaly ν. Recalling Equation 3.20, we can write

$$\mathbf{h} = \mathbf{r} \times \mathbf{v}, \tag{4.1}$$

where, in the orbit-plane coordinate system,

$$\begin{aligned} \mathbf{h} &= \{0, 0, h\} \\ \mathbf{r} &= \{\bar{x}, \bar{y}, 0\} \\ \mathbf{v} &= \{\dot{\bar{x}}, \dot{\bar{y}}, 0\}. \end{aligned} \tag{4.2}$$

Taking the dot product of Equation 4.1 with **h**, we have

$$\mathbf{h} \cdot \mathbf{h} = \mathbf{h} \cdot (\mathbf{r} \times \mathbf{v}),$$

which becomes

$$h^2 = h(\bar{x}\dot{\bar{y}} - \bar{y}\dot{\bar{x}}),$$

so that

$$h = \bar{x}\dot{\bar{y}} - \bar{y}\dot{\bar{x}}. \tag{4.3}$$

4.2. GENERAL RELATIONSHIPS

4.2.1 Angular Momentum and Angular Speed

Equation 4.3 can be used to derive a useful relationship between h and the angular speed $\dot{\nu}$. From the geometry of Figure 4.1, we can write

$$\bar{x} = r \cos \nu \tag{4.4}$$
$$\bar{y} = r \sin \nu . \tag{4.5}$$

Differentiating these two equations with respect to modified time produces

$$\dot{\bar{x}} = \dot{r} \cos \nu - r\dot{\nu} \sin \nu \tag{4.6}$$
$$\dot{\bar{y}} = \dot{r} \sin \nu + r\dot{\nu} \cos \nu . \tag{4.7}$$

Substituting Equations 4.4 through 4.7 into Equation 4.3 and utilizing the trigonometric identity [1]

$$\sin^2 x + \cos^2 x = 1, \tag{4.8}$$

we obtain

$$h = r^2 \dot{\nu} . \tag{4.9}$$

4.2.2 Radial Speed and True Anomaly

We can now use Equation 4.9 to derive an expression for the radial speed $\dot{r}$. Recalling the general equation of a conic, we know that

$$\wp = r(1 + e \cos \nu) , \tag{4.10}$$

where the semiparameter $\wp = h^2/\mu$. Differentiating Equation 4.10 with respect to modified time produces

$$\dot{r}(1 + \cos \nu) - re\dot{\nu} \sin \nu = 0 .$$

Multiplying by r, we get

$$r\dot{r}(1 + \cos \nu) - r^2 e\dot{\nu} \sin \nu = 0 .$$

If we now make use of Equations 4.9 and 4.10, we obtain

$$\dot{r}\wp - he \sin \nu = 0 .$$

Therefore, since $h = \sqrt{\mu\wp}$, the result is

$$\dot{r} = \sqrt{\frac{\mu}{\wp}} e \sin \nu . \tag{4.11}$$

4.2.3 True Anomaly and D

Now, suppose we multiply Equation 4.11 by r and substitute the expression

$$r = \frac{\wp}{1 + e \cos \nu}$$

on the right side. Then we can form the expression

$$\frac{r\dot{r}}{\sqrt{\mu}} = \frac{\sqrt{\wp} e \sin \nu}{1 + e \cos \nu}. \tag{4.12}$$

If we define

$$D = \frac{\mathbf{r} \cdot \mathbf{v}}{\sqrt{\mu}}, \tag{4.13}$$

then, according to Equation 3.10, we have

$$D = \frac{r\dot{r}}{\sqrt{\mu}}, \tag{4.14}$$

and Equation 4.12 becomes

$$D = \frac{\sqrt{\wp} e \sin \nu}{1 + e \cos \nu}. \tag{4.15}$$

The parameter D finds considerable use in orbit computations, particularly in the case of parabolic orbits.

4.2.4 Eccentricity, Semiparameter, and D

Consider now the expression for the vector $\mathbf{e}$ which points from the dynamical center toward the perifocus. According to Equation 3.46,

$$\mathbf{e} = \frac{\mathbf{v} \times \mathbf{h}}{\mu} - \frac{\mathbf{r}}{r},$$

so that

$$\mathbf{e} = \frac{\mathbf{v} \times (\mathbf{r} \times \mathbf{v})}{\mu} - \frac{\mathbf{r}}{r}, \tag{4.16}$$

because $\mathbf{h} = \mathbf{r} \times \mathbf{v}$. Utilizing the vector identity [1]

$$\mathbf{A} \times (\mathbf{B} \times \mathbf{C}) = (\mathbf{A} \cdot \mathbf{C})\mathbf{B} - (\mathbf{A} \cdot \mathbf{B})\mathbf{C},$$

Equation 4.16 can be simplified to yield

$$\mathbf{e} = \left(\frac{v^2}{\mu} - \frac{1}{r}\right) \mathbf{r} - \left(\frac{r\dot{r}}{\mu}\right) \mathbf{v}. \tag{4.17}$$

4.3. RELATIONSHIPS BETWEEN GEOMETRY AND TIME

For the moment, let Equation 4.17 be written as follows:

$$\mathbf{e} = \mathcal{A}\mathbf{r} - \mathcal{B}\mathbf{v}, \tag{4.18}$$

where

$$\mathcal{A} = \frac{v^2}{\mu} - \frac{1}{r}$$

$$\mathcal{B} = \frac{r\dot{r}}{\mu}.$$

Squaring Equation 4.18 by taking the dot product of each side with itself, we obtain

$$e^2 = \mathcal{A}^2(\mathbf{r}\cdot\mathbf{r}) - 2\mathcal{A}\mathcal{B}(\mathbf{r}\cdot\mathbf{v}) + \mathcal{B}^2(\mathbf{v}\cdot\mathbf{v}). \tag{4.19}$$

If we carry out the dot products in Equation 4.19 and replace $\mathcal{A}$ and $\mathcal{B}$ by the terms they represent, the result reduces to

$$e^2 = \left(\frac{v^2}{\mu} - \frac{1}{r}\right)^2 r^2 + D^2\left(\frac{2}{r} - \frac{v^2}{\mu}\right), \tag{4.20}$$

where $D = r\dot{r}/\sqrt{\mu}$. Making use of the vis viva equation of Section 3.4.4, we also have

$$\frac{v^2}{\mu} = \frac{2}{r} - \frac{1}{a}. \tag{4.21}$$

Therefore, Equation 4.20 can be further simplified to yield

$$e^2 = \left(1 - \frac{r}{a}\right)^2 + \frac{D^2}{a}. \tag{4.22}$$

Now, according to Equation 3.43,

$$\wp = a(1 - e^2), \tag{4.23}$$

so that substituting Equation 4.22 for e^2, we obtain

$$\wp = r\left(2 - \frac{r}{a}\right) - D^2. \tag{4.24}$$

4.3 Relationships between Geometry and Time

When Equation 4.3 is applied to the specific geometries of elliptic, hyperbolic, and parabolic orbits, it is possible to derive mathematical relationships between position in the orbit plane and time elapsed from a given epoch for each type of conic. We shall refer to those expressions collectively as Kepler equations, because they serve the same function as the equation derived by Kepler for elliptic orbits [2,3,4].

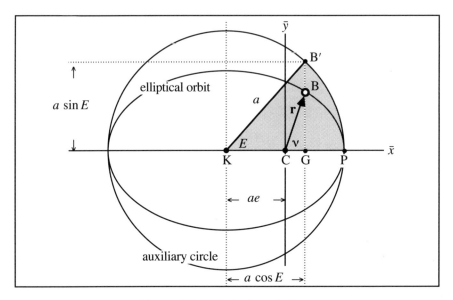

Figure 4.2: Elliptic formulation.

4.3.1 Elliptic Formulation

Consider the geometric construction of Figure 4.2, where an auxiliary circle centered at K circumscribes the actual ellipse of motion about the dynamical center C. As the celestial body B moves along the ellipse, it is followed by a point B′ defined by the projection of B in the y-direction upon the circle. The angle E, which is proportional to the shaded area, is called the *eccentric anomaly* and is measured in the orbital plane from the $\bar{x}$-axis to the line KB′. The distance from B′ to K is always equal to a, the length of the semimajor axis of the ellipse, and the distance between K and C is equal to ae, where e is the eccentricity of the ellipse. The advantage of the auxiliary circle is that by expressing $\bar{x}$ and $\bar{y}$ in terms of the eccentric anomaly, instead of the true anomaly, Equation 4.3 can be reduced to a simple form which is easily integrated.

By carefully examining Figure 4.2, we see that the $\bar{x}$-coordinate of the celestial body is related to the eccentric anomaly by

$$\bar{x} = a(\cos E - e). \tag{4.25}$$

Also, the general conic equation allows us to write

$$\wp = r + e(r \cos \nu). \tag{4.26}$$

Now, by comparing Equations 4.4 and 4.25, we find that

$$r \cos \nu = a \cos E - ae. \tag{4.27}$$

4.3. RELATIONSHIPS BETWEEN GEOMETRY AND TIME

Thus, substituting Equation 4.27 into Equation 4.26 produces

$$\wp = r + ae\cos E - ae^2. \tag{4.28}$$

Finally, if we make use of Equation 4.23, namely

$$\wp = a(1 - e^2),$$

Equation 4.28 can be rearranged to yield an expression for r:

$$r = a(1 - e\cos E). \tag{4.29}$$

We derive the equation for $\bar{y}$ by substituting the expressions for $\bar{x}$ and r into the general relationship

$$r^2 = \bar{x}^2 + \bar{y}^2.$$

The substitutions produce

$$\bar{y}^2 = a^2(1 - e^2)(1 - \cos^2 E).$$

Applying the trigonometric identity of Equation 4.8, we obtain

$$\bar{y}^2 = a^2(1 - e^2)\sin^2 E,$$

which, upon taking the square root, becomes

$$\bar{y} = a\sqrt{1 - e^2}\sin E. \tag{4.30}$$

Expressions for the components of the velocity vector are found by differentiating the equations for $\bar{x}$ and $\bar{y}$. Therefore, from Equation 4.25, we have

$$\dot{\bar{x}} = -a\dot{E}\sin E, \tag{4.31}$$

and, from Equation 4.30, we get

$$\dot{\bar{y}} = a\sqrt{1 - e^2}\,\dot{E}\cos E. \tag{4.32}$$

Now, if we substitute Equations 4.25, 4.30, 4.31, and 4.32 into Equation 4.3, then

$$h = a^2\sqrt{1 - e^2}\,(\cos^2 E - e\cos E + \sin^2 E)\dot{E},$$

which simplifies to

$$\sqrt{\frac{\mu}{a^3}} = (1 - e\cos E)\dot{E} \tag{4.33}$$

when we apply the trigonometric identity of Equation 4.8 and make use of the fact that

$$h = \sqrt{\mu\wp} = \sqrt{\mu a(1 - e^2)}.$$

Finally, since
$$\dot{E} = \frac{dE}{d\tau},$$
Equation 4.33 becomes
$$\sqrt{\frac{\mu}{a^3}}\, d\tau = (1 - e\cos E)\, dE.$$
This equation is easily integrated to produce
$$\sqrt{\frac{\mu}{a^3}}\, \tau = E - e\sin E,$$
where the arbitrary constant of integration is zero because we define the modified time interval τ to be zero when the eccentric anomaly E is zero. Letting T represent the *time of perifocal passage*, then the celestial body's position at any time t can be written
$$n(t - T) = E - e\sin E,$$
where n, known as the *mean motion*, is given by
$$n = k\sqrt{\frac{\mu}{a^3}}. \tag{4.34}$$
It is also convenient to define a quantity M, called the *mean anomaly*, such that
$$M = n(t - T), \tag{4.35}$$
so that we can finally write
$$M = E - e\sin E, \tag{4.36}$$
which is *Kepler's equation*. Although developed for elliptic orbits, Equation 4.36 also holds for the case of circular motion when $e = 0$.

4.3.2 Hyperbolic Formulation

Figure 4.3 illustrates the somewhat unfamiliar geometry required to develop the hyperbolic version of Kepler's equation. The upper portions of the left branches of two hyperbolas are depicted, where the point K is the geometric center of the complete curves. As the celestial body B moves along its hyperbolic orbit about the dynamical center C, it is followed by a point B′ defined by the projection of B in the $\bar{y}$-direction upon an auxiliary hyperbola whose asymptotes radiate from K at 45° angles to the coordinate axes. The orbit's eccentricity and semimajor axis are $e > 1$ and $a < 0$, respectively. The parameter H, which is proportional to the shaded area in the figure, is the *hyperbolic eccentric anomaly*, which is

4.3. RELATIONSHIPS BETWEEN GEOMETRY AND TIME

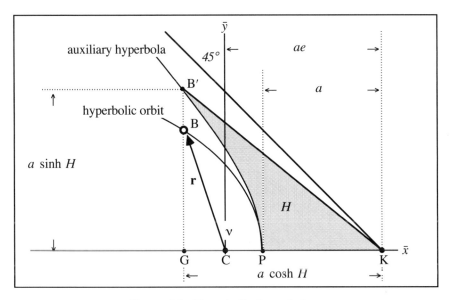

Figure 4.3: Hyperbolic formulation.

used as the argument of the hyperbolic sine (sinh) and cosine (cosh) functions. A positive value of H is associated with an area that is above the $\bar{x}$-axis and a negative value with an area that is below the $\bar{x}$-axis. The term *hyperbolic radian* is sometimes used in connection with H; however, H is not actually an angle but simply a dimensionless real number. Figure 4.4 graphically depicts the general behavior of $\sinh x$ and $\cosh x$ [5].

By carefully examining Figure 4.3 in light of the behavior of $\sinh H$ and $\cosh H$, we can express the $\bar{x}$-coordinate as follows:

$$\bar{x} = a(\cosh H - e). \tag{4.37}$$

Recalling the general equation of a conic, we have

$$\wp = r + e(r \cos \nu),$$

and, since Equations 4.4 and 4.37 show that

$$r \cos \nu = a(\cosh H - e),$$

we can write

$$\wp = r + ae \cosh H - ae^2. \tag{4.38}$$

Now, using Equation 4.23, the above equation can be rearranged to give

$$r = a(1 - e \cosh H). \tag{4.39}$$

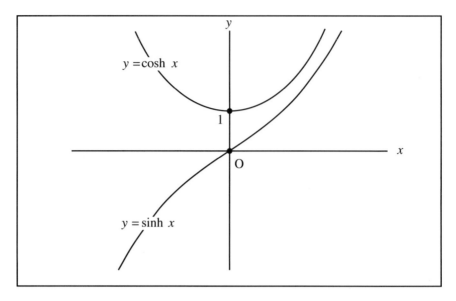

Figure 4.4: Hyperbolic sine and cosine.

We derive the equation for $\bar{y}$ by substituting the expressions for $\bar{x}$ and r into the general relationship
$$r^2 = \bar{x}^2 + \bar{y}^2.$$
The result can be written
$$\bar{y}^2 = a^2(e^2 - 1)(\cosh^2 H - 1).$$
Applying the hyperbolic identity [1]
$$\cosh^2 x - \sinh^2 x = 1, \tag{4.40}$$
we obtain
$$\bar{y}^2 = a^2(e^2 - 1)\sinh^2 H,$$
which, upon taking the square root, becomes either
$$\bar{y} = +a\sqrt{e^2 - 1}\, \sinh H \tag{4.41}$$
or
$$\bar{y} = -a\sqrt{e^2 - 1}\, \sinh H. \tag{4.42}$$
Since $e > 1$ and $a < 0$ for a hyperbola, we choose Equation 4.42 because it yields a positive value for $\bar{y}$ when H is positive.

4.3. RELATIONSHIPS BETWEEN GEOMETRY AND TIME

Expressions for the components of the velocity vector are obtained by differentiating the equations for $\bar{x}$ and $\bar{y}$. Thus, from Equation 4.37, we have

$$\dot{\bar{x}} = a\dot{H} \sinh H, \qquad (4.43)$$

and, from Equation 4.42, we get

$$\dot{\bar{y}} = -a\sqrt{e^2 - 1}\, \dot{H} \cosh H. \qquad (4.44)$$

Now, substituting Equations 4.37, 4.42, 4.43, and 4.44 into Equation 4.3, we can obtain

$$h = a^2\sqrt{e^2 - 1}\,(e \cosh H - \cosh^2 H + \sinh^2 H)\dot{H}.$$

This expression reduces to

$$\sqrt{\frac{\mu}{(-a)^3}} = (e \cosh H - 1)\dot{H} \qquad (4.45)$$

when we apply the hyperbolic identity of Equation 4.40 and make use of the relationship

$$h = \sqrt{\mu p} = \sqrt{\mu(-a)(e^2 - 1)}.$$

Finally, since

$$\dot{H} = \frac{dH}{d\tau},$$

Equation 4.45 can be written in the form

$$\sqrt{\frac{\mu}{(-a)^3}}\, d\tau = (e \cosh H - 1)dH.$$

When this equation is integrated, the result is

$$\sqrt{\frac{\mu}{(-a)^3}}\, \tau = e \sinh H - H. \qquad (4.46)$$

The arbitrary constant of integration is zero because the modified time interval τ and the anomaly H are both zero at the perifocus.

Following the analogy of Kepler's equation for elliptic motion, we define the *mean motion* to be

$$n = k\sqrt{\frac{\mu}{(-a)^3}}. \qquad (4.47)$$

Thus, for any given time t, the *mean anomaly* is again

$$M = n(t - T), \qquad (4.48)$$

where T is the *time of perifocal passage*. Therefore, we can finally write

$$M = e \sinh H - H, \qquad (4.49)$$

which is the hyperbolic equivalent of Kepler's equation.

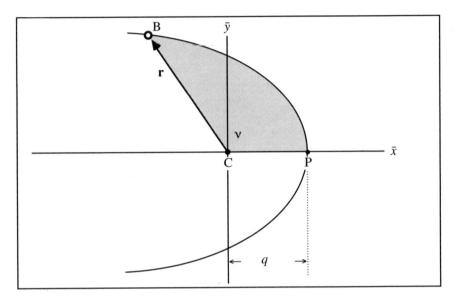

Figure 4.5: Parabolic formulation.

4.3.3 Parabolic Formulation

Figure 4.5 shows celestial body B moving in a parabolic orbit about dynamical center C, which is at the origin of the orbit-plane coordinate system. The angle ν is the true anomaly, and q is the perifocal distance. Since a parabolic orbit is a special case of two-body motion which occurs when the eccentricity is unity, the general conic equation becomes

$$r = \frac{\wp}{1 + \cos \nu}. \tag{4.50}$$

Furthermore, because Equation 3.41 gives

$$\wp = q(1 + e),$$

we find that, for a parabola,

$$\wp = 2q, \tag{4.51}$$

and

$$h = \sqrt{2q\mu}. \tag{4.52}$$

Finally, in the present case where $e = 1$, the fact that Equation 4.23 can be rearranged to yield

$$a = \frac{\wp}{1 - e^2}$$

4.3. RELATIONSHIPS BETWEEN GEOMETRY AND TIME

implies that the semimajor axis of a parabola must be infinite. This condition prevents us from using Equations 4.36 or 4.49 to compute parabolic motion because both those expressions contain the semimajor axis in the denominator of the mean motion term.

The Kepler equation for parabolic orbits can be derived without using an auxiliary curve. We will, however, use the parameter D as a *parabolic eccentric anomaly*. Thus, letting $e = 1$ in Equation 4.15, we obtain

$$D = \frac{\sqrt{\wp} \sin \nu}{1 + \cos \nu}, \qquad (4.53)$$

which, by the trigonometric identity [1]

$$\tan \frac{x}{2} = \frac{\sin x}{1 + \cos x}, \qquad (4.54)$$

becomes

$$D = \sqrt{\wp} \tan \frac{\nu}{2}. \qquad (4.55)$$

The parabolic expression for x is obtained by substituting Equation 4.50 for r in Equation 4.4. Thus,

$$\bar{x} = \frac{\wp \cos \nu}{1 + \cos \nu}. \qquad (4.56)$$

Utilizing the trigonometric identities [1]

$$\cos x = \cos^2 \frac{x}{2} - \sin^2 \frac{x}{2}$$

and

$$\cos x = 2 \cos^2 \frac{x}{2} - 1,$$

Equation 4.56 can be written

$$\bar{x} = \frac{1}{2}\left(\wp - \wp \tan^2 \frac{\nu}{2}\right).$$

Substituting Equation 4.55, the result is

$$\bar{x} = \frac{1}{2}(\wp - D^2). \qquad (4.57)$$

The equation for $\bar{y}$ is more straightforward. Replacing r in Equation 4.5 by Equation 4.50 yields

$$\bar{y} = \frac{\wp \sin \nu}{1 + \cos \nu}.$$

Therefore, according to the trigonometric identity of Equation 4.54, we have

$$\bar{y} = \wp \tan \frac{\nu}{2},$$

which is equivalent to
$$\bar{y} = \sqrt{\wp}D. \tag{4.58}$$

Now, since r is related to $\bar{x}$ and $\bar{y}$ by the familiar general expression
$$r^2 = \bar{x}^2 + \bar{y}^2,$$

we use Equations 4.57 and 4.58 to obtain
$$r^2 = \frac{1}{4}(\wp - D^2)^2 + \wp D^2,$$

which reduces to
$$r = \frac{1}{2}(\wp + D^2). \tag{4.59}$$

The rectangular components of velocity are found by differentiating Equations 4.57 and 4.58 with respect to modified time. Thus,
$$\dot{\bar{x}} = -D\dot{D} \tag{4.60}$$
$$\dot{\bar{y}} = \sqrt{\wp}\dot{D}. \tag{4.61}$$

When we substitute Equations 4.57, 4.58, 4.60, and 4.61 into Equation 4.3, the result is
$$h = \frac{1}{2}(p - D^2)\sqrt{\wp}\dot{D} + \sqrt{\wp}D^2\dot{D}. \tag{4.62}$$

This expression can be simplified by making use of the following relationships:
$$h = \sqrt{\mu\wp}$$
$$\wp = 2q \quad \text{(for a parabola)}$$
$$\dot{D} = \frac{dD}{d\tau}. \tag{4.63}$$

Substituting Equations 4.63 into Equation 4.62 and rearranging, we obtain
$$2\sqrt{\mu}\,d\tau = 2q\,dD + D^2\,dD,$$

which is easily integrated to produce
$$2\sqrt{\mu}\,\tau = 2qD + \frac{D^3}{3}, \tag{4.64}$$

where the arbitrary constant of integration is zero because, by Equation 4.55, $D = 0$ when $\nu = 0$ at $\tau = 0$. If we define the *mean motion* to be
$$n = k\sqrt{\mu}, \tag{4.65}$$

4.4. THE CLASSICAL ELEMENTS FROM POSITION AND VELOCITY

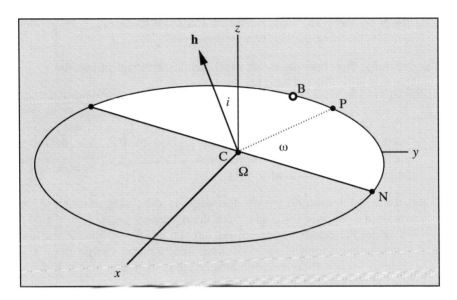

Figure 4.6: Classical geometric elements.

then the *mean anomaly* is
$$M = n(t - T), \qquad (4.66)$$
where T represents the *time of perifocal passage*. Therefore, Equation 4.64 can be written as
$$M = qD + \frac{D^3}{6}, \qquad (4.67)$$
which is the parabolic form of Kepler's equation, also known as *Barker's equation*.

4.4 The Classical Elements from Position and Velocity

The components of **r** and **v** provide a completely general description of orbital motion; however, their vector form does not clearly reveal the orbit's size, shape, and orientation in space. Since it is often helpful to have a geometric perspective such as that depicted in Figure 4.6, we will develop a procedure which will transform **r** and **v** into the following set of parameters known as *classical elements* [2,3,4]:

a **semimajor axis** The conic parameter which is used to define the size of an elliptic or hyperbolic orbit.

q **perifocal distance** The conic parameter which is used instead of the semimajor axis to define the size of a parabolic orbit.

e **eccentricity** The conic parameter which defines the shape of an orbit.

i **inclination** The angle between the $+z$-axis and the angular momentum vector **h**, which is perpendicular to the orbit plane, measured from $0°$ to $180°$. If $i < 90°$, the orbital motion is counterclockwise when viewed from the north side of the fundamental plane (direct motion). If $i > 90°$, the orbital motion is clockwise when viewed from the north side of the fundamental plane (retrograde motion).

Ω **longitude of the ascending node** The angle in the fundamental plane, between the $+x$-axis and a line from the dynamical center C to the point N where the celestial body crosses through the fundamental plane from south to north (ascending node), measured counterclockwise from $0°$ to $360°$ as viewed from the north side of the fundamental plane. If $i = 0$, then Ω is undefined.

ω **argument of the perifocus** The angle in the orbital plane, between the line of the ascending node and a line from the dynamical center C to the perifocus P, measured from $0°$ to $360°$ in the direction of the celestial body's motion. If $e = 0$ or $i = 0$, then ω is undefined.

n **mean motion** A mathematical quantity whose value is the constant angular speed which would be required for the celestial body to complete its orbit in one period.

M **mean anomaly** A mathematical quantity whose value relates the position of the celestial body in the orbit to the elapsed time by means of the Kepler equations. The mean anomaly changes at a uniform rate equal to the mean motion n.

T **time of perifocal passage** The moment when the celestial body passes the perifocus P. This quantity can also be used to relate position along the orbit to the elapsed time by means of the Kepler equations. If $e = 0$, then T is undefined.

4.4.1 Three Fundamental Vectors

We begin the process of determining the classical elements by forming the fundamental vectors **e**, **h**, and **N** illustrated in Figure 4.7. The origin of the inertial rectangular coordinate system is at the dynamical center C, and **I**, **J**, and **K** are unit vectors parallel to the x, y, and z axes, respectively. The coordinate

4.4. THE CLASSICAL ELEMENTS FROM POSITION AND VELOCITY   151

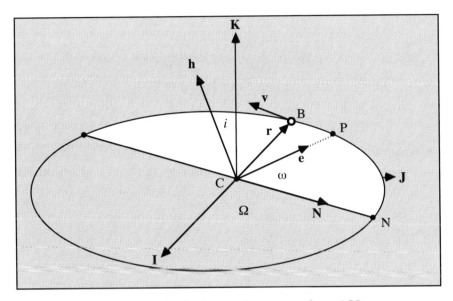

Figure 4.7: The fundamental vectors **e**, **h**, and **N**.

system is aligned so that the $+x$-axis points toward the vernal equinox, and the xy-plane coincides with the fundamental plane of the celestial coordinate system. The fundamental plane will correspond to the equatorial plane if the orbit is geocentric or to the ecliptic plane if the orbit is heliocentric. In the latter case, the vectors **r** and **v** must also be referred to the ecliptic coordinate system by using the equatorial-to-ecliptic transformation described in Section 2.6. Thus, we have for any given time t

$$\mathbf{r} = \{x, y, z\}$$
$$\mathbf{v} = \{\dot{x}, \dot{y}, \dot{z}\},$$

and we can immediately compute

$$r = |\mathbf{r}|$$
$$v^2 = \mathbf{v} \cdot \mathbf{v}$$
$$r\dot{r} = \mathbf{r} \cdot \mathbf{v}. \tag{4.68}$$

Therefore, according to Equation 4.17, the *eccentricity vector* is

$$\mathbf{e} = \left(\frac{v^2}{\mu} - \frac{1}{r}\right)\mathbf{r} - \left(\frac{r\dot{r}}{\mu}\right)\mathbf{v}, \tag{4.69}$$

so that

$$\mathbf{e} = \{e_x, e_y, e_z\}, \tag{4.70}$$

where

$$\begin{aligned} e_x &= \left(\frac{v^2}{\mu} - \frac{1}{r}\right)x - \left(\frac{r\dot{r}}{\mu}\right)\dot{x} \\ e_y &= \left(\frac{v^2}{\mu} - \frac{1}{r}\right)y - \left(\frac{r\dot{r}}{\mu}\right)\dot{y} \\ e_z &= \left(\frac{v^2}{\mu} - \frac{1}{r}\right)z - \left(\frac{r\dot{r}}{\mu}\right)\dot{z}. \end{aligned} \qquad (4.71)$$

The *angular momentum vector* is computed by means of its cross product definition

$$\mathbf{h} = \mathbf{r} \times \mathbf{v}. \qquad (4.72)$$

Therefore,

$$\mathbf{h} = \{h_x, h_y, h_z\}, \qquad (4.73)$$

where

$$\begin{aligned} h_x &= y\dot{z} - z\dot{y} \\ h_y &= z\dot{x} - x\dot{z} \\ h_z &= x\dot{y} - y\dot{x}. \end{aligned} \qquad (4.74)$$

Finally, the *ascending node vector* $\mathbf{N}$ is obtained from the cross product of vectors $\mathbf{K}$ and $\mathbf{h}$. Accordingly,

$$\mathbf{N} = \mathbf{K} \times \mathbf{h}, \qquad (4.75)$$

where

$$\begin{aligned} \mathbf{K} &= \{0, 0, 1\} \\ \mathbf{h} &= \{h_x, h_y, h_z\}. \end{aligned}$$

Thus, taking the vector cross product, we get

$$\mathbf{N} = \{N_x, N_y, N_z\}, \qquad (4.76)$$

where

$$\begin{aligned} N_x &= -h_y \\ N_y &= +h_x \\ N_z &= 0. \end{aligned} \qquad (4.77)$$

4.4.2 The Conic Parameters

The conic parameters a, e, and q are easily found by using relationships already discussed. According to the vis-viva equation, we have

$$\frac{1}{a} = \frac{2}{r} - \frac{v^2}{\mu}, \qquad (4.78)$$

so the semimajor axis is determined. The eccentricity is obtained form Equation 4.69. Thus,

$$e = |\mathbf{e}|. \qquad (4.79)$$

We find the perifocal distance by employing Equation 4.72 to compute

$$h = |\mathbf{h}|, \qquad (4.80)$$

so that, by Equation 3.40, we have

$$\wp = \frac{h^2}{\mu}, \qquad (4.81)$$

and Equation 3.41 yields

$$q = \frac{\wp}{1+e}. \qquad (4.82)$$

4.4.3 The Orientation Angles

Returning to Figure 4.7, we see that the orientation angles i, Ω, and ω can all be determined from various dot products between fundamental vectors $\mathbf{e}$, $\mathbf{h}$, $\mathbf{N}$ and unit vectors $\mathbf{I}$, $\mathbf{J}$, $\mathbf{K}$. Accordingly, the *angle of inclination* is computed from the dot product

$$\mathbf{K} \cdot \mathbf{h} = |\mathbf{K}||\mathbf{h}|\cos i, \qquad (4.83)$$

where

$$\mathbf{K} = \{0, 0, 1\}$$
$$\mathbf{h} = \{h_x, h_y, h_z\}.$$

Therefore, we find that

$$\mathbf{K} \cdot \mathbf{h} = h_z$$
$$|\mathbf{K}| = 1$$
$$|\mathbf{h}| = h,$$

so Equation 4.83 simplifies to

$$\cos i = \frac{h_z}{h}, \qquad (4.84)$$

and the inclination is determined.

In the case of the *longitude of the ascending node*, we begin with the dot product
$$\mathbf{I} \cdot \mathbf{N} = |\mathbf{I}||\mathbf{N}| \cos \Omega, \tag{4.85}$$
where
$$\mathbf{I} = \{1, 0, 0\}$$
$$\mathbf{N} = \{N_x, N_y, N_z\}.$$
Thus,
$$\mathbf{I} \cdot \mathbf{N} = N_x$$
$$|\mathbf{I}| = 1$$
$$|\mathbf{N}| = N,$$
so Equation 4.85 reduces to
$$\cos \Omega = \frac{N_x}{N}, \tag{4.86}$$
where $\Omega > 180°$ if $N_y < 0$. Therefore, the longitude of the ascending node is determined.

Finally, the *argument of the perifocus* is obtained from the dot product
$$\mathbf{N} \cdot \mathbf{e} = |\mathbf{N}||\mathbf{e}| \cos \omega, \tag{4.87}$$
where
$$\mathbf{N} = \{N_x, N_y, N_z\}$$
$$\mathbf{e} = \{e_x, e_y, e_z\}.$$
Now, since
$$|\mathbf{N}| = N$$
$$|\mathbf{e}| = e,$$
Equation 4.87 can be rewritten as
$$\cos \omega = \frac{\mathbf{N} \cdot \mathbf{e}}{Ne}, \tag{4.88}$$
where $\omega > 180°$ if $e_z < 0$. Thus, the argument of the perifocus is determined. This element is frequently replaced by the *longitude of the perifocus*,
$$\varpi = \Omega + \omega,$$
so that
$$\omega = \varpi - \Omega.$$

4.4. THE CLASSICAL ELEMENTS FROM POSITION AND VELOCITY

4.4.4 The Mean Anomaly

According to Equations 4.4 and 4.5, the celestial body's position with reference to the orbit-plane coordinate system is given by

$$\bar{x} = r \cos \nu \qquad (4.89)$$
$$\bar{y} = r \sin \nu. \qquad (4.90)$$

Rearranging the conic equation to obtain

$$r \cos \nu = \frac{\wp - r}{e}$$

permits us to write Equation 4.89 in the form

$$\bar{x} = \frac{\wp - r}{e}. \qquad (4.91)$$

Now, recalling Equation 4.11, we have

$$\dot{r} = \sqrt{\frac{\mu}{\wp}} e \sin \nu,$$

so that, multiplying through by r, we get

$$r \sin \nu = \frac{r\dot{r}}{e} \sqrt{\frac{\wp}{\mu}}.$$

Substituting the above expression in Equation 4.90, the result is

$$\bar{y} = \frac{r\dot{r}}{e} \sqrt{\frac{\wp}{\mu}}. \qquad (4.92)$$

Now, if the eccentricity determined by Equation 4.79 is less than unity, the mean anomaly must be computed using the elliptic equations of motion. We begin with Equations 4.25 and 4.30, namely

$$\bar{x} = a(\cos E - e) \qquad (4.93)$$
$$\bar{y} = b \sin E, \qquad (4.94)$$

where, for convenience, we have let

$$b = a\sqrt{1 - e^2}. \qquad (4.95)$$

Since $\bar{x}$ and $\bar{y}$ are known quantities given by Equations 4.91 and 4.92, we can write

$$\cos E = \frac{\bar{x}}{a} + e \qquad (4.96)$$

$$\sin E = \frac{\bar{y}}{b}, \qquad (4.97)$$

which determine the eccentric anomaly without ambiguity. The elliptic mean anomaly and mean motion can now be computed in radian measure from

$$M = E - e \sin E \tag{4.98}$$

$$n = k\sqrt{\frac{\mu}{a^3}}. \tag{4.99}$$

If desired, the period of the elliptic orbit can be found from Equation 3.52.

If the eccentricity of the orbit is greater than unity, the mean anomaly must be determined using the hyperbolic equations of motion. According to Equation 4.42, we write

$$\sinh H = \frac{\bar{y}}{b}, \tag{4.100}$$

where $\bar{y}$ is given by Equation 4.92, and

$$b = -a\sqrt{e^2 - 1}. \tag{4.101}$$

Since the hyperbolic eccentric anomaly is determined, the hyperbolic mean anomaly and mean motion can be computed from

$$M = e \sinh H - H \tag{4.102}$$

$$n = k\sqrt{\frac{\mu}{-a^3}}. \tag{4.103}$$

Finally, if the eccentricity of the orbit is equal to unity (within some reasonable tolerance), we first compute the parabolic eccentric anomaly from Equation 4.14,

$$D = \frac{r\dot{r}}{\sqrt{\mu}}. \tag{4.104}$$

Then, the parabolic mean anomaly and mean motion can be found immediately from

$$M = qD + \frac{D^3}{6} \tag{4.105}$$

$$n = k\sqrt{\mu}. \tag{4.106}$$

An element known as the *mean longitude L* is sometimes used in place of the mean anomaly. It is related to M by the expression

$$L = \varpi + M,$$

so that

$$M = L - \varpi,$$

or

$$M = L - \omega - \Omega.$$

4.5. POSITION AND VELOCITY FROM THE CLASSICAL ELEMENTS

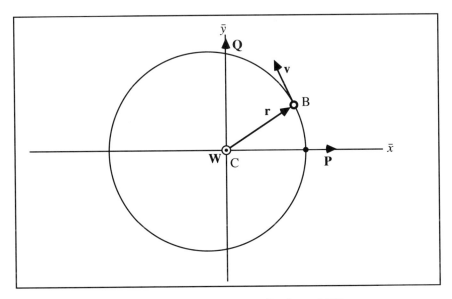

Figure 4.8: The unit vectors **P**, **Q**, and **W**.

4.4.5 The Time of Perifocal Passage

The time of perifocal passage is determined from the definition of the mean anomaly,

$$M = n(t - T). \tag{4.107}$$

This expression yields

$$T = t - \frac{M}{n}. \tag{4.108}$$

Therefore, in order to determine T, we first compute M and n from the expressions appropriate to the conic section which describes the orbit.

4.5 Position and Velocity from the Classical Elements

Occasionally, situations arise where it is necessary to convert a set of classical elements into the elements of position and velocity at a given epoch time. This transformation is facilitated by defining a set of mutually perpendicular unit vectors **P**, **Q**, and **W** in the orbit-plane coordinate system. As shown in Figure 4.8, **P** is directed along the $\bar{x}$-axis toward the perifocus, **Q** lies along the $\bar{y}$-axis, and **W** is perpendicular to the orbit plane. Thus,

$$\mathbf{W} = \mathbf{P} \times \mathbf{Q}. \tag{4.109}$$

Therefore, as celestial body B orbits dynamical center C, its position and velocity may be described in terms of the unit vectors as follows:

$$\begin{aligned} \mathbf{r} &= \bar{x}\mathbf{P} + \bar{y}\mathbf{Q} \\ \mathbf{v} &= \dot{\bar{x}}\mathbf{P} + \dot{\bar{y}}\mathbf{Q} \, . \end{aligned} \quad (4.110)$$

Now, if we express all the quantities on the right sides of these equations in terms of the classical elements, the vector elements $\mathbf{r}$ and $\mathbf{v}$ can be computed. We begin by transforming the scalar components using expressions which are appropriate to the conic section which describes the orbit.

4.5.1 The Scalar Components of Elliptic Motion

In the case of elliptic motion, we assume the following set of classical elements:

$$\{a, e, M, i, \Omega, \omega\},$$

where $e < 1$. The mean and eccentric anomalies are related by the elliptic Kepler equation, namely

$$M = E - e \sin E, \qquad (4.111)$$

which can be written

$$f = E - e \sin E - M. \qquad (4.112)$$

Differentiating f with respect to E, we obtain

$$\frac{df}{dE} = 1 - e \cos E. \qquad (4.113)$$

If we choose $E = M$ as a first approximation for the eccentric anomaly, Equations 4.112 and 4.113 can be solved for an accurate value of E by successive iterations using the Newton-Raphson method described in Section 7.2.2. This procedure is straightforward because a given value of the mean anomaly determines a unique value of the eccentric anomaly. When a satisfactory value of E has been found, we use Equation 4.29 to compute

$$r = a(1 - e \cos E) \qquad (4.114)$$

and rewrite Equation 4.33 to obtain

$$\sqrt{\frac{\mu}{a}} = a(1 - e \cos E)\dot{E}. \qquad (4.115)$$

Substituting Equation 4.114 into 4.115, the result can be arranged to yield

$$\dot{E} = \frac{1}{r}\sqrt{\frac{\mu}{a}}. \qquad (4.116)$$

4.5. POSITION AND VELOCITY FROM THE CLASSICAL ELEMENTS

We now use Equations 4.25, 4.30, 4.31, and 4.32 to compute the scalar components of position and velocity:

$$\begin{aligned} \bar{x} &= a(\cos E - e) \\ \bar{y} &= b \sin E \\ \dot{\bar{x}} &= -a\dot{E} \sin E \\ \dot{\bar{y}} &= b\dot{E} \cos E, \end{aligned} \quad (4.117)$$

where $b = a\sqrt{1 - e^2}$.

4.5.2 The Scalar Components of Hyperbolic Motion

When the orbital motion is hyperbolic, we begin with the classical element set

$$\{a, e, T, i, \Omega, \omega\},$$

where $a < 0$ and $e > 1$. At any given time t, we can compute $M = n(t - T)$. Then, following a procedure similar to that used for elliptic motion, we express Equation 4.49 as the function

$$f = e \sinh H - H - M, \quad (4.118)$$

and differentiate f with respect to H to obtain

$$\frac{df}{dH} = e \cosh H - 1. \quad (4.119)$$

If the celestial body is near perifocus, we can choose $H = M$ as a first approximation for the hyperbolic eccentric anomaly and solve Equations 4.118 and 4.119 by the Newton-Raphson method. This yields the unique value of H which corresponds to the given value of M. Next, we use Equation 4.39 to obtain

$$r = -a(e \cosh H - 1), \quad (4.120)$$

and rewrite Equation 4.45 in the form

$$\sqrt{\frac{\mu}{-a}} = \sqrt{a^2} (e \cosh H - 1) \dot{H}.$$

Choosing $\sqrt{a^2} = -a$ so that $\dot{H}$ remains positive, we have

$$\sqrt{\frac{\mu}{-a}} = -a(e \cosh H - 1) \dot{H}. \quad (4.121)$$

Therefore, substituting Equation 4.120 into Equation 4.121 yields

$$\dot{H} = \frac{1}{r}\sqrt{\frac{\mu}{-a}}. \qquad (4.122)$$

Thus, the scalar components of hyperbolic motion follow from Equations 4.37, 4.42, 4.43, and 4.44:

$$\begin{aligned}
\bar{x} &= a(\cosh H - e) \\
\bar{y} &= b\sinh H \\
\dot{\bar{x}} &= a\dot{H}\sinh H \\
\dot{\bar{y}} &= b\dot{H}\cosh H,
\end{aligned} \qquad (4.123)$$

where $b = -a\sqrt{e^2 - 1}$.

4.5.3 The Scalar Components of Parabolic Motion

The special case of parabolic motion can be described by the classical element set

$$\{q, e, T, i, \Omega, \omega\},$$

where $e = 1$. At any given time t, we have $M = n(t - T)$, and Barker's equation can be written as the function

$$f = qD + \frac{D^3}{6} - M. \qquad (4.124)$$

Differentiating this equation with respect to D yields

$$\frac{df}{dD} = q + \frac{D^2}{2}. \qquad (4.125)$$

Equations 4.124 and 4.125 are solved for D by the Newton-Raphson method. Starting with the initial approximation $D = M$, the procedure will converge to a value of D which is unique to the given mean anomaly. Taking advantage of the fact that $\wp = 2q$ for a parabola, Equation 4.59 becomes

$$r = q + \frac{D^2}{2}. \qquad (4.126)$$

Recalling Equation 4.64, we have the relationship

$$\sqrt{\mu}\,\tau = qD + \frac{D^3}{6},$$

4.5. POSITION AND VELOCITY FROM THE CLASSICAL ELEMENTS

which, when differentiated with respect to modified time, yields

$$\sqrt{\mu} = \left(q + \frac{D^2}{2}\right)\dot{D}. \qquad (4.127)$$

Substituting Equation 4.126 for the expression in parentheses, we get

$$\dot{D} = \frac{1}{r}\sqrt{\mu}. \qquad (4.128)$$

Finally, the scalar components of parabolic motion are determined from Equations 4.57, 4.58, 4.60, and 4.61:

$$\begin{aligned}
\bar{x} &= q - \frac{D^2}{2} \\
\bar{y} &= \sqrt{2q}D \\
\dot{\bar{x}} &= -D\dot{D} \\
\dot{\bar{y}} &= \sqrt{2q}\dot{D},
\end{aligned} \qquad (4.129)$$

where the semiparameter $\wp$ has been replaced by its parabolic equivalent, $2q$.

4.5.4 The Unit Vector Components of Motion

Now that we have developed equations which express the scalar components of position and velocity in terms of the classical elements, all that remains is to accomplish this for the vector components of $\mathbf{P}$ and $\mathbf{Q}$. It will then be possible to use Equations 4.110 to compute the vector elements $\mathbf{r}$ and $\mathbf{v}$ when the classical elements are given.

Consider the geometric relationships depicted in Figure 4.9. The $\{\mathbf{P}, \mathbf{Q}, \mathbf{W}\}$ unit vector system can be obtained by successive rotations of the $\{\mathbf{I}, \mathbf{J}, \mathbf{K}\}$ unit vector system through angles Ω, i, and ω. We proceed as follows:

Step 1 Rotate the $\{\mathbf{I}, \mathbf{J}, \mathbf{K}\}$ system about $\mathbf{K}$ through the angle Ω, as shown in Figure 4.10. The result is an $\{\mathbf{I}', \mathbf{J}', \mathbf{K}'\}$ system, where

$$\begin{aligned}
\mathbf{I}' &= +\mathbf{I}\cos\Omega + \mathbf{J}\sin\Omega \\
\mathbf{J}' &= -\mathbf{I}\sin\Omega + \mathbf{J}\cos\Omega \\
\mathbf{K}' &= +\mathbf{K}.
\end{aligned} \qquad (4.130)$$

Step 2 Rotate the $\{\mathbf{I}', \mathbf{J}', \mathbf{K}'\}$ system about $\mathbf{I}'$ through the angle i, as shown in Figure 4.11. The result is an $\{\mathbf{I}'', \mathbf{J}'', \mathbf{K}''\}$ system, where

$$\begin{aligned}
\mathbf{I}'' &= +\mathbf{I}' \\
\mathbf{J}'' &= +\mathbf{J}'\cos i + \mathbf{K}'\sin i \\
\mathbf{K}'' &= -\mathbf{J}'\sin i + \mathbf{K}'\cos i.
\end{aligned} \qquad (4.131)$$

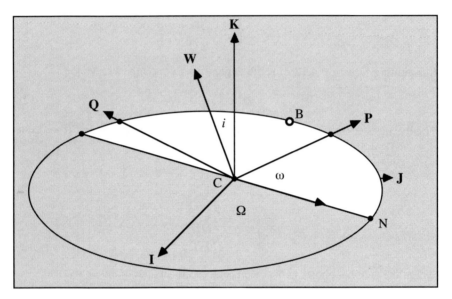

Figure 4.9: The rotation angles Ω, i, and ω.

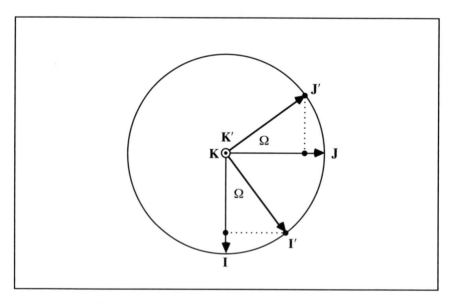

Figure 4.10: Rotation about **K** through angle Ω.

4.5. POSITION AND VELOCITY FROM THE CLASSICAL ELEMENTS 163

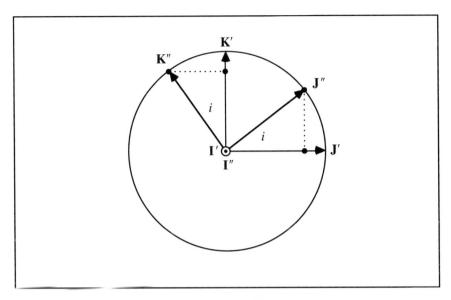

Figure 4.11: Rotation about $\mathbf{I}'$ through angle i.

Step 3 Rotate the $\{\mathbf{I}'', \mathbf{J}'', \mathbf{K}''\}$ system about $\mathbf{K}''$ through the angle ω, as shown in Figure 4.12. The result is the $\{\mathbf{P}, \mathbf{Q}, \mathbf{W}\}$ system, where

$$\begin{aligned} \mathbf{P} &= +\mathbf{I}''\cos\omega + \mathbf{J}''\sin\omega \\ \mathbf{Q} &= -\mathbf{I}''\sin\omega + \mathbf{J}''\cos\omega \\ \mathbf{W} &= +\mathbf{K}''. \end{aligned} \qquad (4.132)$$

Step 4 Substitute Equations 4.130 into Equations 4.131 to obtain the following relationships:

$$\begin{aligned} \mathbf{I}'' &= +\mathbf{I}\cos\Omega + \mathbf{J}\sin\Omega \\ \mathbf{J}'' &= -\mathbf{I}(\sin\Omega\cos i) + \mathbf{J}(\cos\Omega\cos i) + \mathbf{K}\sin i \\ \mathbf{K}'' &= +\mathbf{I}(\sin\Omega\sin i) - \mathbf{J}(\cos\Omega\sin i) + \mathbf{K}\cos i. \end{aligned} \qquad (4.133)$$

Step 5 Substitute Equations 4.133 into Equations 4.132, factor, and rearrange to obtain the final forms of the unit vectors:

$$\begin{aligned} \mathbf{P} = \;&\mathbf{I}(+\cos\omega\cos\Omega - \sin\omega\sin\Omega\cos i) + \\ &\mathbf{J}(+\cos\omega\sin\Omega + \sin\omega\cos\Omega\cos i) + \\ &\mathbf{K}(+\sin\omega\sin i) \end{aligned} \qquad (4.134)$$

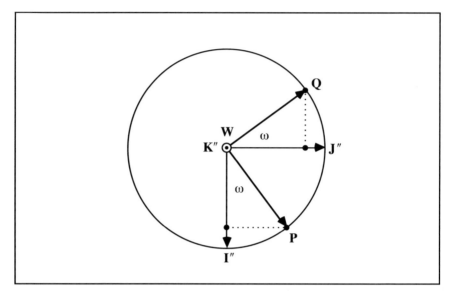

Figure 4.12: Rotation about **K″** through angle ω.

$$\begin{aligned}
\mathbf{Q} &= \mathbf{I}(-\sin\omega\cos\Omega - \cos\omega\sin\Omega\cos i) + \\
&\quad \mathbf{J}(-\sin\omega\sin\Omega + \cos\omega\cos\Omega\cos i) + \\
&\quad \mathbf{K}(+\cos\omega\sin i) \qquad\qquad\qquad\qquad (4.135)
\end{aligned}$$

$$\begin{aligned}
\mathbf{W} &= \mathbf{I}(+\sin\Omega\sin i) + \\
&\quad \mathbf{J}(-\cos\Omega\sin i) + \\
&\quad \mathbf{K}(+\cos i) \,. \qquad\qquad\qquad\qquad\qquad (4.136)
\end{aligned}$$

Equations 4.134, 4.135, and 4.136 express the x, y, and z components of the unit vectors **P**, **Q**, and **W** in terms of the classical elements. In the case of geocentric orbits, the orientation angles i, Ω, and ω are referred to the equatorial coordinate system, so the corresponding unit vectors are also referred to the equator. Thus, Equations 4.110 yield equatorial values for **r** and **v**. However, when the orbit is heliocentric, i, Ω, and ω are measured with respect to the ecliptic coordinate system, and the resulting values of **r** and **v** must be reduced to the equator by the ecliptic-to-equatorial transformation given in Section 2.6.

4.6 Computer Programs

The two computer programs presented in this section perform the computations described in Sections 4.4 and 4.5. They work equally well for geocentric and heliocentric orbits. Although these programs are not essential for the solution of any other problems in this book, their use provides great insight into the workings of the two-body orbit.

4.6.1 Program CLASSEL

Program CLASSEL computes the classical elements from the components of position and velocity.

Program Algorithm

Define	Line Number
FNVS(X,Y,Z): vector squaring function	1070
FNMG(X,Y,Z): vector magnitude function	1080
FNDP(X1,Y1,Z1,X2,Y2,Z2): dot product function	1090
FNASN(X): inverse sine function	1100
FNACN(X): inverse cosine function	1110
FNGSH(X): inverse hyperbolic sine function	1120

Given	Line Number
angle to radian conversion factor	1150
value of π	1160
ε: obliquity of the ecliptic (set $\varepsilon = 0$ for geocentric orbits)	1200
name of the object for information only	1210
equinox for information only	1210
k: gravitational constant	1210

μ: combined mass	1220
t: epoch time	1230
x, y, z: position elements	1240
$\dot{x}, \dot{y}, \dot{z}$: velocity elements	1250
Compute	Line Number
ε in radians	1500
x referred to the ecliptic: Equation 2.39	1520
y referred to the ecliptic: Equation 2.40	1530
z referred to the ecliptic: Equation 2.41	1540
$\dot{x}$ referred to the ecliptic: Equation 2.39	1560
$\dot{y}$ referred to the ecliptic: Equation 2.40	1570
$\dot{z}$ referred to the ecliptic: Equation 2.41	1580
r: Equations 4.68	1620
v^2: Equations 4.68	1630
$r\dot{r}$: Equations 4.68	1640
e: Equations 4.71	1670
h: Equations 4.74	1700-1720
N: Equations 4.77	1740-1760
$1/a$: Equation 4.78	1800
e: Equation 4.79	1810

4.6. COMPUTER PROGRAMS 167

$\wp$: Equation 4.81	1820
q: Equation 4.82	1830
$h = |\mathbf{h}|$	1870
i: Equation 4.84	1880
$N = |\mathbf{N}|$	1900
Ω: Equation 4.86	1910
proper quadrant for Ω	1920
$Ne = \mathbf{N} \cdot \mathbf{e}$	1940
ω: Equation 4.88	1950
proper quadrant for ω	1960
$\bar{x}$: Equation 4.91	2000
$\bar{y}$: Equation 4.92	2010
test for parabolic motion	2060
test for hyperbolic motion	2070
test for elliptic motion	2080
In the case of elliptic motion, compute	
a	2140
b: Equation 4.95	2150
$\cos E$: Equation 4.96	2160
$\sin E$: Equation 4.97	2170
E in the proper quadrant	2190

M: Equation 4.98	2210
M in degrees	2220
n: Equation 4.99	2230
T: Equation 4.35	2240
P: Equation 3.52	2250

In the case of hyperbolic motion, compute

a	2320
b: Equation 4.101	2330
$\sinh H$: Equation 4.100	2340
H	2350
M: Equation 4.102	2360
n: Equation 4.103	2370
T: Equation 4.48	2380

In the case of parabolic motion, compute

q	2450
D: Equation 4.104	2460
M: Equation 4.105	2470
n: Equation 4.106	2480
T: Equation 4.66	2490

End.

4.6. COMPUTER PROGRAMS

Program Listing

```
1000 CLS
1010 PRINT"# CLASSEL # CLASSICAL ELEMENTS"
1020 PRINT"# FROM POSITION AND VELOCITY"
1030 REM
1040 DEFDBL A-Z
1050 DEFINT K
1060 REM
1070 DEF FNVS(X,Y,Z)=X*X+Y*Y+Z*Z
1080 DEF FNMG(X,Y,Z)=SQR(X*X+Y*Y+Z*Z)
1090 DEF FNDP(X1,Y1,Z1,X2,Y2,Z2)=X1*X2+Y1*Y2+Z1*Z2
1100 DEF FNASN(X)=ATN(X/SQR(-X*X+1))
1110 DEF FNACN(X)=-ATN(X/SQR(-X*X+1))+1.5707963263#
1120 DEF FNGSH(X)=LOG(X+SQR(X*X+1))
1130 REM
1140 REM
1150 Q1=.0174532925#
1160 PI=3.1415926536#
1170 G$="#######.#######"
1180 S$="#########.#####"
1190 REM
1200 READ EC
1210 READ N$, E$, K#
1220 READ M
1230 READ T0
1240 READ R0(1),R0(2),R0(3)
1250 READ V0(1),V0(2),V0(3)
1260 REM
1270 LINE INPUT"";A$
1280 CLS
1290 PRINT"CLASSICAL ELEMENTS"
1300 PRINT"FROM"
1310 PRINT"POSITION AND VELOCITY "
1320 PRINT N$
1330 PRINT E$
1340 PRINT"K";TAB(7);:PRINT USING G$;K#
1350 PRINT"EC";TAB(7);:PRINT USING G$;EC
1360 PRINT
1370 PRINT"T(0)";TAB(7);:PRINT USING S$;T0
1380 PRINT
1390 PRINT"POSITION AND VELOCITY"
```

```
1400 PRINT
1410 PRINT"R(K)";TAB(7);
1420 PRINT USING G$;R0(1),R0(2),R0(3)
1430 PRINT"V(K)";TAB(7);
1440 PRINT USING G$;V0(1),V0(2),V0(3)
1450 REM
1460 REM
1470 REM
1480 REM *** REDUCTION TO ECLIPTIC IF REQUIRED ***
1490 REM
1500 EC=EC*Q1
1510 REM
1520 R(1)=R0(1)
1530 R(2)=R0(2)*COS(EC)+R0(3)*SIN(EC)
1540 R(3)=R0(3)*COS(EC)-R0(2)*SIN(EC)
1550 REM
1560 V(1)=V0(1)
1570 V(2)=V0(2)*COS(EC)+V0(3)*SIN(EC)
1580 V(3)=V0(3)*COS(EC)-V0(2)*SIN(EC)
1590 REM
1600 REM *** VECTORS E(K), H(K), AND N(K) ***
1610 REM
1620 R=FNMG(R(1),R(2),R(3))
1630 V2=FNVS(V(1),V(2),V(3))
1640 RV=FNDP(R(1),R(2),R(3),V(1),V(2),V(3))
1650 REM
1660 FOR K=1 TO 3
1670    E(K)=(V2/M-1/R)*R(K)-(RV/M)*V(K)
1680 NEXT K
1690 REM
1700 H(1)=R(2)*V(3)-R(3)*V(2)
1710 H(2)=R(3)*V(1)-R(1)*V(3)
1720 H(3)=R(1)*V(2)-R(2)*V(1)
1730 REM
1740 NN(1)=-H(2)
1750 NN(2)=+H(1)
1760 NN(3)= 0
1770 REM
1780 REM *** ELEMENTS A, E, AND Q ***
1790 REM
1800 AI=2/R-V2/M
1810 E=FNMG(E(1),E(2),E(3))
```

4.6. COMPUTER PROGRAMS

```
1820 SP=FNVS(H(1),H(2),H(3))/M
1830 Q=SP/(1+E)
1840 REM
1850 REM *** ELEMENTS I, OO, AND W ***
1860 REM
1870 H=FNMG(H(1),H(2),H(3))
1880 I=FNACN(H(3)/H)/Q1
1890 REM
1900 NN=FNMG(NN(1),NN(2),NN(3))
1910 OO=FNACN(NN(1)/NN)/Q1
1920 IF NN(2)<0 THEN OO=360-OO
1930 REM
1940 NE=FNDP(NN(1),NN(2),NN(3),E(1),E(2),E(3))
1950 W=FNACN(NE/(NN*E))/Q1
1960 IF E(3)<0 THEN W=360-W
1970 REM
1980 REM *** ELEMENTS MM AND TT ***
1990 REM
2000 XB=(SP-R)/E
2010 YB=RV*SQR(SP/M)/E
2020 PE=0
2030 REM
2040 REM *** DETERMINE CONIC SECTION ***
2050 REM
2060 IF ABS(1-E)<.001 THEN 2430
2070 IF E>1 THEN 2300
2080 IF E<1 THEN 2120
2090 REM
2100 REM *** FOR ELLIPTIC MOTION ***
2110 REM
2120 AQ$="A"
2130 MT$="MM"
2140 AQ=1/AI
2150 B=(1/AI)*SQR(1-E*E)
2160 CX=XB*AI+E
2170 SX=YB/B
2180 REM
2190 GOSUB 11010 REM ARC/SUB
2200 REM
2210 MM=X-E*SX
2220 MT=MM/Q1
2230 N=K#*AI*SQR(M*AI)
```

```
2240 TP=T0-MM/N
2250 PE=(2*PI/K#)*SQR(1/(M*AI*AI*AI))
2260 GOTO 2510
2270 REM
2280 REM *** FOR HYPERBOLIC MOTION
2290 REM
2300 AQ$="A"
2310 MT$="TT"
2320 AQ=1/AI
2330 B=(-1/AI)*SQR(E*E-1)
2340 SX=YB/B
2350 X=FNGSH(SX)
2360 MM=E*SX-X
2370 N=K#*(-AI)*SQR(M*(-AI))
2380 MT=T0-MM/N
2390 GOTO 2510
2400 REM
2410 REM *** FOR PARABOLIC MOTION
2420 REM
2430 AQ$="Q"
2440 MT$="TT"
2450 AQ=Q
2460 DD=RV/SQR(M)
2470 MM=Q*DD+DD*DD*DD/6
2480 N=K#*SQR(M)
2490 MT=T0-MM/N
2500 REM
2510 LINE INPUT"";A$
2520 CLS
2530 PRINT"CLASSICAL ELEMENTS"
2540 PRINT
2550 PRINT AQ$;TAB(7);:PRINT USING G$;AQ
2560 PRINT"E";TAB(7);:PRINT USING G$;E
2570 PRINT MT$;TAB(7);:PRINT USING S$;MT
2580 PRINT"I";TAB(7);:PRINT USING S$;I
2590 PRINT"OO";TAB(7);:PRINT USING S$;OO
2600 PRINT"W";TAB(7);:PRINT USING S$;W
2610 PRINT
2620 REM
2630 IF PE=0 THEN 2680 ELSE 2650
2640 REM
2650 PRINT"TT";TAB(7);:PRINT USING S$;TP
```

4.6. COMPUTER PROGRAMS

173

```
2660 PRINT"PERIOD";TAB(7);:PRINT USING S$;PE
2670 REM
2680 END
11000 STOP
11010 REM # ARC/SUB # COMPUTES X FROM SIN(X) AND COS(X)
11020 IF ABS(SX)<=.707107 THEN X=FNASN(ABS(SX))
11030 IF ABS(CX)<=.707107 THEN X=FNACN(ABS(CX))
11040 IF CX>=0 AND SX>=0 THEN X=X
11050 IF CX<0 AND SX>=0 THEN X=180*Q1-X
11060 IF CX<0 AND SX<0 THEN X=180*Q1+X
11070 IF CX>=0 AND SX<0 THEN X=360*Q1-X
11080 RETURN
20000 DATA   23.439291#
20010 DATA   "MARS", "J2000.0", 0.017202099#
20020 DATA   1.000000323#
20030 DATA   2446280.5#
20040 DATA  -0.8008462#,+1.2418782#,+0.5936583#
20050 DATA  -0.6523424#,-0.3449870#,-0.1405741#
20060 REM
30000 DATA   23.439291#
30010 DATA   "COMET X", "J2000.0", 0.017202099#
30020 DATA   1
30030 DATA   2446400.5#
30040 DATA  +0.9322759#,+1.4977398#,+0.6355047#
30050 DATA  +0.0733900#,-1.0062202#,-0.2356457#
30060 REM
40000 DATA   23.439291#
40010 DATA   "COMET Y","J2000.0",0.017202099#
40020 DATA   1
40030 DATA   2446400.5#
40040 DATA  +0.9322075#,+1.4974398#,+0.6353796#
40050 DATA  +0.0736077#,-1.0031691#,-0.2344978#
40060 REM
50000 DATA   0.0#
50010 DATA   "GEOS","J2000.0",0.07436680#
50020 DATA   1
50030 DATA   95.0#
50040 DATA  +1.0825318#,+0.0000000#,+0.6250000#
50050 DATA  +0.0000000#,+1.0954451#,+0.0000000#
```

4.6.2 Program POSVEL

Program POSVEL computes the position and velocity elements from the classical elements.

Program Algorithm

Define	Line Number
FNSH(X): hyperbolic sine function	1070
FNCH(X): hyperbolic cosine function	1080

Given	Line Number
angle to radian conversion factor	1110
ε: obliquity of the ecliptic (set $\varepsilon = 0$ for geocentric orbits)	1170
name of the object for information only	1180
equinox for information only	1180
k: gravitational constant	1180
μ: combined mass	1190
t: epoch time	1200
elements a(or q), e, M(or T)	1210
elements i, Ω, ω	1220

Compute	Line Number
ε in radians	1510
i in radians	1520
Ω in radians	1530
ω in radians	1540

4.6. COMPUTER PROGRAMS

test for parabolic motion	1600
test for hyperbolic motion	1610
test for elliptic motion	1620
In the case of elliptic motion, compute	
M in radians	1660
initial approximation for E	1670
f: Equation 4.112	1680
test the magnitude of f	1710
df/dE: Equation 4.113	1720
improved value of E by the Newton-Raphson method	1730
r: Equation 4.114	1750
$\dot{E}$: Equation 4.116	1760
b: Equation 4.95	1770
$\bar{x}$: Equations 4.117	1780
$\bar{y}$: Equations 4.117	1790
$\dot{\bar{x}}$: Equations 4.117	1800
$\dot{\bar{y}}$: Equations 4.117	1810
In the case of hyperbolic motion, compute	
n: Equation 4.103	1860
M: Equation 4.107	1870

initial approximation for H	1880
f: Equation 4.118	1890
test the magnitude of f	1920
df/dH: Equation 4.119	1930
improved value of H by the Newton-Raphson method	1940
r: Equation 4.120	1960
$\dot{H}$: Equation 4.122	1970
b: Equation 4.101	1980
$\bar{x}$: Equations 4.123	1990
$\bar{y}$: Equations 4.123	2000
$\dot{\bar{x}}$: Equations 4.123	2010
$\dot{\bar{y}}$: Equations 4.123	2020

In the case of parabolic motion, compute

n: Equation 4.106	2070
M: Equation 4.107	2080
initial approximation for D	2090
f: Equation 4.124	2100
test the magnitude of f	2130
df/dD: Equation 4.125	2140
improved value of D by the Newton-Raphson method	2150
r: Equation 4.126	2170

4.6. COMPUTER PROGRAMS

$\dot{D}$: Equation 4.128 — 2180

$\bar{x}$: Equations 4.129 — 2190

$\bar{y}$: Equations 4.129 — 2200

$\dot{\bar{x}}$: Equations 4.129 — 2210

$\dot{\bar{y}}$: Equations 4.129 — 2220

for the unit vector **P**, compute

P_x: Equation 4.134 — 2260

P_y: Equation 4.134 — 2270

P_z: Equation 4.134 — 2280

for the unit vector **Q**, compute

Q_x: Equation 4.135 — 2300

Q_y: Equation 4.135 — 2310

Q_z: Equation 4.135 — 2320

r: Equations 4.110 — 2370

v: Equations 4.110 — 2380

for heliocentric orbits, compute

x referred to the equator: Equation 2.42 — 2430

y referred to the equator: Equation 2.43 — 2440

z referred to the equator: Equation 2.44 — 2450

$\dot{x}$ referred to the equator: Equation 2.42 — 2470

$\dot{y}$ referred to the equator: Equation 2.43   2480

$\dot{z}$ referred to the equator: Equation 2.44   2490

End.

4.6. COMPUTER PROGRAMS

Program Listing

```
1000 CLS
1010 PRINT"# POSVEL # POSITION AND VELOCITY"
1020 PRINT"# FROM CLASSICAL ELEMENTS"
1030 REM
1040 DEFDBL A-Z
1050 DEFINT K
1060 REM
1070 DEF FNSH(X)=(EXP(X)-EXP(-X))/2
1080 DEF FNCH(X)=(EXP(X)+EXP(-X))/2
1090 REM
1100 REM
1110 Q1=.0174532925#
1120 G$="#######.#######"
1130 S$="#########.#####"
1140 AQ$="A"
1150 MT$="MM"
1160 REM
1170 READ EC
1180 READ N$, E$, K#
1190 READ M
1200 READ T0
1210 READ AQ,E,MT
1220 READ I,O0,W
1230 REM
1240 IF AQ<0 THEN MT$="TT"
1250 IF ABS(1-E)<.001# THEN AQ$="Q"
1260 IF AQ$="Q" THEN MT$="TT"
1270 REM
1280 LINE INPUT"";A$
1290 CLS
1300 PRINT"POSITION AND VELOCITY"
1310 PRINT"FROM"
1320 PRINT"CLASSICAL ELEMENTS"
1330 PRINT N$
1340 PRINT E$
1350 PRINT"K";TAB(7);:PRINT USING G$;K#
1360 PRINT"EC";TAB(7);:PRINT USING G$;EC
1370 PRINT
1380 PRINT"T(0)";TAB(7);:PRINT USING S$;T0
1390 PRINT
```

```
1400 PRINT"CLASSICAL ELEMENTS"
1410 PRINT
1420 PRINT AQ$;TAB(7);:PRINT USING G$;AQ
1430 PRINT"E";TAB(7);:PRINT USING G$;E
1440 PRINT MT$;TAB(7);:PRINT USING S$;MT
1450 PRINT"I";TAB(7);:PRINT USING S$;I
1460 PRINT"OO";TAB(7);:PRINT USING S$;OO
1470 PRINT"W";TAB(7);:PRINT USING S$;W;
1480 REM
1490 LINE INPUT"";A$
1500 CLS
1510 EC=EC*Q1
1520 I=I*Q1
1530 OO=OO*Q1
1540 W=W*Q1
1550 REM
1560 REM
1570 REM
1580 PRINT"NEWTON-RAPHSON SOLUTION"
1590 PRINT
1600 IF AQ$="Q" THEN 2070
1610 IF E>1 THEN 1860
1620 IF E<1 THEN 1660
1630 REM
1640 REM *** ELLIPTIC MOTION ***
1650 REM
1660 MM=MT*Q1
1670 EE=MM
1680    F=EE-E*SIN(EE)-MM
1690    PRINT"F =";
1700    PRINT USING G$;F
1710    IF ABS(F)<.0000001# THEN 1750
1720    DF=1-E*COS(EE)
1730    EE=EE-F/DF
1740    GOTO 1680
1750 R=AQ*(1-E*COS(EE))
1760 EP=SQR(M/AQ)/R
1770 B=AQ*SQR(1-E*E)
1780 XB=AQ*(COS(EE)-E)
1790 YB=B*SIN(EE)
1800 XP=-AQ*EP*SIN(EE)
1810 YP=+B*EP*COS(EE)
```

4.6. COMPUTER PROGRAMS

```
1820 GOTO 2260
1830 REM
1840 REM *** HYPERBOLIC MOTION ***
1850 REM
1860 N=K#*(-1/AQ)*SQR(M*(-1/AQ))
1870 MM=N*(T0-MT)
1880 HH=MM
1890   F=E*FNSH(HH)-HH-MM
1900   PRINT"F =";
1910   PRINT USING G$;F
1920   IF ABS(F)<.0000001# THEN 1960
1930   DF=E*FNCH(HH)-1
1940   HH=HH-F/DF
1950   GOTO 1890
1960 R=AQ*(1-E*FNCH(HH))
1970 HP=SQR(M/(-AQ))/R
1980 B=(-AQ)*SQR(E*E-1)
1990 XB=AQ*(FNCH(HH)-E)
2000 YB=B*FNSH(HH)
2010 XP=AQ*HP*FNSH(HH)
2020 YP=B*HP*FNCH(HH)
2030 GOTO 2260
2040 REM
2050 REM *** PARABOLIC MOTION ***
2060 REM
2070 N=K#*SQR(M)
2080 MM=N*(T0-MT)
2090 DD=MM
2100   F=AQ*DD+DD*DD*DD/6-MM
2110   PRINT"F =";
2120   PRINT USING G$;F
2130   IF ABS(F)<.0000001# THEN 2170
2140   DF=AQ+DD*DD/2
2150   DD=DD-F/DF
2160   GOTO 2100
2170 R=AQ+DD*DD/2
2180 DP=SQR(M)/R
2190 XB=AQ-DD*DD/2
2200 YB=DD*SQR(2*AQ)
2210 XP=-DD*DP
2220 YP=DP*SQR(2*AQ)
2230 REM
```

```
2240 REM *** UNIT VECTORS PP AND QQ ***
2250 REM
2260 PP(1)=+COS(W)*COS(OO)-SIN(W)*SIN(OO)*COS(I)
2270 PP(2)=+COS(W)*SIN(OO)+SIN(W)*COS(OO)*COS(I)
2280 PP(3)=+SIN(W)*SIN(I)
2290 REM
2300 QQ(1)=-SIN(W)*COS(OO)-COS(W)*SIN(OO)*COS(I)
2310 QQ(2)=-SIN(W)*SIN(OO)+COS(W)*COS(OO)*COS(I)
2320 QQ(3)=+COS(W)*SIN(I)
2330 REM
2340 REM *** POSITION AND VELOCITY ***
2350 REM
2360 FOR K=1 TO 3
2370    R(K)=XB*PP(K)+YB*QQ(K)
2380    V(K)=XP*PP(K)+YP*QQ(K)
2390 NEXT K
2400 REM
2410 REM *** REDUCTION TO EQUATOR IF REQUIRED ***
2420 REM
2430 RO(1)=R(1)
2440 RO(2)=R(2)*COS(EC)-R(3)*SIN(EC)
2450 RO(3)=R(3)*COS(EC)+R(2)*SIN(EC)
2460 REM
2470 VO(1)=V(1)
2480 VO(2)=V(2)*COS(EC)-V(3)*SIN(EC)
2490 VO(3)=V(3)*COS(EC)+V(2)*SIN(EC)
2500 REM
2510 LINE INPUT"";A$
2520 CLS
2530 PRINT"POSITION AND VELOCITY"
2540 PRINT
2550 PRINT"R(K)";TAB(7);
2560 PRINT USING G$;RO(1),RO(2),RO(3)
2570 PRINT"V(K)";TAB(7);
2580 PRINT USING G$;VO(1),VO(2),VO(3)
2590 END
20000 DATA   23.439291#
20010 DATA   "PALLAS", "J2000.0", 0.017202099#
20020 DATA   1
20030 DATA   2446400.5#
20040 DATA   2.7720#,0.2337#,334.594#
20050 DATA   34.795#,173.346#,309.909#
```

4.6. COMPUTER PROGRAMS

```
20060 REM
30000 DATA  0.0#
30010 DATA  "RECON1", "J2000.0", 0.07436680#
30020 DATA  1
30030 DATA  1200.0#
30040 DATA  4.0#,0.5#,0.0#
30050 DATA  30.0#,180.0#,270.0#
30060 REM
40000 DATA  0.0#
40010 DATA  "RECON2", "J2000.0", 0.07436680#
40020 DATA  1
40030 DATA  1220.0#
40040 DATA  -2.0#,2.0#,1200.0#
40050 DATA  30.0#,90.0#,90.0#
40060 REM
50000 DATA  0.0#
50010 DATA  "RECON3", "J2000.0", 0.07436680#
50020 DATA  1
50030 DATA  1180.0#
50040 DATA  2.0#,1.0#,1200.0#
50050 DATA  150.0#,270.0#,270.0#
```

4.7 Numerical Examples

4.7.1 Classical Elements for Mars

Problem

Compute a set of heliocentric classical elements for Mars referred to the mean ecliptic and equinox of J2000.0. Assume the following data for the epoch time 1985 August 3 [6]:

- $\varepsilon = 23.439291$
- $k = 0.017202099$
- $\mu = 1.000000323$
- $t_0 = 2446280.5$
- $\mathbf{r}_0 = \{-0.8888462, +1.2418782, +0.5936583\}$
- $\dot{\mathbf{r}}_0 = \{-0.6523424, -0.3449870, -0.1405741\}$

Solution

Use the given information to write data lines 20000 to 20050 as shown at the end of program CLASSEL. Run the program.

Results

```
CLASSICAL ELEMENTS
FROM
POSITION AND VELOCITY
MARS
J2000.0
K              0.0172021
EC            23.4392910

T(0)    2446280.50000

POSITION AND VELOCITY

R(K)      -0.8888462     1.2418782     0.5936583
V(K)      -0.6523424    -0.3449870    -0.1405741
```

4.7. NUMERICAL EXAMPLES

CLASSICAL ELEMENTS

A	1.5237210
E	0.0933060
MM	140.69324
I	1.85078
OO	49.60220
W	286.38354
TT	2446012.01053
PERIOD	686.99965

Discussion of Results

As expected, the orbit of Mars is an ellipse. Additionally, we can compute the longitude of the perifocus

$$\varpi = \Omega + \omega$$
$$\varpi = 49°.60220 + 286°.38354$$
$$\varpi = 335°.98574$$

and the mean longitude

$$L = M + \varpi$$
$$L = 140°.69324 + 335°.98574$$
$$L = 116°.67898 \,.$$

The corresponding values of the Martian elements published in Reference 6 are listed below for comparison:

- $t_0 = 2446280.5$
- $a = 1.5237214$
- $e = 0.0933058$
- $L = 116.67897$
- $i = 1.85078$
- $\Omega = 49.6025$
- $\varpi = 335.9859$

4.7.2 Classical Elements for Comet X

Problem

Compute the heliocentric classical elements for Comet X referred to the standard equinox of J2000.0 given the following data:

- $\varepsilon = 23.439291$
- $k = 0.017202099$
- $\mu = 1$
- $t_0 = 2446400.5$
- $\mathbf{r}_0 = \{+0.9322759, +1.4977398, +0.6355047\}$
- $\dot{\mathbf{r}}_0 = \{+0.0733900, -1.0062202, -0.2356457\}$

Solution

Use what is known to write data lines 30000 to 30050 as shown at the end of program CLASSEL. Run the program.

Results

```
CLASSICAL ELEMENTS
FROM
POSITION AND VELOCITY
COMET X
J2000.0
K            0.0172021
EC          23.4392910

T(0)    2446400.50000

POSITION AND VELOCITY

R(K)      0.9322759      1.4977398      0.6355047
V(K)      0.0733900     -1.0062202     -0.2356457
```

4.7. NUMERICAL EXAMPLES

CLASSICAL ELEMENTS

A	-146.6481544
E	1.0042573
TT	2446495.94869
I	162.60338
OO	58.94733
W	108.04813

Discussion of Results

The orbit of Comet X is found to be a hyperbola with an eccentricity nearly equal to unity. Notice that its motion is retrograde because its orbital inclination is greater than 90 degrees.

4.7.3 Classical Elements for Comet Y

Problem

Compute a set of heliocentric classical elements for Comet Y referred to the standard equinox of J2000.0. Assume the following:

- $\varepsilon = 23.439291$
- $k = 0.017202099$
- $\mu = 1$
- $t_0 = 2446400.5$
- $\mathbf{r}_0 = \{+0.9322075, +1.4974398, +0.6353796\}$
- $\dot{\mathbf{r}}_0 = \{+0.0736077, -1.0031691, -0.2344978\}$

Solution

Use the given data to write lines 40000 to 40050 as shown at the end of program CLASSEL. Run the program.

Results

```
CLASSICAL ELEMENTS
FROM
POSITION AND VELOCITY
COMET Y
J2000.0
K              0.0172021
EC            23.4392910

T(0)    2446400.50000

POSITION AND VELOCITY

R(K)      0.9322075      1.4974398      0.6353796
V(K)      0.0736077     -1.0031691     -0.2344978
```

4.7. NUMERICAL EXAMPLES

CLASSICAL ELEMENTS

Q	0.6225937
E	1.0000005
TT	2446496.17920
I	162.57637
OO	58.94649
W	108.32786

Discussion of Results

The orbit of Comet Y is a parabola because its eccentricity is essentially equal to unity. Its orbital motion is also retrograde. One would strongly suspect that Comet X and Comet Y are the same celestial body and that the differences in their classical elements are due to uncertainties in the assumed positions and velocities.

4.7.4 Classical Elements for GEOS

Problem

Position and velocity elements are listed below for the earth satellite GEOS. Compute a set of geocentric classical elements for GEOS with respect to the standard equinox J2000.0.

- $\varepsilon = 0$
- $k = 0.07436680$
- $\mu = 1$
- $t_0 = 95$
- $\mathbf{r}_0 = \{+1.0825318, 0, +0.6250000\}$
- $\dot{\mathbf{r}}_0 = \{0, +1.0954451, 0\}$

Solution

Use the given information to write data lines 50000 to 50050 as shown at the end of program CLASSEL. Run the program.

Results

```
CLASSICAL ELEMENTS
FROM
POSITION AND VELOCITY
GEOS
J2000.0
K              0.0743668
EC             0.0000000

T(0)          95.00000

POSITION AND VELOCITY

R(K)           1.0825318        0.0000000        0.6250000
V(K)           0.0000000        1.0954451        0.0000000
```

4.7. NUMERICAL EXAMPLES

CLASSICAL ELEMENTS

A	2.5000001
E	0.5000000
MM	0.00000
I	30.00000
OO	270.00000
W	90.00000
TT	95.00000
PERIOD	333.97258

Discussion of Results

The orbit of satellite GEOS is an ellipse, and its motion is direct. The epoch of the elements corresponds to the moment of perifocal passage.

4.7.5 Position and Velocity Elements for Pallas

Problem

A set of classical elliptic elements for the heliocentric orbit of the minor planet Pallas is given below [6]. Compute a corresponding set of heliocentric position and velocity elements for Pallas with respect to the mean equator and equinox of J2000.0.

- $\varepsilon = 23.439291$
- $k = 0.017202099$
- $\mu = 1$
- $t_0 = 2446400.5$
- $a = 2.7720$
- $e = 0.2337$
- $M = 334.594$
- $i = 34.795$
- $\Omega = 173.346$
- $\omega = 309.909$

Solution

Use the given information to write data lines 20000 to 20050 as shown at the end of program POSVEL. Run the program.

Results

```
POSITION AND VELOCITY
FROM
CLASSICAL ELEMENTS
PALLAS
J2000.0
K              0.0172021
EC            23.4392910

T(0)      2446400.50000
```

4.7. NUMERICAL EXAMPLES

CLASSICAL ELEMENTS

A	2.7720000
E	0.2337000
MM	334.59400
I	34.79500
OO	173.34600
W	309.90900

POSITION AND VELOCITY

| R(K) | 0.2440368 | 2.1678371 | -0.4447231 |
| V(K) | -0.7314521 | -0.0041239 | 0.0502226 |

194   CHAPTER 4. ORBIT GEOMETRY

4.7.6   Position and Velocity Elements for Recon 1

Problem

Classical elements for the geocentric elliptic orbit of space probe Recon 1 are listed below for the epoch which corresponds to the time of perifocal passage. Compute a set of geocentric position and velocity elements referred to the standard equinox J2000.0.

- $\varepsilon = 0$
- $k = 0.07436680$
- $\mu = 1$
- $t_0 = 1200$
- $a = 4$
- $e = 0.5$
- $M = 0$
- $i = 30$
- $\Omega = 180$
- $\omega = 270$

Solution

Use the given information to write data lines 30000 to 30050 as shown at the end of program POSVEL. Run the program.

Results

```
POSITION AND VELOCITY
FROM
CLASSICAL ELEMENTS
RECON1
J2000.0
K               0.0743668
EC              0.0000000

T(0)         1200.00000
```

4.7. NUMERICAL EXAMPLES

CLASSICAL ELEMENTS

A	4.0000000
E	0.5000000
MM	0.00000
I	30.00000
OO	180.00000
W	270.00000

POSITION AND VELOCITY

R(K)	0.0000000	1.7320508	-1.0000000
V(K)	-0.8660254	0.0000000	-0.0000000

4.7.7 Position and Velocity Elements for Recon 2

Problem

A set of classical elements for the geocentric hyperbolic orbit of space probe Recon 2 is given below. Compute a set of geocentric position and velocity elements for Recon 2 with respect to the standard equinox J2000.0.

- $\varepsilon = 0$
- $k = 0.07436680$
- $\mu = 1$
- $t_0 = 1220$
- $a = -2$
- $e = 2$
- $T = 1200$
- $i = 30$
- $\Omega = 90$
- $\omega = 90$

Solution

Use the given data to write lines 40000 to 40050 as shown at the end of program POSVEL. Run the program.

Results

```
POSITION AND VELOCITY
FROM
CLASSICAL ELEMENTS
RECON2
J2000.0
K               0.0743668
EC              0.0000000

T(0)         1220.00000
```

4.7. NUMERICAL EXAMPLES

CLASSICAL ELEMENTS

A	-2.0000000
E	2.0000000
TT	1200.00000
I	30.00000
OO	90.00000
W	90.00000

POSITION AND VELOCITY

R(K)	-1.5226403	-1.7541632	0.8790968
V(K)	0.2497130	-1.1055028	-0.1441718

4.7.8 Position and Velocity Elements for Recon 3

Problem

Classical elements for the geocentric parabolic orbit of space probe Recon 3 are listed below. Compute a set of geocentric position and velocity elements for the standard equinox J2000.0.

- $\varepsilon = 0$
- $k = 0.07436680$
- $\mu = 1$
- $t_0 = 1180$
- $q = 2$
- $e = 1$
- $T = 1200$
- $i = 150$
- $\Omega = 270$
- $\omega = 270$

Solution

Use the given information to write data lines 50000 to 50050 as shown at the end of program POSVEL. Run the program.

Results

```
POSITION AND VELOCITY
FROM
CLASSICAL ELEMENTS
RECON3
J2000.0
K              0.0743668
EC             0.0000000

T(0)           1180.00000
```

4.7. NUMERICAL EXAMPLES

CLASSICAL ELEMENTS

Q	2.0000000
E	1.0000000
TT	1200.00000
I	150.00000
OO	270.00000
W	270.00000

POSITION AND VELOCITY

R(K)	1.5116672	1.4268205	−0.8727615
V(K)	0.2740464	−0.8871237	−0.1582208

References

[1] CRC., *Standard Mathematical Tables*, CRC Press, Inc., 1974.

[2] Escobal, *Methods of Orbit Determination*, Krieger Publishing Co., 1976.

[3] Baker and Makemson, *An Introduction to Astrodynamics*, Academic Press, 1967.

[4] Bate, Mueller, and White, *Fundamentals of Astrodynamics*, Dover Publications Inc., 1971.

[5] Thomas, *Calculus and Analytic Geometry*, Addison-Wesley Publishing Company, Inc., 1960.

[6] *The Astronomical Almanac 1985*, U.S. Government Printing Office, 1984.

Chapter 5

Ephemeris Generation

5.1 Introduction

In the previous chapter we derived three simple equations which relate position in the orbital plane to the time elapsed from the moment of perifocal passage:

- $M = E - e \sin E$ for elliptic motion
- $M = e \sinh H - H$ for hyperbolic motion
- $M = qD + D^3/6$ for parabolic motion

However, since these expressions are not in the most convenient form for our application, we shall modify them to obtain three other relationships which express motion in terms of time elapsed from an arbitrary epoch. This is done with a view toward using the modified forms with closed f and g expressions and as a starting point for a universal formulation which is equally applicable to all conic sections.

5.2 The Differenced Kepler Equations

5.2.1 Elliptic Formulation

Consider the case of elliptic motion where, at some given epoch time t_0, there exist corresponding values of the eccentric anomaly E_0 and mean anomaly M_0. Then we can write

$$M_0 = E_0 - e \sin E_0. \tag{5.1}$$

Subtracting Equation 5.1 from the more general form yields

$$M - M_0 = E - E_0 - e \sin E + e \sin E_0, \tag{5.2}$$

where, according to their definitions,

$$M - M_0 = n(t - t_0)$$

$$n = k\sqrt{\frac{\mu}{a^3}}. \tag{5.3}$$

Now, there is a trigonometric identity which states [1]

$$\sin(x + y) = \sin x \cos y + \cos x \sin y. \tag{5.4}$$

Therefore, if we write the $\sin E$ term in the following form

$$\sin E = \sin(E - E_0 + E_0),$$

then we can apply the identity of Equation 5.4 to obtain

$$\sin E = \sin(E - E_0) \cos E_0 + \cos(E - E_0) \sin E_0. \tag{5.5}$$

Thus, substituting Equation 5.5 into Equation 5.2, the result is

$$M - M_0 = (E - E_0) - (e \cos E_0) \sin(E - E_0) - \\ (e \sin E_0) \cos(E - E_0) + (e \sin E_0). \tag{5.6}$$

Equation 5.6 can be written in a more practical form by equating the three trigonometric terms in parentheses to quantities which are easy to compute when the position and velocity elements are known for the epoch. Equations 4.14, 4.29, and 4.116 provide the following relationships:

$$D = \frac{r\dot{r}}{\sqrt{\mu}} \tag{5.7}$$

$$r = a(1 - e \cos E) \tag{5.8}$$

$$\dot{E} = \frac{1}{r}\sqrt{\frac{\mu}{a}}. \tag{5.9}$$

If we rearrange Equation 5.8, we can write

$$e \cos E = 1 - \frac{r}{a}. \tag{5.10}$$

Therefore, at the epoch time t_0 we also have

$$e \cos E_0 = 1 - \frac{r_0}{a}. \tag{5.11}$$

Differentiating Equation 5.10 with respect to modified time yields

$$(e \sin E)\dot{E} = \frac{\dot{r}}{a},$$

5.2. THE DIFFERENCED KEPLER EQUATIONS

and substituting Equation 5.9 for $\dot{E}$ produces

$$e \sin E = \frac{r\dot{r}}{\sqrt{\mu}} \sqrt{\frac{1}{a}}.$$

When the first term on the right side of this expression is replaced by Equation 5.7, we obtain

$$e \sin E = D \sqrt{\frac{1}{a}}. \quad (5.12)$$

Thus, at the epoch t_0 Equation 5.12 becomes

$$e \sin E_0 = D_0 \sqrt{\frac{1}{a}}. \quad (5.13)$$

Now, if we define

$$\begin{aligned} C_0 &= e \cos E_0 \\ S_0 &= e \sin E_0, \end{aligned} \quad (5.14)$$

then, by Equations 5.11 and 5.13, we have

$$C_0 = 1 - \frac{r_0}{a} \quad (5.15)$$

$$S_0 = D_0 \sqrt{\frac{1}{a}}, \quad (5.16)$$

where, in terms of the position and velocity elements,

$$\begin{aligned} r_0 &= |\mathbf{r}_0| \\ D_0 &= \frac{\mathbf{r}_0 \cdot \mathbf{v}_0}{\sqrt{\mu}} \\ \frac{1}{a} &= \frac{2}{r_0} - \frac{\mathbf{v}_0 \cdot \mathbf{v}_0}{\mu}. \end{aligned} \quad (5.17)$$

Finally, if we let

$$\begin{aligned} W &= M - M_0 \\ G &= E - E_0, \end{aligned} \quad (5.18)$$

then Equation 5.6 can be written in the simpler form

$$W = G - C_0 \sin G - S_0 \cos G + S_0. \quad (5.19)$$

5.2.2 Hyperbolic Formulation

The differenced Kepler equation for hyperbolic motion is derived by a process analogous to that used for elliptic motion. Therefore, at a given epoch time t_0, we have

$$M_0 = e \sinh H_0 - H_0, \quad (5.20)$$

which, when subtracted from the general hyperbolic form, yields

$$M - M_0 = e \sinh H - e \sinh H_0 - H + H_0, \quad (5.21)$$

where, according to their definitions,

$$M - M_0 = n(t - t_0)$$

$$n = k\sqrt{\frac{\mu}{(-a)^3}}. \quad (5.22)$$

Using the hyperbolic identity which states [1]

$$\sinh(x + y) = \sinh x \cosh y + \cosh x \sinh y, \quad (5.23)$$

and applying it to the relationship

$$\sinh H = \sinh(H - H_0 + H_0),$$

we obtain

$$\sinh H = \sinh(H - H_0) \cosh H_0 + \cosh(H - H_0) \sinh H_0. \quad (5.24)$$

Thus, substituting Equation 5.24 into Equation 5.21, the result is

$$M - M_0 = -(H - H_0) + (e \cosh H_0) \sinh(H - H_0) + \\ (e \sinh H_0) \cosh(H - H_0) - (e \sinh H_0). \quad (5.25)$$

Equation 5.25 can be written in its more practical form by employing Equations 4.14, 4.39, and 4.122:

$$D = \frac{r\dot{r}}{\sqrt{\mu}} \quad (5.26)$$

$$r = a(1 - e \cosh H) \quad (5.27)$$

$$\dot{H} = \frac{1}{r}\sqrt{\frac{\mu}{(-a)}}. \quad (5.28)$$

5.2. THE DIFFERENCED KEPLER EQUATIONS

Rearranging Equation 5.27, we can write

$$e \cosh H = 1 - \frac{r}{a}. \tag{5.29}$$

Therefore, at the epoch time t_0 we also have

$$e \cosh H_0 = 1 - \frac{r_0}{a}. \tag{5.30}$$

If we differentiate Equation 5.29 with respect to modified time, then

$$(e \sinh H)\dot{H} = \frac{\dot{r}}{(-a)}.$$

Substituting Equation 5.28 for $\dot{H}$ in the above equation produces

$$e \sinh H = \frac{r\dot{r}}{\sqrt{\mu}} \sqrt{\frac{1}{(-a)}}.$$

When the first term on the right side of this expression is replaced by Equation 5.26, we obtain

$$e \sinh H = D \sqrt{\frac{1}{(-a)}}. \tag{5.31}$$

Thus, at the epoch t_0 Equation 5.31 becomes

$$e \sinh H_0 = D_0 \sqrt{\frac{1}{(-a)}}. \tag{5.32}$$

Consequently, if we define

$$\begin{aligned} C_0 &= e \cosh H_0 \\ S_0 &= e \sinh H_0, \end{aligned} \tag{5.33}$$

then, by Equations 5.30 and 5.32,

$$C_0 = 1 - \frac{r_0}{a} \tag{5.34}$$

$$S_0 = D_0 \sqrt{\frac{1}{(-a)}}, \tag{5.35}$$

where, in terms of the position and velocity elements,

$$r_0 = |\mathbf{r}_0|$$

$$D_0 = \frac{\mathbf{r}_0 \cdot \mathbf{v}_0}{\sqrt{\mu}}$$

$$\frac{1}{a} = \frac{2}{r_0} - \frac{\mathbf{v}_0 \cdot \mathbf{v}_0}{\mu}. \tag{5.36}$$

Finally, if we let

$$W = M - M_0$$
$$G = H - H_0, \qquad (5.37)$$

then Equation 5.25 becomes

$$W = -G + C_0 \sinh G + S_0 \cosh G - S_0. \qquad (5.38)$$

5.2.3 Parabolic Formulation

The differenced equation for parabolic motion is obtained from Barker's equation. Thus, at a given epoch t_0 we have

$$M_0 = qD_0 + \frac{D_0^3}{6}. \qquad (5.39)$$

If Equation 5.39 is subtracted from its general parabolic form, the result can be written

$$6(M - M_0) = 6q(D - D_0) + D^3 - D_0^3, \qquad (5.40)$$

where

$$M - M_0 = n(t - t_0)$$
$$n = k\sqrt{\mu}. \qquad (5.41)$$

Now, if we add and subtract appropriate terms on the right side of Equation 5.40 in order to complete the square and the cube, we can manipulate the resulting expression to finally yield

$$6(M - M_0) = (D - D_0)^3 + 3D_0(D - D_0)^2 +$$
$$6q(D - D_0) + 3D_0^2(D - D_0), \qquad (5.42)$$

which can be factored again and rewritten

$$6(M - M_0) = (D - D_0)^3 + 3D_0(D - D_0)^2 +$$
$$6\left(q + \frac{D_0^2}{2}\right)(D - D_0). \qquad (5.43)$$

Furthermore, according to Equation 4.126,

$$r_0 = q + \frac{D_0^2}{2}, \qquad (5.44)$$

so Equation 5.43 can be simplified to the form

$$6W = G^3 + 3D_0 G^2 + 6r_0 G, \qquad (5.45)$$

5.3. THE CLOSED F AND G EXPRESSIONS

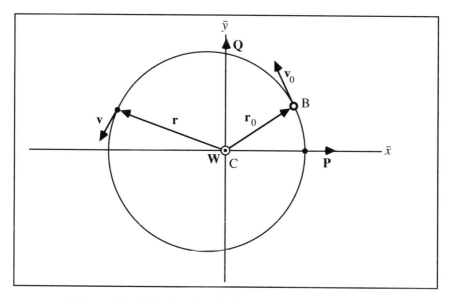

Figure 5.1: Orbital motion during the time interval $t - t_0$.

where

$$W = M - M_0$$
$$G = D - D_0$$
$$r_0 = |\mathbf{r}_0|$$
$$D_0 = \frac{\mathbf{r}_0 \cdot \mathbf{v}_0}{\sqrt{\mu}}. \qquad (5.46)$$

5.3 The Closed f and g Expressions

Closed expressions for the f and g series can be developed for each conic section. Since these closed forms do not suffer from series truncation error, they maintain their accuracy when the computed positions are separated by long intervals of time [2].

Consider the situation shown in Figure 5.1, where the position and motion of celestial body B are referred to the orbit-plane coordinate system at an epoch time t_0. Using $\mathbf{r}_0$ and $\mathbf{v}_0$ as the orbital elements, the position and velocity of B at some other time t is described by the expressions

$$\mathbf{r} = f\mathbf{r}_0 + g\mathbf{v}_0 \qquad (5.47)$$
$$\mathbf{v} = \dot{f}\mathbf{r}_0 + \dot{g}\mathbf{v}_0. \qquad (5.48)$$

Equation 5.47 can be solved for f by taking its vector cross product with $\mathbf{v}_0$. Thus,

$$\mathbf{r} \times \mathbf{v}_0 = f(\mathbf{r}_0 \times \mathbf{v}_0) + g(\mathbf{v}_0 \times \mathbf{v}_0),$$

which becomes

$$\mathbf{r} \times \mathbf{v}_0 = f\mathbf{h} \qquad (5.49)$$

because crossing any vector with itself yields the null vector, and

$$\mathbf{r}_0 \times \mathbf{v}_0 = \mathbf{h}$$

by definition. Now, if $\mathbf{W}$ is the unit vector perpendicular to the orbit-plane and h is the magnitude of the angular momentum vector, then Equation 5.49 can be written

$$(\bar{x}\dot{\bar{y}}_0 - \bar{y}\dot{\bar{x}}_0)\mathbf{W} = fh\mathbf{W}.$$

Equating the scalar coefficients of $\mathbf{W}$, we have the general equation

$$f = \frac{\bar{x}\dot{\bar{y}}_0 - \bar{y}\dot{\bar{x}}_0}{h}. \qquad (5.50)$$

The expression for g is derived by crossing Equation 5.48 with $\mathbf{r}_0$. So that by following a process similar to that used to derive the expression for f, we obtain the general form

$$g = \frac{\bar{x}_0\bar{y} - \bar{y}_0\bar{x}}{h}. \qquad (5.51)$$

5.3.1 Elliptic Motion

According to the formulation of orbital motion described in Sections 4.3.1 and 4.5.1, we have the following relationships:

$$\bar{x} = a(\cos E - e) \qquad (5.52)$$
$$\bar{y} = b\sin E \qquad (5.53)$$
$$\dot{\bar{x}} = -a\dot{E}\sin E \qquad (5.54)$$
$$\dot{\bar{y}} = b\dot{E}\cos E, \qquad (5.55)$$

where

$$b = a\sqrt{1-e^2} \qquad (5.56)$$
$$\dot{E} = \frac{1}{r}\sqrt{\frac{\mu}{a}} \qquad (5.57)$$
$$r = a(1 - e\cos E). \qquad (5.58)$$

5.3. THE CLOSED F AND G EXPRESSIONS

If we substitute Equations 5.52 through 5.55, evaluated at t and t_0, into Equations 5.50 and 5.51, the resulting expressions for f and g can be written

$$f = \frac{ab\dot{E}_0}{h}[(\cos E - e)\cos E_0 + \sin E \sin E_0]$$

$$g = \frac{ab}{h}[(\cos E_0 - e)\sin E - \sin E_0(\cos E - e)], \quad (5.59)$$

where, according to Equations 3.40 and 3.43,

$$h = \sqrt{\mu a(1-e^2)}. \quad (5.60)$$

Utilizing Equations 5.56, 5.57, and 5.60, Equations 5.59 can be written

$$f = \frac{a}{r_0}[(\cos E - e)\cos E_0 + \sin E \sin E_0]$$

$$g = \sqrt{\frac{a^3}{\mu}}[(\cos E_0 - e)\sin E - \sin E_0(\cos E - e)] \quad (5.61)$$

Employing the trigonometric identities [1]

$$\sin(x+y) = \sin x \cos y + \cos x \sin y$$
$$\cos(x+y) = \cos x \cos y - \sin x \sin y, \quad (5.62)$$

we can replace $\sin E$ and $\cos E$ in Equations 5.61 by the expressions

$$\sin(G + E_0) = \sin G \cos E_0 + \cos G \sin E_0$$
$$\cos(G + E_0) = \cos G \cos E_0 - \sin G \sin E_0, \quad (5.63)$$

where

$$G = E - E_0. \quad (5.64)$$

When this substitution is made, Equations 5.61 become

$$f = \frac{a}{r_0}(\cos G - e \cos E_0)$$

$$g = \sqrt{\frac{a^3}{\mu}}[\sin G(1 - e\cos E_0) + (1 - \cos G)(e\sin E_0)]. \quad (5.65)$$

Recalling Equations 5.11 and 5.13, Equations 5.65 can be further reduced to yield

$$f = 1 - \frac{a}{r_0}(1 - \cos G)$$

$$g = \sqrt{\frac{1}{\mu}}[r_0(\sqrt{a}\sin G) + D_0 a(1 - \cos G)]. \quad (5.66)$$

If we define

$$C = a(1 - \cos G)$$
$$S = \sqrt{a} \sin G, \qquad (5.67)$$

then the closed f and g expressions can be written in the compact form

$$f = 1 - \frac{C}{r_0}$$

$$g = \sqrt{\frac{1}{\mu}}(r_0 S + D_0 C). \qquad (5.68)$$

Before proceeding to develop closed expressions for the derivatives of f and g, we must derive an equation for the magnitude of the radius vector at time t. If we replace $\cos E$ in Equation 5.58 with the corresponding trigonometric identity from Equations 5.63, we can write

$$r = a[1 - \cos G(e \cos E_0) + \sin G(e \sin E_0)]. \qquad (5.69)$$

Substituting Equations 5.11 and 5.13 for the appropriate terms above, the result can be arranged to obtain

$$r = a(1 - \cos G) + r_0(\cos G) + D_0(\sqrt{a} \sin G), \qquad (5.70)$$

Finally, if Equations 5.67 are used to replace the terms in parentheses above, Equation 5.70 can be simplified to

$$r = r_0 + C\left(1 - \frac{r_0}{a}\right) + D_0 S. \qquad (5.71)$$

The closed expressions for $\dot{f}$ and $\dot{g}$, required by Equation 5.48, are derived by differentiating Equations 5.68 with respect to modified time. Thus, we obtain

$$\dot{f} = -\frac{\dot{C}}{r_0}$$

$$\dot{g} = \sqrt{\frac{1}{\mu}}(r_0 \dot{S} + D_0 \dot{C}). \qquad (5.72)$$

The next step is to find relationships which express $\dot{C}$ and $\dot{S}$ in terms of quantities which can be computed from the vector elements. Beginning with the definition

$$G = E - E_0,$$

5.3. THE CLOSED F AND G EXPRESSIONS

differentiation yields

$$\dot{G} = \dot{E},$$

since E_0 is a constant. Therefore, according to Equation 5.57,

$$\dot{G} = \frac{1}{r}\sqrt{\frac{\mu}{a}}. \tag{5.73}$$

Turning to the definition for C, we have

$$\dot{C} = a\dot{G}\sin G,$$

which, according to Equation 5.73, can be written

$$\dot{C} = \frac{\sqrt{\mu}}{r}\sqrt{a}\sin G. \tag{5.74}$$

Making use of Equations 5.67, we obtain

$$\dot{C} = \frac{\sqrt{\mu}}{r}S \tag{5.75}$$

Following a similar process for S, we have

$$\dot{S} = \sqrt{a}\,\dot{G}\cos G,$$

which, by Equation 5.73, is

$$\dot{S} = \frac{\sqrt{\mu}}{r}\cos G. \tag{5.76}$$

Again, employing Equations 5.67, we find that

$$\dot{S} = \frac{\sqrt{\mu}}{r}\left(1 - \frac{C}{a}\right). \tag{5.77}$$

Now, if Equations 5.75 and 5.77 are substituted for the appropriate terms in expressions for $\dot{f}$ and $\dot{g}$, the results can be written

$$\dot{f} = -\frac{\sqrt{\mu}}{rr_0}S \tag{5.78}$$

$$\dot{g} = \frac{1}{r}\left\{\left[r_0 + C\left(1 - \frac{r_0}{a}\right) + D_0 S\right] - C\right\}.$$

Finally, if we replace the term in square brackets by Equation 5.71, the expression for $\dot{g}$ reduces to the simple relationship

$$\dot{g} = 1 - \frac{C}{r}. \tag{5.79}$$

In order to use the closed f and g expressions to compute elliptic motion over the time interval $t - t_0$, we first solve the differenced Kepler equation for G by applying the Newton-Raphson method to the function

$$f(G) = G - C_0 \sin G - S_0 \cos G + S_0 - W \qquad (5.80)$$

and its derivative

$$\frac{df(G)}{dG} = 1 - C_0 \cos G + S_0 \sin G. \qquad (5.81)$$

Once G has been used to calculate C and S, the values of f, g, $\dot{f}$, and $\dot{g}$ are determined by their respective closed expressions, and the position and velocity at t are computed from Equations 5.47 and 5.48.

5.3.2 Hyperbolic Motion

The closed f and g expressions for hyperbolic orbital motion are obtained by a process very similar to that used to derive the closed elliptic expressions. Beginning from the hyperbolic formulation described in Sections 4.3.2 and 4.5.2, we have

$$\bar{x} = a(\cosh H - e) \qquad (5.82)$$
$$\bar{y} = b \sinh H \qquad (5.83)$$
$$\dot{\bar{x}} = a\dot{H} \sinh H \qquad (5.84)$$
$$\dot{\bar{y}} = b\dot{H} \cosh H, \qquad (5.85)$$

where

$$b = -a\sqrt{e^2 - 1} \qquad (5.86)$$

$$\dot{H} = \frac{1}{r}\sqrt{\frac{\mu}{(-a)}} \qquad (5.87)$$

$$r = a(1 - e \cosh H). \qquad (5.88)$$

Using these relationships along with Equation 5.60, the expressions for f and g can be transformed to yield the following:

$$f = \frac{a}{r_0}[(\cosh H - e) \cosh H_0 - \sinh H \sinh H_0]$$

$$g = \sqrt{\frac{(-a)^3}{\mu}} \left[(\cosh H_0 - e) \sinh H - \sinh H_0 (\cosh H - e) \right].$$

If we employ the hyperbolic identities [1]

$$\sinh(x + y) = \sinh x \cosh y + \cosh x \sinh y$$
$$\cosh(x + y) = \cosh x \cosh y + \sinh x \sinh y, \qquad (5.89)$$

5.3. THE CLOSED F AND G EXPRESSIONS

we can replace $\sinh H$ and $\cosh H$ by the equivalent expressions

$$\sinh(G + H_0) = \sinh G \cosh H_0 + \cosh G \sinh H_0$$
$$\cosh(G + H_0) = \cosh G \cosh H_0 + \sinh G \sinh H_0, \qquad (5.90)$$

where

$$G = H - H_0. \qquad (5.91)$$

When this substitution is made, the f and g equations reduce to

$$f = \frac{a}{r_0}(\cosh G - e \cosh H_0)$$

$$g = \sqrt{\frac{(-a)^3}{\mu}}\,[\sinh G(1 - e\cosh H_0) + (1 - \cosh G)(e\sinh H_0)]. \qquad (5.92)$$

Making use of Equations 5.30 and 5.32, Equations 5.92 can be written

$$f = 1 - \frac{a}{r_0}(1 - \cosh G)$$

$$g = \sqrt{\frac{1}{\mu}}\,[r_0(\sqrt{(-a)}\sinh G) + D_0 a(1 - \cosh G)]. \qquad (5.93)$$

If we define

$$C = a(1 - \cosh G)$$
$$S = \sqrt{(-a)}\sinh G, \qquad (5.94)$$

then the closed hyperbolic f and g expressions finally reduce to

$$f = 1 - \frac{C}{r_0}$$

$$g = \sqrt{\frac{1}{\mu}}\,(r_0 S + D_0 C), \qquad (5.95)$$

which are the same as their elliptic counterparts. Likewise, the expression for the magnitude of the radius vector at time t can be written

$$r = r_0 + C\left(1 - \frac{r_0}{a}\right) + D_0 S. \qquad (5.96)$$

The closed hyperbolic expressions for $\dot{f}$ and $\dot{g}$ are derived by differentiating Equations 5.95 with respect to modified time. Thus,

$$\dot{f} = -\frac{\dot{C}}{r_0}$$

$$\dot{g} = \sqrt{\frac{1}{\mu}}\,(r_0\dot{S} + D_0\dot{C}). \qquad (5.97)$$

Following the same procedure used to derived the elliptic relationships, we obtain

$$\dot{G} = \frac{1}{r}\sqrt{\frac{\mu}{(-a)}} \qquad (5.98)$$

$$\dot{C} = \frac{\sqrt{\mu}}{r}S \qquad (5.99)$$

$$\dot{S} = \frac{\sqrt{\mu}}{r}\left(1 - \frac{C}{a}\right). \qquad (5.100)$$

If we now substitute Equation 5.99 and 5.100 into the expressions for $\dot{f}$ and $\dot{g}$, the final result is

$$\dot{f} = -\frac{\sqrt{\mu}}{rr_0}S$$

$$\dot{g} = 1 - \frac{C}{r}. \qquad (5.101)$$

We compute hyperbolic motion over the time interval $t - t_0$ by first solving the differenced hyperbolic Kepler equation for G using the Newton-Raphson method. To accomplish this, we use

$$f(G) = -G + C_0 \sinh G + S_0 \cosh G - S_0 - W \qquad (5.102)$$

and its derivative

$$\frac{df(G)}{dG} = -1 + C_0 \cosh G + S_0 \sinh G. \qquad (5.103)$$

Once G has been used to find C and S, the values of f, g, $\dot{f}$, and $\dot{g}$ are calculated from their respective closed hyperbolic expressions, and the position and velocity at t are given by Equations 5.47 and 5.48.

5.3.3 Parabolic Motion

The parabolic formulation of orbital motion described in Sections 4.3.3 and 4.5.3 provide the following relationships:

$$\bar{x} = \frac{1}{2}(\wp - D^2) \qquad (5.104)$$

$$\bar{y} = \sqrt{\wp}D \qquad (5.105)$$

$$\dot{\bar{x}} = -D\dot{D} \qquad (5.106)$$

$$\dot{\bar{y}} = \sqrt{\wp}\dot{D}, \qquad (5.107)$$

5.3. THE CLOSED F AND G EXPRESSIONS

where

$$\dot{D} = \frac{1}{r}\sqrt{\mu} \qquad (5.108)$$

$$r = \frac{1}{2}(\wp + D^2) \qquad (5.109)$$

$$h = \sqrt{\mu\wp}. \qquad (5.110)$$

If we substitute Equations 5.104 through 5.110, evaluated at t and t_0, into Equations 5.50 and 5.51, we obtain the following after some algebraic manipulation:

$$f = 1 - \frac{1}{2r_0}(D - D_0)^2$$

$$g = \sqrt{\frac{1}{\mu}}\left[r_0(D - D_0) + \frac{D_0}{2}(D - D_0)^2\right]. \qquad (5.111)$$

Making use of the parabolic definition

$$G = D - D_0, \qquad (5.112)$$

Equations 5.111 take the more compact forms

$$f = 1 - \frac{1}{2r_0}G^2$$

$$g = \sqrt{\frac{1}{\mu}}\left[r_0 G + \frac{D_0}{2}G^2\right],$$

which are beginning to look familiar. If we now define

$$C = \frac{G^2}{2} \qquad (5.113)$$

$$S = G, \qquad (5.114)$$

then we obtain closed parabolic f and g expressions

$$f = 1 - \frac{C}{r_0}$$

$$g = \sqrt{\frac{1}{\mu}}(r_0 S + D_0 C), \qquad (5.115)$$

which have the same form as those obtained previously for elliptic and hyperbolic motion.

Consider now the parabolic expression for the magnitude of the radius vector given by Equation 5.109, which can be written

$$\wp = 2r - D^2. \tag{5.116}$$

Additionally, since $\wp$ is a constant of the orbit, it is also true that

$$\wp = 2r_0 - D_0^2. \tag{5.117}$$

Substituting Equation 5.117 into Equation 5.109, we obtain

$$r = r_0 + \frac{1}{2}(D^2 - D_0^2). \tag{5.118}$$

Completing the square for the term in parentheses, Equation 5.118 can be made to yield

$$r = r_0 + \frac{1}{2}\left[(D - D_0)^2 + 2D_0(D - D_0)\right], \tag{5.119}$$

which, by the definitions of Equations 5.112 through 5.114, becomes

$$r = r_0 + C + D_0 S. \tag{5.120}$$

The closed parabolic expressions for $\dot{f}$ and $\dot{g}$ are obtained by differentiating Equations 5.115 with respect to modified time. Thus,

$$\begin{aligned} \dot{f} &= -\frac{\dot{C}}{r_0} \\ \dot{g} &= \sqrt{\frac{1}{\mu}}(r_0\dot{S} + D_0\dot{C}). \end{aligned} \tag{5.121}$$

Differentiating Equations 5.112 through 5.114 with respect to modified time and using the relationship given by Equation 5.108, the result can be transformed to yield

$$\dot{C} = \frac{\sqrt{\mu}}{r}S \tag{5.122}$$

$$\dot{S} = \frac{\sqrt{\mu}}{r}. \tag{5.123}$$

Finally, when Equations 5.120, 5.122, and 5.123 are substituted into Equations 5.121, we obtain the same forms common to elliptic and hyperbolic motion, namely,

$$\begin{aligned} \dot{f} &= -\frac{\sqrt{\mu}}{rr_0}S \\ \dot{g} &= 1 - \frac{C}{r}. \end{aligned} \tag{5.124}$$

5.4. THE UNIVERSAL FORMULATION

Parabolic motion during the time interval $t - t_0$ is computed by first solving the differenced form of Barker's equation for G by means of the Newton-Raphson method. Thus, we write Equation 5.45 in the form

$$f(G) = G^3 + 3D_0 G^2 + 6r_0 G - 6W. \tag{5.125}$$

Differentiating with respect to G, we obtain

$$\frac{df(G)}{dG} = 3G^2 + 6D_0 G + 6r_0. \tag{5.126}$$

Once G has been used to calculate C and S, these values are used in the general closed expressions to determine the values of f, g, $\dot{f}$, and $\dot{g}$. Finally, the new position and velocity vectors at time t are computed from Equations 5.47 and 5.48.

5.4 The Universal Formulation

The closed f and g expressions are very convenient for computing orbital motion when the value of the eccentricity clearly indicates that the orbit is an ellipse, hyperbola, or parabola. Unfortunately, the closed elliptic and hyperbolic formulations begin to yield inaccurate results as the eccentricity approaches unity in the ambiguous case of a nearly parabolic orbit. Therefore, it is often advantageous to avoid the need to switch formulas when changing from one conic section to another. Fortunately, general expressions have been developed which retain their accuracy for all values of the eccentricity; however, in order to obtain this *universal formulation*, we shall abandon the closed functions derived in the previous sections and use series expansions [3,4,5].

5.4.1 The Coefficients C, S, and U

According to Equations 5.67, the coefficients C and S are given by the trigonometric relationships

$$C = a(1 - \cos G) \tag{5.127}$$
$$S = \sqrt{a} \sin G. \tag{5.128}$$

These equations can be converted into series expansions by utilizing the following general identities [1]:

$$\cos x = 1 - \frac{x^2}{2!} + \frac{x^4}{4!} - \frac{x^6}{6!} + \cdots \tag{5.129}$$

$$\sin x = x - \frac{x^3}{3!} + \frac{x^5}{5!} - \frac{x^7}{7!} + \cdots \tag{5.130}$$

where, for any integer $n > 0$,

$$n! = n(n-1)(n-2)\cdots(1). \tag{5.131}$$

Therefore, if for convenience we let

$$B_n = \frac{1}{n!}, \tag{5.132}$$

then Equations 5.127 and 5.128 can be written

$$C = a(B_2 G^2 - B_4 G^4 + B_6 G^6 - B_8 G^8 + \cdots) \tag{5.133}$$
$$S = \sqrt{a}(G - B_3 G^3 + B_5 G^5 - B_7 G^7 + \cdots). \tag{5.134}$$

We now make a crucial change of variable which will cause the value of the semimajor axis to appear only in the denominator so that the terms containing $1/a$ will vanish when the semimajor axis becomes infinite. Let

$$X = \sqrt{a}\, G, \tag{5.135}$$

so that

$$G = \frac{X}{\sqrt{a}}. \tag{5.136}$$

Substituting Equation 5.136 for G in Equations 5.133 and 5.134, the result is

$$C = B_2 X^2 - B_4 \frac{X^4}{a} + B_6 \frac{X^6}{a^2} - \cdots \tag{5.137}$$

$$S = X - B_3 \frac{X^3}{a} + B_5 \frac{X^5}{a^2} - \cdots \tag{5.138}$$

If we define a new coefficient U such that

$$U = B_3 X^3 - B_5 \frac{X^5}{a} + \cdots \tag{5.139}$$

then we may also write

$$S = X - \frac{U}{a}. \tag{5.140}$$

The derivatives of C and U with respect to X will be required for the solution of the universal Kepler equation. They are easily found by differentiating Equations 5.137 and 5.139. Thus, we obtain

$$\frac{dC}{dX} = 2B_2 X - 4B_4 \frac{X^3}{a} + \cdots \tag{5.141}$$

$$\frac{dU}{dX} = 3B_3 X^2 - 5B_5 \frac{X^4}{a} + \cdots \tag{5.142}$$

5.4. THE UNIVERSAL FORMULATION

When we take advantage of the values of the B-coefficients given by Equation 5.132, the above expressions for the derivatives can be simplified to yield

$$\frac{dC}{dX} = X - B_3 \frac{X^3}{a} + \cdots \tag{5.143}$$

$$\frac{dU}{dX} = B_2 X^2 - B_4 \frac{X^4}{a} + \cdots \tag{5.144}$$

Therefore, comparing the above expressions with Equations 5.137 and 5.138, we can finally write

$$\frac{dC}{dX} = S \tag{5.145}$$

$$\frac{dU}{dX} = C. \tag{5.146}$$

5.4.2 The Equations of Motion

Now that we have series expansions in X for the coefficients C, S, and U, we can use them to produce a universal formulation of the f and g equations of motion. We begin by combining Equations 5.3, 5.18, and 5.19 to produce the following expression for Kepler's equation:

$$n(t - t_0) = G - C_0 \sin G - S_0 \cos G + S_0, \tag{5.147}$$

where

$$n = k\sqrt{\frac{\mu}{a^3}}. \tag{5.148}$$

If we redefine W to be

$$W = k\sqrt{\mu}\,(t - t_0), \tag{5.149}$$

then Equation 5.147 can be written

$$W = a\sqrt{a}\,G - aC_0\sqrt{a}\sin G + \sqrt{a}\,S_0 a(1 - \cos G). \tag{5.150}$$

Substituting Equations 5.127 and 5.128 for the appropriate terms in Equation 5.150, we obtain

$$W = a\sqrt{a}\,G - aC_0 S + \sqrt{a}\,S_0 C. \tag{5.151}$$

Implementing the same change to the variable X introduced previously, the result is

$$W = aX - aC_0 S + \sqrt{a}\,S_0 C. \tag{5.152}$$

Now, when Equations 5.15, 5.16, and 5.140 are substituted into the above, the result can be arranged to yield

$$W = r_0 X + C_0 U + D_0 C, \qquad (5.153)$$

which is the *universal Kepler's equation*.

The remaining universal equations of motion follow immediately by using the series expansions for C and S in the f and g formulation derived in Section 5.3.1. Thus,

$$f = 1 - \frac{C}{r_0} \qquad (5.154)$$

$$g = \sqrt{\frac{1}{\mu}} (r_0 S + D_0 C) \qquad (5.155)$$

$$r = r_0 + C_0 C + D_0 S \qquad (5.156)$$

$$\dot{f} = -\frac{\sqrt{\mu}}{r r_0} S \qquad (5.157)$$

$$\dot{g} = 1 - \frac{C}{r}, \qquad (5.158)$$

where

$$r_0 = |\mathbf{r}_0| \qquad (5.159)$$

$$D_0 = \frac{\mathbf{r}_0 \cdot \mathbf{v}_0}{\sqrt{\mu}} \qquad (5.160)$$

$$\frac{1}{a} = \frac{2}{r_0} - \frac{\mathbf{v}_0 \cdot \mathbf{v}_0}{\mu} \qquad (5.161)$$

$$C_0 = 1 - \frac{r_0}{a}. \qquad (5.162)$$

In order to use the universal f and g expressions to compute orbital motion over the time interval $t - t_0$, we must first solve the universal Kepler equation for X by applying the Newton-Raphson method to the function

$$f(X) = r_0 X + C_0 U + D_0 C - W \qquad (5.163)$$

and its derivative

$$\frac{df(X)}{dX} = r_0 + C_0 \frac{dU}{dX} + D_0 \frac{dC}{dX}, \qquad (5.164)$$

which can be written as simply

$$\frac{df(X)}{dX} = r_0 + C_0 C + D_0 S \qquad (5.165)$$

5.5. THE EPHEMERIS

when Equations 5.145 and 5.146 are substituted for the derivatives of C and U. Once X has been used to calculate C and S, the values of f, g, $\dot{f}$, and $\dot{g}$ are determined from their universal expressions, and the position and velocity at time t are computed from Equations 5.47 and 5.48.

The universal formulation employed in this text uses series expansions having a limited number of terms. This means that, in some cases, the accuracy of the results will degrade if the length of the time interval between computed positions is extended too far; however, these series converge so rapidly that the step-size can always be much greater than is possible by numerical integration methods. Alternatively, it is possible to compute the series by means of a recurrence process which includes a test that terminates the computation when the required accuracy is achieved [5].

5.5 The Ephemeris

Given a celestial body's position and velocity at a particular epoch time, it is not difficult to compute a sequence of right ascension and declination coordinates at a series of other convenient times. Consider again the fundamental vector equation introduced in Section 2.7:

$$\mathbf{p} = \mathbf{r} + \mathbf{R}. \tag{5.166}$$

As illustrated in Figure 5.2, $\mathbf{r}$ is the radius vector from the dynamical center of motion to the orbiting celestial body, $\mathbf{R}$ is the vector from the observer to the dynamical center, and $\mathbf{p}$ is the vector which defines the position of the orbiting body with respect to the observer. Now, for any given time t, the vector $\mathbf{r}$ can be computed from the orbital elements by numerical integration, the closed f and g series, or a universal formulation. Furthermore, we may assume that the vector $\mathbf{R}$ is also known because the daily geocentric rectangular coordinates of the Sun tabulated in Section C of *The Astronomical Almanac* can be used for heliocentric orbits. The components of $\mathbf{R}$ can also be computed for both geocentric and heliocentric orbits by the procedures described in Chapter 8. Therefore, $\mathbf{p}$ can be determined and used to compute the unit vector $\mathbf{L}$ from

$$\mathbf{L} = \frac{\mathbf{p}}{|\mathbf{p}|}. \tag{5.167}$$

Finally, since

$$\mathbf{L} = \{\cos \delta \cos \alpha, \cos \delta \sin \alpha, \sin \delta\}, \tag{5.168}$$

the scalar components of $\mathbf{L}$ are

$$L_x = \cos \delta \cos \alpha \tag{5.169}$$
$$L_y = \cos \delta \sin \alpha \tag{5.170}$$
$$L_z = \sin \delta, \tag{5.171}$$

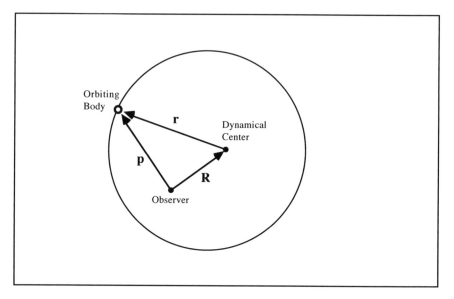

Figure 5.2: The fundamental vector triangle.

and, since $|\mathbf{L}| = 1$, Equations 2.20 through 2.23 reduce to

$$\sin \delta = L_z \tag{5.172}$$

$$\cos \delta = \sqrt{1 - L_z^2} \tag{5.173}$$

$$\cos \alpha = \frac{L_x}{\cos \delta} \tag{5.174}$$

$$\sin \alpha = \frac{L_y}{\cos \delta}, \tag{5.175}$$

which permit α and δ to be found for time t.

In the case of geocentric orbits, we can usually assume that the effects of light-time are negligible. Therefore, the vector $\mathbf{p}$ given by Equation 5.166 can be used immediately to compute $\mathbf{L}$ by Equation 5.167. However, when a heliocentric orbit is being computed, the effects of light-time are normally taken into account because the light which reaches the observer at a given time t had to leave the celestial body at a significantly earlier time t_c. Thus, it appears to the observer to come from the direction of a slightly different orbital position $\mathbf{r}_c$. If we now let $p = |\mathbf{p}|$ represent the distance which the light must travel between the point where it leaves the celestial body and the point where it reaches the

5.5. THE EPHEMERIS

observer, then

$$t_c = t - \frac{p}{c}, \qquad (5.176)$$

where $c = 173.1446$ au/day, so that

$$\frac{1}{c} = 0.005775519 \text{ day/au}. \qquad (5.177)$$

In practice, since $\mathbf{r}_c$ is initially unknown, an approximate value of p is first calculated from Equation 5.166 and used in Equation 5.176 to find t_c. The vector $\mathbf{r}_c$ is then computed for time t_c and used to find an improved value for $\mathbf{p}$ from

$$\mathbf{p} = \mathbf{r}_c + \mathbf{R}. \qquad (5.178)$$

Therefore, we now have as before

$$\mathbf{L} = \frac{\mathbf{p}}{|\mathbf{p}|}, \qquad (5.179)$$

and Equations 5.169 through 5.175 yield α and δ at time t.

224   CHAPTER 5.   EPHEMERIS GENERATION

5.6 Computer Programs

5.6.1 Program SEARCH

Program SEARCH computes geometric ephemerides for geocentric orbits or astrometric ephemerides for heliocentric orbits using the closed f and g expressions.

Program Algorithm

Define   Line Number

    FNVS(X,Y,Z): vector squaring function   1070

    FNMG(X,Y,Z): vector magnitude function   1080

    FNDP(X1,Y1,Z1,X2,Y2,Z2): dot product function   1090

    FNASN(X): inverse sine function   1100

    FNACN(X): inverse cosine function   1110

    FNSH(X): hyperbolic sine function   1120

    FNCH(X): hyperbolic cosine function   1130

Given   Line Number

    angle to radian conversion factor   1180

    $1/c$: light-time constant (set to zero for geocentric orbits)   1270

    name of the object for information only   1290

    equinox for information only   1290

    k: gravitational constant   1290

    μ: combined mass   1300

    t_0: epoch time   1310

5.6. COMPUTER PROGRAMS

$\mathbf{r}_0$: position elements	1320
$\mathbf{v}_0$: velocity elements	1330
number of ephemeris positions	1350
t_i: time for each ephemeris position	1370
$\mathbf{R}_i$: position of the dynamical center for each t_i	1380

Compute	Line Number
r_0: Equations 5.17	1400
D_0: Equations 5.17	1410
$1/a$: Equations 5.17	1420
e: Equation 4.22	1430
For each ephemeris position:	1730
test for parabolic motion	16030
test for hyperbolic motion	16040
test for elliptic motion	16050
In the case of elliptic motion,	
C_0: Equation 5.15	16100
S_0: Equation 5.16	16110
n: Equation 5.3	16120
W: Equations 5.18	16130
G by the Newton-Raphson method	16140-16190
C: Equations 5.67	16200

S: Equations 5.67	16210
In the case of hyperbolic motion,	
$\quad C_0$: Equation 5.34	16270
$\quad S_0$: Equation 5.35	16280
$\quad n$: Equation 5.22	16290
$\quad W$: Equations 5.37	16300
$\quad G$ by the Newton-Raphson method	16310-16360
$\quad C$: Equations 5.94	16370
$\quad S$: Equations 5.94	16380
In the case of parabolic motion,	
$\quad C_0 = 1$	16440
$\quad n$: Equation 5.41	16450
$\quad W$: Equations 5.46	16460
$\quad G$ by the Newton-Raphson method	16470-16520
$\quad C$: Equation 5.113	16530
$\quad S$: Equation 5.114	16540
f: general closed expression	1770
g: general closed expression	1780
$\mathbf{r}$: Equation 5.47	1800
$\mathbf{p}$: Equation 5.166	1810

5.6. COMPUTER PROGRAMS

$p = \|\mathbf{p}\|$	1830
light-time correction for heliocentric orbits	1870
test for improvement of the light-time correction	1880
t_c: Equation 5.176	1890
$\mathbf{r}_c$ for time t_c	1900
L: Equation 5.167	1950
$\cos\delta$: Equation 5.173	1970
$\cos\alpha$: Equation 5.174	1980
$\sin\alpha$: Equation 5.175	1990
α in radians by subroutine ARC/SUB	2010
α in hours	2030
δ in degrees by Equation 5.172	2040
RA in hours, minutes, and seconds	2060-2080
DEC in degrees, minutes, and seconds	2100-2120

End.

Program Listing

```
1000 CLS
1010 PRINT"# SEARCH # SEARCH EPHEMERIS"
1020 PRINT"# BY CLOSED F&G EXPRESSIONS"
1030 REM
1040 DEFDBL A-Z
1050 DEFINT I,K,N
1060 REM
1070 DEF FNVS(X,Y,Z)=X*X+Y*Y+Z*Z
1080 DEF FNMG(X,Y,Z)=SQR(X*X+Y*Y+Z*Z)
1090 DEF FNDP(X1,Y1,Z1,X2,Y2,Z2)=X1*X2+Y1*Y2+Z1*Z2
1100 DEF FNASN(X)=ATN(X/SQR(-X*X+1))
1110 DEF FNACN(X)=-ATN(X/SQR(-X*X+1))+1.5707963263#
1120 DEF FNSH(X)=(EXP(X)-EXP(-X))/2
1130 DEF FNCH(X)=(EXP(X)+EXP(-X))/2
1140 REM
1150 DIM T(20),TA(20)
1160 DIM R(20,3),V(20,3),RR(20,3),P(20,3),LL(20,3)
1170 REM
1180 Q1=.017453293#
1190 G$="####.#######"
1200 A$="####.#####"
1210 D$="######.#####"
1220 AM$="##"
1230 AC$="####.#"
1240 DM$="######"
1250 DC$="######"
1260 REM
1270 READ AB
1280 REM
1290 READ N$, E$, K#
1300 READ M
1310 READ T(0)
1320 READ R(0,1),R(0,2),R(0,3)
1330 READ V(0,1),V(0,2),V(0,3)
1340 REM
1350 READ NP
1360 FOR I=1 TO NP
1370    READ TA(I)
1380    READ RR(I,1),RR(I,2),RR(I,3)
1390 NEXT I
```

5.6. COMPUTER PROGRAMS

```
1400 R0=FNMG(R(0,1),R(0,2),R(0,3))
1410 D0=FNDP(R(0,1),R(0,2),R(0,3),
     V(0,1),V(0,2),V(0,3))/SQR(M)
1420 AI=2/R0-FNVS(V(0,1),V(0,2),V(0,3))/M
1430 E=SQR((1-R0*AI)*(1-R0*AI)+AI*D0*D0)
1440 REM
1450 FOR I=1 TO NP
1460   T(I)=TA(I)
1470 NEXT I
1480 REM
1490 LINE INPUT"";L$
1500 CLS
1510 PRINT"SEARCH EPHEMERIS"
1520 PRINT"MOTION BY CLOSED F&G EXPRESSIONS"
1530 PRINT N$
1540 PRINT E$
1550 PRINT
1560 PRINT"ELEMENTS"
1570 PRINT
1580 PRINT"T(0)";TAB(9);:PRINT USING A$;T(0)
1590 PRINT"R(0,K)";TAB(9);
1600 PRINT USING G$;R(0,1),R(0,2),R(0,3)
1610 PRINT"V(0,K)";TAB(9);
1620 PRINT USING G$;V(0,1),V(0,2),V(0,3)
1630 PRINT
1640 LINE INPUT"";L$
1650 CLS
1660 PRINT "EPHEMERIS"
1670 PRINT
1680 PRINT TAB(4)"T(I)";
1690 PRINT TAB(16)"A(I)";TAB(28)"D(I)";TAB(38)"P(I)"
1700 REM
1710 REM
1720 REM
1730 FOR I=1 TO NP
1740   REM
1750   GOSUB 16010 REM CLOSED/SUB
1760   REM
1770   F=1-CC/R0
1780   G=(R0*SS+D0*CC)/SQR(M)
1790   FOR K=1 TO 3
1800     R(I,K)=F*R(0,K)+G*V(0,K)
```

```
1810    P(I,K)=R(I,K)+RR(I,K)
1820    NEXT K
1830    P=FNMG(P(I,1),P(I,2),P(I,3))
1840    REM
1850    REM *** LIGHT-TIME ***
1860    REM
1870    AP=AB*P
1880    IF T(I)-(TA(I)-AP)<.00001 THEN 1940
1890    T(I)=TA(I)-AP
1900    GOTO 1750
1910    REM
1920    REM *** RA AND DEC ***
1930    REM
1940    FOR K=1 TO 3
1950       LL(I,K)=P(I,K)/P
1960    NEXT K
1970    CD=SQR(1-LL(I,3)*LL(I,3))
1980    CX=LL(I,1)/CD
1990    SX=LL(I,2)/CD
2000    REM
2010    GOSUB 11010 REM ARC/SUB
2020    REM
2030    A=X/(Q1*15)
2040    D=FNASN(LL(I,3))/Q1
2050    REM
2060    AH=FIX(A)
2070    AM=FIX((A-AH)*60)
2080    AC=(A-AH-AM/60)*3600
2090    REM
2100    DD=FIX(D)
2110    DM=FIX((D-DD)*60)
2120    DC=(D-DD-DM/60)*3600
2130    REM
2140    PRINT USING A$;TA(I);
2150    PRINT USING A$;A;
2160    PRINT USING D$;D;
2170    PRINT USING G$;P
2180    REM
2190    PRINT TAB(13);
2200    PRINT USING AM$;AM;
2210    PRINT USING AC$;AC;
2220    PRINT USING DM$;DM;
```

5.6. COMPUTER PROGRAMS

```
2230    PRINT USING DC$;DC;
2240    LINE INPUT"";L$
2250 NEXT I
2260 PRINT CS$
2270 END
11000 STOP
11010 REM # ARC/SUB # COMPUTES X FROM SIN(X) AND COS(X)
11020 IF ABS(SX)<=.707107 THEN X=FNASN(ABS(SX))
11030 IF ABS(CX)<=.707107 THEN X=FNACN(ABS(CX))
11040 IF CX>=0 AND SX>=0 THEN X=X
11050 IF CX<0 AND SX>=0 THEN X=180*Q1-X
11060 IF CX<0 AND SX<0 THEN X=180*Q1+X
11070 IF CX>=0 AND SX<0 THEN X=360*Q1-X
11080 RETURN
16000 STOP
16010 REM # CLOSED/SUB # CLOSED F&G EXPRESSIONS
16020 REM
16030 IF ABS(1-E)<.0001 THEN 16430
16040 IF E>1 THEN 16260
16050 IF E<1 THEN 16090
16060 REM
16070 REM *** ELLIPSE ***
16080 REM
16090 CS$="ELLIPSE"
16100 C0=1-R0*AI
16110 S0=D0*SQR(AI)
16120 N#=K#*SQR(M)*SQR(AI*AI*AI)
16130 WW=N#*(T(I)-T(0))
16140 GG=WW
16150    FG=GG-C0*SIN(GG)-S0*COS(GG)+S0-WW
16160    IF ABS(FG)<.0000001# THEN 16200
16170    DF=1-C0*COS(GG)+S0*SIN(GG)
16180    GG=GG-FG/DF
16190    GOTO 16150
16200 CC=(1-COS(GG))/AI
16210 SS=SIN(GG)/SQR(AI)
16220 RETURN
16230 REM
16240 REM *** HYPERBOLA ***
16250 REM
16260 CS$="HYPERBOLA"
16270 C0=1-R0*AI
```

```
16280 S0=D0*SQR(-AI)
16290 N#=K#*SQR(M)*SQR(-AI*AI*AI)
16300 WW=N#*(T(I)-T(0))
16310 GG=WW
16320    FG=-GG+C0*FNSH(GG)+S0*FNCH(GG)-S0-WW
16330    IF ABS(FG)<.0000001# THEN 16370
16340    DF=-1+C0*FNCH(GG)+S0*FNSH(GG)
16350    GG=GG-FG/DF
16360    GOTO 16320
16370 CC=(1-FNCH(GG))/AI
16380 SS=FNSH(GG)/SQR(-AI)
16390 RETURN
16400 REM
16410 REM *** PARABOLA ***
16420 REM
16430 CS$="PARABOLA"
16440 C0=1
16450 N#=K#*SQR(M)
16460 WW=N#*(T(I)-T(0))
16470 GG=WW
16480    FG=GG*GG*GG+3*D0*GG*GG+6*R0*GG-6*WW
16490    IF ABS(FG)<.0000001# THEN 16530
16500    DF=3*GG*GG+6*D0*GG+6*R0
16510    GG=GG-FG/DF
16520    GOTO 16480
16530 CC=(GG*GG)/2
16540 SS=GG
16550 RETURN
20000 DATA 0
20010 'DATA 0.005775519#
20020 REM
30000 DATA   "GEOS", "J2000.0", 0.07436680#
30010 DATA   1
30020 DATA   95.0#
30030 DATA +1.0825318#,+0.0000000#,+0.6250000#
30040 DATA +0.0000000#,+1.0954451#,+0.0000000#
30050 REM
30060 DATA   7
30070 REM
31000 REM # RR-VECTOR #
31010 DATA   90.0#,-0.7675204#,+0.2004763#,-0.6068605#
31020 DATA   92.0#,-0.7692452#,+0.1937452#,-0.6068629#
```

5.6. COMPUTER PROGRAMS

```
31030 DATA   94.0#,-0.7709111#,+0.1869993#,-0.6068652#
31040 DATA   96.0#,-0.7725179#,+0.1802390#,-0.6068674#
31050 DATA   98.0#,-0.7740654#,+0.1734650#,-0.6068696#
31060 DATA  100.0#,-0.7755536#,+0.1666776#,-0.6068716#
31070 DATA  102.0#,-0.7769824#,+0.1598775#,-0.6068736#
31080 REM
31090 REM
40000 DATA   "PALLAS", "J2000.0", 0.017202099#
40010 DATA   1
40020 DATA   6400.5#
40030 DATA +0.2440368#,+2.1678371#,-0.4447231#
40040 DATA -0.7314521#,-0.0041239#,+0.0502226#
40050 REM
41000 DATA   "COMET X", "J2000.0", 0.017202099#
41010 DATA   1
41020 DATA   6400.5#
41030 DATA +0.0322759#,+1.4977398#,+0.6355047#
41040 DATA +0.0733900#,-1.0062202#,-0.2356457#
41050 REM
41060 DATA   7
41070 REM
45000 REM # RR-VECTOR #
45010 DATA 6070.5#,+0.2513477#,-0.8721056#,-0.3781384#
45020 DATA 6130.5#,+0.9608238#,-0.2266742#,-0.0982791#
45030 DATA 6190.5#,+0.7173926#,+0.6504346#,+0.2820253#
45040 DATA 6250.5#,-0.2145064#,+0.9117811#,+0.3953369#
45050 DATA 6310.5#,-0.9461372#,+0.3214489#,+0.1393729#
45060 DATA 6370.5#,-0.7744047#,-0.5695064#,-0.2469331#
45070 DATA 6430.5#,+0.1614512#,-0.8898986#,-0.3858482#
```

5.6.2 Program RADEC

Program RADEC computes a geometric position for geocentric orbits or an astrometric position for heliocentric orbits using the universal formulation.

Program Algorithm

Define	Line Number
FNVS(X,Y,Z): vector squaring function	1070
FNMG(X,Y,Z): vector magnitude function	1080
FNDP(X1,Y1,Z1,X2,Y2,Z2): dot product function	1090
FNASN(X): inverse sine function	1100
FNACN(X): inverse cosine function	1110

Given	Line Number
angle to radian conversion factor	1160
$1/c$: light-time constant (set to zero for geocentric orbits)	1260
name of the object for information only	1280
equinox for information only	1280
k: gravitational constant	1280
μ: combined mass	1290
t_0: epoch time	1300
$\mathbf{r}_0$: position elements	1310
$\mathbf{v}_0$: velocity elements	1320
t: given time for the RA and DEC position	1340
$\mathbf{R}$: position of the dynamical center for time t	1360

5.6. COMPUTER PROGRAMS 235

number of computation steps	1460
Compute	Line Number
B_n: Equation 5.132	1590-1620
step-size in normal time units	1650
For each ephemeris position:	1680
time at each step	1720
r_0: Equations 5.159	11040
D_0: Equations 5.160	11050
$1/a$: Equations 5.161	11060
C_0: Equation 5.162	11070
W: Equation 5.149	11080
X by the Newton-Raphson method	11090-11200
X^2	11100
X^2/a	11110
X^3	11120
C: Equation 5.137	11130
U: Equation 5.139	11140
S: Equation 5.140	11150
$f(X)$: Equation 5.163	11160
test for solution	11170

$df(X)/dX$: Equations 5.165	11180
improved value for X	11190
f: Equation 5.154	11210
g: Equation 5.155	11220
r: Equation 5.156	11230
$\dot{f}$: Equation 5.157	11240
$\dot{g}$: Equation 5.158	11250
r: Equation 5.47	1770
v: Equation 5.48	1780
$r = |\mathbf{r}|$	1810
p: Equation 5.166	1970
$p = |\mathbf{p}|$	2000
light-time correction for heliocentric orbits	2020
test for improvement of the light-time correction	2030
t_c: Equation 5.176	2040
light-time corrected step-size to compute $\mathbf{r}_c$	2050
$\mathbf{r}_c$ for time t_c	2070-2110
compute improved value of t_c if required	2130
L: Equation 5.167	2180
$\cos \delta$: Equation 5.173	2200
$\cos \alpha$: Equation 5.174	2210

5.6. COMPUTER PROGRAMS

$\sin\alpha$: Equation 5.175	2220
α in radians by subroutine ARC/SUB	2240
α in hours	2260
δ in degrees by Equation 5.172	2270
RA in hours, minutes, and seconds	2290-2310
DEC in degrees, minutes, and seconds	2330-2350

End.

Program Listing

```
1000 CLS
1010 PRINT"# RADEC # RA AND DEC"
1020 PRINT"# BY UNIVERSAL VARIABLES"
1030 PRINT
1040 DEFDBL A-Z
1050 DEFINT I,J,K,N
1060 REM
1070 DEF FNVS(X,Y,Z)=X*X+Y*Y+Z*Z
1080 DEF FNMG(X,Y,Z)=SQR(X*X+Y*Y+Z*Z)
1090 DEF FNDP(X1,Y1,Z1,X2,Y2,Z2)=X1*X2+Y1*Y2+Z1*Z2
1100 DEF FNASN(X)=ATN(X/SQR(-X*X+1))
1110 DEF FNACN(X)=-ATN(X/SQR(-X*X+1))+1.5707963263#
1120 REM
1130 DIM T(100),B(19)
1140 DIM R(100,3),V(100,3)
1150 REM
1160 Q1=.017453293#
1170 G$="####.#######"
1180 A$="####.#####"
1190 D$="######.#####"
1200 P$="######.#######"
1210 AM$="##"
1220 AC$="####.#"
1230 DM$="######"
1240 DC$="######"
1250 REM
1260 READ AB
1270 REM
1280 READ N$, E$, K#
1290 READ M
1300 READ T(0)
1310 READ R(0,1),R(0,2),R(0,3)
1320 READ V(0,1),V(0,2),V(0,3)
1330 REM
1340 READ TF
1350 REM
1360 READ RR(1),RR(2),RR(3)
1370 REM
1380 PRINT N$
1390 PRINT"INITIAL EPOCH:";TAB(15);
```

5.6. COMPUTER PROGRAMS

```
1400 PRINT USING A$;T(0)
1410 PRINT"FINAL EPOCH:";TAB(15);
1420 PRINT USING A$;TF
1430 PRINT"ELAPSED TIME:";TAB(15);
1440 PRINT USING A$;(TF-T(0))
1450 PRINT
1460 INPUT">>>NUMBER OF STEPS";NS
1470 CLS
1480 PRINT"RA AND DEC"
1490 PRINT"MOTION BY UNIVERSAL VARIABLES"
1500 PRINT N$
1510 PRINT E$
1520 PRINT
1530 PRINT TAB(4)"T(I)";
1540 PRINT TAB(18)"X";TAB(30)"Y";TAB(42)"Z";
1550 PRINT TAB(54)"R"
1560 REM
1570 REM
1580 REM
1590 B(1)=1
1600 FOR J=2 TO 19
1610    B(J)=B(J-1)/J
1620 NEXT J
1630 REM
1640 TN=TF
1650 DT=(TN-T(0))/NS
1660 REM
1670 REM
1680 FOR I=0 TO NS
1690    REM
1700    IF I=0 THEN 1810
1710    REM
1720    T(I)=T(I-1)+DT
1730    REM
1740    GOSUB 11010 REM UNIVERSAL/SUB
1750    REM
1760    FOR K=1 TO 3
1770      R(I,K)=F*R(I-1,K)+G*V(I-1,K)
1780      V(I,K)=FP*R(I-1,K)+GP*V(I-1,K)
1790    NEXT K
1800    REM
1810    R=FNMG(R(I,1),R(I,2),R(I,3))
```

```
1820    REM
1830    PRINT USING A$;T(I);
1840    PRINT USING G$;R(I,1),R(I,2),R(I,3),R;
1850    REM
1860    LINE INPUT"";L$
1870 NEXT I
1880 REM
1890 I=I-1
1900 REM
1910 PRINT TAB(11);
1920 PRINT USING G$;V(I,1),V(I,2),V(I,3)
1930 REM
1940 REM *** LIGHT-TIME ***
1950 REM
1960    FOR K=1 TO 3
1970       P(K)=R(I,K)+RR(K)
1980    NEXT K
1990    REM
2000    P=FNMG(P(1),P(2),P(3))
2010    REM
2020    AP=AB*P
2030    IF T(I)-(TF-AP)<.00001 THEN 2170
2040    T(I)=TF-AP
2050    DT=T(I)-T(I-1)
2060    REM
2070    GOSUB 11010 REM UNIVERSAL/SUB
2080    REM
2090    FOR K=1 TO 3
2100       R(I,K)=F*R(I-1,K)+G*V(I-1,K)
2110    NEXT K
2120    REM
2130    GOTO 1960
2140 REM
2150 REM *** RA AND DEC ***
2160 REM
2170 FOR K=1 TO 3
2180    LL(K)=P(K)/P
2190 NEXT K
2200 CD=SQR(1-LL(3)*LL(3))
2210 CX=LL(1)/CD
2220 SX=LL(2)/CD
2230 REM
```

5.6. COMPUTER PROGRAMS

```
2240 GOSUB 15010 REM ARC/SUB
2250 REM
2260 A=X/(15*Q1)
2270 D=FNASN(LL(3))/Q1
2280 REM
2290 AH=FIX(A)
2300 AM=FIX((A-AH)*60)
2310 AC=(A-AH-AM/60)*3600
2320 REM
2330 DD=FIX(D)
2340 DM=FIX((D-DD)*60)
2350 DC=(D-DD-DM/60)*3600
2360 PRINT
2370 PRINT TAB(4)"T(I)";
2380 PRINT TAB(16)"A(I)";TAB(28)"D(I)";TAB(40)"P(I)"
2390 REM
2400 PRINT USING A$;TF;
2410 PRINT USING A$;A;
2420 PRINT USING D$;D;
2430 PRINT USING P$;P
2440 REM
2450 PRINT TAB(13);
2460 PRINT USING AM$;AM;
2470 PRINT USING AC$;AC;
2480 PRINT USING DM$;DM;
2490 PRINT USING DC$;DC
2500 PRINT
2510 END
11000 STOP
11010 REM # UNIVERSAL/SUB # UNIVERSAL VARIABLES
11020 REM
11030 REM
11040 R0=FNMG(R(I-1,1),R(I-1,2),R(I-1,3))
11050 D0=FNDP(R(I-1,1),R(I-1,2),R(I-1,3),
      V(I-1,1),V(I-1,2),V(I-1,3))/SQR(M)
11060 AI=2/R0-FNVS(V(I-1,1),V(I-1,2),V(I-1,3))/M
11070 C0=1-R0*AI
11080 WW=K#*SQR(M)*DT
11090 XX=WW/R0
11100    X2=XX*XX
11110    XA=X2*AI
11120    X3=X2*XX
```

242   CHAPTER 5. EPHEMERIS GENERATION

```
11130    CC=X2*(B(2)-XA*(B(4)-XA*(B(6)-XA*(B(8)-XA*(B(10)-
         XA*(B(12)-XA*(B(14)-XA*(B(16)-XA*(B(18))))))))))
11140    UU=X3*(B(3)-XA*(B(5)-XA*(B(7)-XA*(B(9)-XA*(B(11)-
         XA*(B(13)-XA*(B(15)-XA*(B(17)-XA*(B(19))))))))))
11150    SS=XX-UU*AI
11160    FX=R0*XX+C0*UU+D0*CC-WW
11170    IF ABS(FX)<.00000005# THEN 11210
11180    DF=R0+C0*CC+D0*SS
11190    XX=XX-FX/DF
11200    GOTO 11100
11210 F=1-CC/R0
11220 G=(R0*SS+D0*CC)/SQR(M)
11230 R=R0+C0*CC+D0*SS
11240 FP=-SQR(M)*SS/(R*R0)
11250 GP=1-CC/R
11260 RETURN
15000 STOP
15010 REM # ARC/SUB # COMPUTES X FROM SIN(X) AND COS(X)
15020 IF ABS(SX)<=.707107 THEN X=FNASN(ABS(SX))
15030 IF ABS(CX)<=.707107 THEN X=FNACN(ABS(CX))
15040 IF CX>=0 AND SX>=0 THEN X=X
15050 IF CX<0 AND SX>=0 THEN X=180*Q1-X
15060 IF CX<0 AND SX<0 THEN X=180*Q1+X
15070 IF CX>=0 AND SX<0 THEN X=360*Q1-X
15080 RETURN
20000 'DATA 0
20010 DATA 0.005775519#
20020 REM
30000 DATA   "COMET X", "J2000.0",  0.017202099#
30010 DATA  1
30020 DATA  6400.5#
30030 DATA +0.9322759#,+1.4977398#,+0.6355047#
30040 DATA +0.0733900#,-1.0062202#,-0.2356457#
30050 REM
45000 REM # RR-VECTOR #
45010 DATA 6070.5#,+0.2513477#,-0.8721056#,-0.3781384#
```

5.7 Numerical Examples

5.7.1 Ephemeris for GEOS

Problem

The position and velocity elements for the Earth satellite GEOS are listed below for the standard equinox J2000.0:

- $k = 0.07436680$
- $\mu = 1$
- $t_0 = 95$
- $\mathbf{r}_0 = \{+1.0825318, 0, +0.6250000\}$
- $\dot{\mathbf{r}}_0 = \{0, +1.0954451, 0\}$

Compute an ephemeris for GEOS using the topocentric coordinates of the geocenter listed in the following table:

	GEOCENTER TOPOCENTRIC COORDINATES		
t	X	Y	Z
90.0	−0.7675204	0.2004763	−0.6068605
92.0	−0.7692452	0.1937452	−0.6068629
94.0	−0.7709111	0.1869993	−0.6068652
96.0	−0.7725179	0.1802390	−0.6068674
98.0	−0.7740654	0.1734650	−0.6068696
100.0	−0.7755536	0.1666776	−0.6068716
102.0	−0.7769824	0.1598775	−0.6068736

Solution

Use the given information to write data lines 20000 and 30000 through 31070 as shown at the end of program SEARCH. Run the program.

Results

```
SEARCH EPHEMERIS
MOTION BY CLOSED F&G EXPRESSIONS
GEOS
J2000.0

ELEMENTS

T(0)       95.00000
R(0,K)     1.0825318    0.0000000    0.6250000
V(0,K)     0.0000000    1.0954451    0.0000000

EPHEMERIS

    T(I)         A(I)         D(I)        P(I)
  90.00000    21.59369    -0.61169    0.3431292
              35  37.3    -36  -42
  92.00000    23.37302     1.92681    0.3038187
              22  22.9     55   37
  94.00000     1.25344     3.01454    0.3280219
              15  12.4      0   52
  96.00000     2.68705     2.44159    0.4048796
              41  13.4     26   30
  98.00000     3.64902     1.14557    0.5106123
              38  56.5      8   44
 100.00000     4.31271    -0.33430    0.6297441
              18  45.7    -20   -3
 102.00000     4.80278    -1.82516    0.7545481
              48  10.0    -49  -31
ELLIPSE
```

5.7. NUMERICAL EXAMPLES

5.7.2 Ephemeris for Pallas

Problem

The position and velocity elements for the minor planet Pallas are listed below for the standard equinox J2000.0:

- $k = 0.017202099$
- $\mu = 1$
- $t_0 = 6400.5$
- $\mathbf{r}_0 = \{+0.2440368, +2.1678371, -0.4447231\}$
- $\dot{\mathbf{r}}_0 = \{-0.7314521, -0.0041239, +0.0502226\}$

Compute an ephemeris for Pallas using the geocentric coordinates of the Sun listed in the following table [6]:

SUN
GEOCENTRIC COORDINATES

t	X	Y	Z
6070.5	0.2513477	−0.8721056	−0.3781384
6130.5	0.9608238	−0.2266742	−0.0982791
6190.5	0.7173926	0.6504346	0.2820253
6250.5	−0.2145064	0.9117811	0.3953369
6310.5	−0.9461372	0.3214489	0.1393729
6370.5	−0.7744047	−0.5695064	−0.2469331
6430.5	0.1614512	−0.8898986	−0.3858482

Solution

Use the given information to write data lines 20010, 40000 through 40040, and 41060 through 45070 as shown at the end of program SEARCH. Run the program.

Results

```
SEARCH EPHEMERIS
MOTION BY CLOSED F&G EXPRESSIONS
PALLAS
J2000.0

ELEMENTS

T(0)     6400.50000
R(0,K)      0.2440368    2.1678371   -0.4447231
V(0,K)     -0.7314521   -0.0041239    0.0502226

EPHEMERIS

   T(I)          A(I)          D(I)         P(I)
6070.50000    23.14612     -10.41108    3.3549830
               8   46.0    -24   -40
6130.50000     0.46252      -6.38246    3.7550266
              27   45.1    -22   -57
6190.50000     2.03325      -1.79648    3.6130638
               1   59.7    -47   -47
6250.50000     3.74169      -1.02508    3.0530529
              44   30.1     -1   -30
6310.50000     5.38573      -8.28627    2.3338871
              23    8.6    -17   -11
6370.50000     6.37979     -24.22820    1.7356034
              22   47.2    -13   -42
6430.50000     5.91842     -32.60771    1.4832553
              55    6.3    -36   -28
ELLIPSE
```

5.7. NUMERICAL EXAMPLES

Discussion of Results

The astrometric ephemeris for Pallas listed in the table below was taken from *The Astronomical Almanac* [6]. It represents the motion of the minor planet during the same 360-day period used in this numerical example. A comparison of this ephemeris with the results of program SEARCH shows that during this interval of time, a two-body orbit agrees fairly well with the perturbed motion which takes into account the influences of the other bodies in the solar system.

PALLAS
ALMANAC POSITIONS

t	α	δ	p
	h m s	° ′ ″	
6070.5	23 08 45.2	−10 24 37	3.355
6130.5	0 27 44.9	− 6 22 54	3.755
6190.5	2 01 59.7	− 1 47 45	3.613
6250.5	3 44 30.1	− 1 01 29	3.053
6310.5	5 23 08.6	− 8 17 10	2.334
6370.5	6 22 47.1	−24 13 41	1.736
6430.5	5 55 06.2	−32 36 27	1.483

5.7.3 Ephemeris for Comet X

Problem

The position and velocity elements for the Comet X are listed below for the standard equinox J2000.0:

- $k = 0.017202099$
- $\mu = 1$
- $t_0 = 6400.5$
- $\mathbf{r}_0 = \{+0.9322759, +1.4977398, +0.6355047\}$
- $\dot{\mathbf{r}}_0 = \{+0.0733900, -1.0062202, -0.2356457\}$

Compute an ephemeris for Comet X using the geocentric coordinates of the Sun employed in the previous numerical example, namely [6]:

	SUN		
	GEOCENTRIC COORDINATES		
t	X	Y	Z
6070.5	0.2513477	−0.8721056	−0.3781384
6130.5	0.9608238	−0.2266742	−0.0982791
6190.5	0.7173926	0.6504346	0.2820253
6250.5	−0.2145064	0.9117811	0.3953369
6310.5	−0.9461372	0.3214489	0.1393729
6370.5	−0.7744047	−0.5695064	−0.2469331
6430.5	0.1614512	−0.8898986	−0.3858482

Solution

Use the given information to write data lines 20010, 41000 through 41040, and 41060 through 45070 as shown at the end of program SEARCH. Run the program.

5.7. NUMERICAL EXAMPLES

Results

```
SEARCH EPHEMERIS
MOTION BY CLOSED F&G EXPRESSIONS
COMET X
J2000.0

ELEMENTS

T(0)     6400.50000
R(0,K)    0.9322759    1.4977398    0.6355047
V(0,K)    0.0733900   -1.0062202   -0.2356457

EPHEMERIS

   T(I)          A(I)         D(I)        P(I)
6070.50000     5.84719      12.21383    4.7612022
              50   49.9     12    50
6130.50000     5.12348      13.86498    4.9732119
               7   24.5     51    54
6190.50000     5.19625      16.16836    5.2652978
              11   46.5     10     6
6250.50000     5.72884      17.84424    4.7620471
              43   43.8     50    39
6310.50000     6.26918      18.45120    3.3551362
              16    9.0     27     4
6370.50000     5.71846      19.31346    1.5093709
              43    6.4     18    48
6430.50000     0.18782       5.80094    1.1137195
              11   16.2     48     3
HYPERBOLA
```

5.7.4 Right Ascension and Declination of Comet X

Problem

Use the information below to compute the right ascension and declination of Comet X for 0^h TDT on 1985 Jan 5. All positions are measured with respect to the equinox J2000.0.

- $k = 0.017202099$
- $\mu = 1$
- $t_0 = 2446400.5$
- $\mathbf{r}_0 = \{+0.9322759, +1.4977398, +0.6355047\}$
- $\dot{\mathbf{r}}_0 = \{+0.0733900, -1.0062202, -0.2356457\}$
- $t = 2446070.5$
- $\mathbf{R} = \{+0.2513477, -0.8721056, -0.3781384\}$

Solution

Use the given information to write data lines 20010 through 45010 as shown at the end of program RADEC. Run the program and answer the prompt as follows:

```
COMET X
INITIAL EPOCH:6400.50000
FINAL EPOCH:   6070.50000
ELAPSED TIME:  -330.00000

>>>NUMBER OF STEPS? 1
```

5.7. NUMERICAL EXAMPLES

Results

```
RA AND DEC
MOTION BY UNIVERSAL VARIABLES
COMET X
J2000.0
```

T(I)	X	Y	Z	R
6400.50000	0.9322759	1.4977398	0.6355047	1.8751611
6070.50000	-0.0651432	5.5215499	1.3853802	5.6930691
	0.1963883	-0.5583680	-0.0882054	

T(I)	A(I)	D(I)	P(I)
6070.50000	5.84719	12.21383	4.7612022
	50 49.9	12 50	

Discussion of Results

As we would expect, the coordinates computed by program RADEC agree with those found for the same date using program SEARCH in the previous numerical example.

References

[1] CRC, *Standard Mathematical Tables*, CRC Press Inc., 1974.

[2] Escobal, *Methods of Orbit Determination*, Krieger Publishing Co., 1976.

[3] Baker and Makemson, *An Introduction to Astrodynamics*, Academic Press, 1967.

[4] Baker, *Astrodynamics: Applications and Advanced Topics*, Academic Press, 1967.

[5] Danby, *Fundamentals of Celestial Mechanics*, Willmann-Bell, 1988.

[6] *The Astronomical Almanac 1985*, U.S. Government Printing Office, 1984.

Chapter 6

Special Perturbations

6.1 Introduction

The long-term motion of a well known celestial body can be computed from a set of analytical expressions which take into account perturbations and yield the position of the body over great intervals of past or future time. This approach is called the method of *general perturbations*, and such theories permit the most detailed understanding of a celestial body's motion. In contrast to this, the method of *special perturbations* is normally used to predict orbital motion over a relatively limited span of time by numerical methods. Although this procedure may not be as useful for rigorous studies of an object's motion, it is usually much simpler to accomplish and follows directly from first principles. These fundamental principles were considered in Chapter 1 during the derivation of Equation 1.26, namely,

$$\ddot{\mathbf{r}} = -\frac{\mu \mathbf{r}}{r^3} + \sum_{q=1}^{n} m_q \left(\frac{\mathbf{p}_q}{p_q^3} - \frac{\mathbf{r}_q}{r_q^3} \right), \tag{6.1}$$

where the first term on the right side represents the two-body acceleration, and the summation terms are the attractions of the n perturbing masses m_q.

The special perturbation procedures described in this introductory treatment are based on the general features of the methods of *Cowell* and *Encke*. The resulting computer programs utilize several algorithms from previous chapters such as the Runge-Kutta method, closed f and g expressions, and universal variables. Since the treatment is introductory, the application is restricted to the case of heliocentric orbits because the tables published in *The Astronomical Almanac* afford the best opportunity to demonstrate the method. The interested reader should consult References 1 through 7 for additional information about special perturbations methods, various numerical procedures, and the effects of accumulated round-off error.

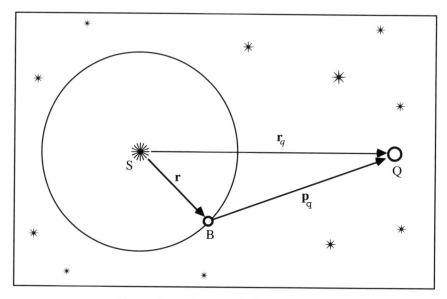

Figure 6.1: The perturbation geometry.

6.2 Direct and Indirect Attractions

The subject of perturbations was briefly discussed in Section 3.2 in order to justify their exclusion from the two-body equation of motion. From that point until now, all orbital computations have been based on the simplifying assumption that the effects of the attractions of other celestial bodies have not been significant during the time periods covered by the various calculations. However, while the gravitational perturbations of objects orbiting the Sun are normally small, their cumulative effect can be significant when the orbital motion is followed for several years.

Consider the situation depicted in Figure 6.1, where the orbit of a celestial body B about the Sun S is perturbed by the gravitational attraction of a third body Q whose mass m_q and position $\mathbf{r}_q$ are known with sufficient accuracy from published data. The interaction of these three bodies is similar to that which occurs in the numerical example of Section 1.7. In this case we can drop the summation notation, and Equation 6.1 reduces to

$$\ddot{\mathbf{r}} = -\frac{\mu \mathbf{r}}{r^3} + m_q \left(\frac{\mathbf{p}_q}{p_q^3} - \frac{\mathbf{r}_q}{r_q^3} \right), \tag{6.2}$$

which can also be written

$$\ddot{\mathbf{r}} = -\frac{\mu \mathbf{r}}{r^3} + \frac{m_q \mathbf{p}_q}{p_q^3} - \frac{m_q \mathbf{r}_q}{r_q^3}. \tag{6.3}$$

6.2. DIRECT AND INDIRECT ATTRACTIONS

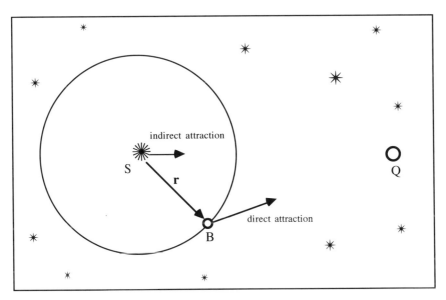

Figure 6.2: The direct and indirect attractions.

The first term on the right side of Equation 6.3 is, of course, the relative acceleration caused by the combined mass of the Sun and the orbiting body. The second term represents the action of m_q on the orbiting body, and the third term represents the action of m_q on the Sun. These latter two quantities are called the *direct* and *indirect attractions*, respectively. In other words,

$$\text{direct attraction} = \frac{m_q \mathbf{p}_q}{p_q^3}$$

$$\text{indirect attraction} = \frac{m_q \mathbf{r}_q}{r_q^3}.$$

These actions are illustrated in Figure 6.2, where, in the case shown, the direct attraction is greater than the indirect attraction because the perturbing body is closer to the the orbiting body than to the Sun. Notice also that the Sun carries the origin of the position vector $\mathbf{r}$ as it is attracted toward body Q. Therefore, as was pointed out in Section 3.2, the observed perturbation of $\mathbf{r}$ results from a net attraction

$$\mathbf{a}_q = m_q \left(\frac{\mathbf{p}_q}{p_q^3} - \frac{\mathbf{r}_q}{r_q^3} \right),$$

which is the difference between the direct and indirect attractions produced by mass m_q. Finally, it is important to keep in mind that this net attraction is in fact an *acceleration*, not a force or a perturbation. The perturbation results when

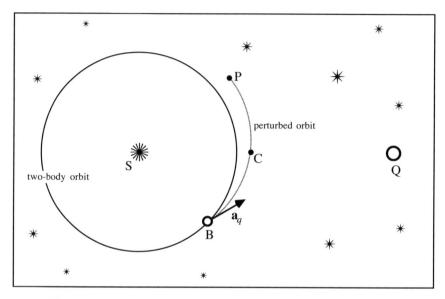

Figure 6.3: The perturbation resulting from the net acceleration.

the net acceleration acts over time to change the velocity of the orbiting body with respect to the Sun, and the new velocity, in time, moves the orbiting body to its perturbed position. An example of this action is illustrated in Figure 6.3, where the greatest perturbation of B does not occur near C, when it is closest to Q, but at some point P farther along the orbit where its increased velocity has had its full effect.

Regardless of the special perturbation method used, the numerical integration requires one to know each perturbing planetary mass m_q as well as the position $\mathbf{r}_q$ of each perturber at the beginning of every step of the computation. When these are available, the distance $\mathbf{p}_q$ can be found from the relationship

$$\mathbf{p}_q = \mathbf{r}_q - \mathbf{r}, \tag{6.4}$$

which follows from the geometry of Figure 6.1. Fortunately, the necessary data can be obtained from Reference 8 or *The Astronomical Almanac*.

The computer programs described in this chapter use the position and velocity elements of the major planets published in *The Astronomical Almanac* to compute the positions of all these perturbing bodies by means of closed elliptic f and g expressions. Although these functions produce two-body orbits for the major planets, the differences between the actual and computed positions are not significant for our purposes because the perturbing attractions are small quantities. This procedure will usually work for periods up to several years unless

6.3 The Method of Cowell

The method of Cowell computes orbital motion by a direct numerical integration of all the terms on the right side of Equation 6.1. The perturbed position and velocity found at the end of each integration step are used as the orbital elements to begin the next step. Program COWELL, listed in Section 6.6.2, is an example of Cowell's method using a fifth-order Runge-Kutta method similar to that described in Section 3.5.3. In this case, however, the algorithm is much more complex.

Consider the situation where an orbiting celestial body's position $\mathbf{r}_0$ and velocity $\mathbf{v}_0$ are known for a given epoch t_0, along with the positions $\mathbf{r}_q$ and velocities $\mathbf{v}_q$ of n perturbing masses m_q. Let Equation 6.1 be rewritten as follows:

$$\ddot{\mathbf{r}} = -(1+m)\frac{\mathbf{r}}{r^3} + \mathbf{a}_t, \tag{6.5}$$

where m is the mass of the orbiting body, $\mathbf{a}_t$ is the total perturbing attraction given by

$$\mathbf{a}_t = \sum_{q=1}^{n} m_q \left(\frac{\mathbf{p}_q}{p_q^3} - \frac{\mathbf{r}_q}{r_q^3} \right), \tag{6.6}$$

and

$$\mathbf{p}_q = \mathbf{r}_q - \mathbf{r}. \tag{6.7}$$

Since the positions $\mathbf{r}_q$ and velocities $\mathbf{v}_q$ of the perturbing bodies can be determined for any epoch by means of closed f and g expressions or other analytical procedures, we may assume that all the quantities on the right side of Equation 6.5 are known at the beginning of each integration step.

In order to apply the Runge-Kutta method to Equation 6.5, we must express this second-order differential equation in terms of two first-order differential equations as follows:

$$\frac{d\mathbf{r}}{d\tau} = f(\mathbf{v}) \tag{6.8}$$

$$\frac{d\mathbf{v}}{d\tau} = g(m, \mathbf{r}, r, \mathbf{a}_t), \tag{6.9}$$

where

$$f(\mathbf{v}) = \mathbf{v}$$
$$g(m, \mathbf{r}, r, \mathbf{a}_t) = -(1+m)\frac{\mathbf{r}}{r^3} + \mathbf{a}_t.$$

However, the integration of Equations 6.8 and 6.9 also makes it necessary to move the perturbers along their orbits during the integration step so that their changing attractions can be evaluated. The calculation of these quantities for each of the six *internal* steps of the Runge-Kutta method may be included as part of the numerical integration using only two-body accelerations. Thus, we also have for each perturbing body

$$\frac{d\mathbf{r}_q}{d\tau} = \mathbf{v}_q \tag{6.10}$$

$$\frac{d\mathbf{v}_q}{d\tau} = -(1 + m_q)\frac{\mathbf{r}_q}{r_q^3}. \tag{6.11}$$

The Runge-Kutta special perturbation routine integrates the functions of Equations 6.8 through 6.11 by means of auxiliary vectors having the following general forms:

$$\mathbf{F}_j = hf(\mathbf{v})_j \tag{6.12}$$
$$\mathbf{G}_j = hg(m, \mathbf{r}, r, \mathbf{a}_t)_j, \tag{6.13}$$

where h is the modified time interval of the step and $j = 1$ to 6. Finally, the vector increments for the celestial body's orbital motion are found from

$$\delta\mathbf{r} = \frac{1}{90}(7\mathbf{F}_1 + 32\mathbf{F}_3 + 12\mathbf{F}_4 + 32\mathbf{F}_5 + 7\mathbf{F}_6) \tag{6.14}$$

$$\delta\mathbf{v} = \frac{1}{90}(7\mathbf{G}_1 + 32\mathbf{G}_3 + 12\mathbf{G}_4 + 32\mathbf{G}_5 + 7\mathbf{G}_6), \tag{6.15}$$

and the body's new position and velocity are given by

$$\mathbf{r} = \mathbf{r}_0 + \delta\mathbf{r} \tag{6.16}$$
$$\mathbf{v} = \mathbf{v}_0 + \delta\mathbf{v}, \tag{6.17}$$

which become the starting point for the next integration step.

6.4 The Method of Encke

The method of Encke takes advantage of the fact that only perturbing attractions must be integrated numerically, since two-body motion can be computed by closed f and g expressions, universal variables, or any other convenient device. Therefore, in Encke's method, only the difference between a two-body *reference orbit* and the perturbed orbit is integrated. This approach permits a larger stepsize than Cowell's method; however, the computation becomes somewhat more

6.4. THE METHOD OF ENCKE

complicated. Furthermore, since the method of Encke becomes less efficient as the difference between the reference and perturbed orbits grows larger, it is periodically necessary to adjust the elements of the reference orbit to take into account the accumulated perturbations. This procedure is called *rectification*.

Consider again the total acceleration of an orbiting body given by Equation 6.5, that is,

$$\ddot{\mathbf{r}} = -(1+m)\frac{\mathbf{r}}{r^3} + \mathbf{a}_t. \tag{6.18}$$

As before, the last term is the total perturbing attraction

$$\mathbf{a}_t = \sum_{q=1}^{n} m_q \left(\frac{\mathbf{p}_q}{p_q^3} - \frac{\mathbf{r}_q}{r_q^3} \right), \tag{6.19}$$

where

$$\mathbf{p}_q = \mathbf{r}_q - \mathbf{r}. \tag{6.20}$$

Now, if we let the two-body acceleration along the reference orbit be represented by

$$\ddot{\mathbf{r}}_\mu = -(1+m)\frac{\mathbf{r}_\mu}{r_\mu^3} \tag{6.21}$$

and subtract this expression from Equation 6.18, the result is

$$\ddot{\mathbf{r}} - \ddot{\mathbf{r}}_\mu = (1+m)\left(\frac{\mathbf{r}_\mu}{r_\mu^3} - \frac{\mathbf{r}}{r^3}\right) + \mathbf{a}_t,$$

or

$$\delta\ddot{\mathbf{r}} = (1+m)\left(\frac{\mathbf{r}_\mu}{r_\mu^3} - \frac{\mathbf{r}}{r^3}\right) + \mathbf{a}_t. \tag{6.22}$$

Equation 6.22 represents the differential acceleration which leads to the perturbed motion.

Program ENCKE, listed in Section 6.6.3, demonstrates a special perturbation method which is similar to Encke's in that a reference orbit is computed by means of universal variables, and only the perturbing acceleration of Equation 6.22 is integrated numerically. However, the procedure is also similar to Cowell's method because the position and velocity elements are continuously updated at the end of each step and used as the basis for the next step. This procedure avoids some of the complexities found in the classical method of Encke. In any event, the result can be expressed as follows:

$$\mathbf{r} = \mathbf{r}_\mu + \delta\mathbf{r} \tag{6.23}$$
$$\mathbf{v} = \mathbf{v}_\mu + \delta\mathbf{v}, \tag{6.24}$$

where $\mathbf{r}_\mu$ and $\mathbf{v}_\mu$ are the position and velocity vectors calculated for the end of the computation step from the two-body acceleration by universal variables,

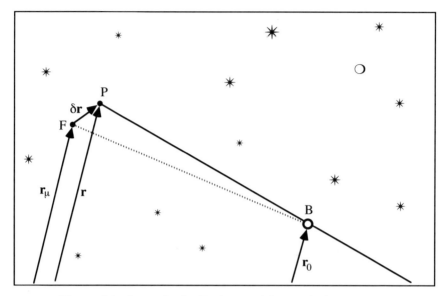

Figure 6.4: A step in the Encke special perturbation method.

and the increments $\delta\mathbf{r}$ and $\delta\mathbf{v}$ are the perturbations of the position and velocity vectors computed by numerical integration of the differential acceleration.

Figure 6.4 depicts a small portion of the orbit to illustrate the relationship between the vectors of Equation 6.23. At the beginning of the computation step, celestial body B is located at position $\mathbf{r}_0$, where the two-body reference orbit and perturbed orbit coincide. In the absence of perturbations, B would travel along the reference orbit during the time interval of the computation to arrive at point F, where the position vector is $\mathbf{r}_\mu$. However, when perturbations are taken into account, the altered motion of B will carry it to point P, where the position vector $\mathbf{r}$ is the sum of $\mathbf{r}_\mu$ and the increment $\delta\mathbf{r}$. The next step in the orbit computation will use $\mathbf{r}$ and the corresponding velocity vector $\mathbf{v}$ from Equations 6.23 and 6.24 as the new elements $\mathbf{r}_0$ and $\mathbf{v}_0$, respectively. Thus, the process continues all around the orbit.

The principal functions which must be numerically integrated to determine the vectors $\delta\mathbf{r}$ and $\delta\mathbf{v}$ are the following:

$$\frac{d}{d\tau}(\delta\mathbf{r}) = f(\delta\mathbf{v}) \tag{6.25}$$

$$\frac{d}{d\tau}(\delta\mathbf{v}) = g(m, \mathbf{r}_\mu, r_\mu, \mathbf{r}, r, \mathbf{a}_t), \tag{6.26}$$

6.4. THE METHOD OF ENCKE

where

$$f(\delta \mathbf{v}) = \delta \mathbf{v}$$
$$g(m, \mathbf{r}_\mu, r_\mu, \mathbf{r}, r, \mathbf{a}_t) = (1+m)\left(\frac{\mathbf{r}_\mu}{r_\mu^3} - \frac{\mathbf{r}}{r^3}\right) + \mathbf{a}_t.$$

The vectors $\mathbf{r}_\mu$ and $\mathbf{v}_\mu$ are known at the beginning of each integration step because they are equal to the position $\mathbf{r}$ and velocity $\mathbf{v}$ computed at the end of the previous step. Thus, the difference between $\mathbf{r}_\mu/r_\mu^3$ and $\mathbf{r}/r^3$ remains small during each integration step. The calculation of these quantities during the *internal* steps of the Runge-Kutta method may be accomplished by numerical integration. The functions to be integrated are the following:

$$\frac{d\mathbf{r}_\mu}{d\tau} = \mathbf{v}_\mu \tag{6.27}$$

$$\frac{d\mathbf{v}_\mu}{d\tau} = -(1+m)\frac{\mathbf{r}_\mu}{r_\mu^3} \tag{6.28}$$

$$\frac{d\mathbf{r}}{d\tau} = \mathbf{v} \tag{6.29}$$

$$\frac{d\mathbf{v}}{d\tau} = -(1+m)\frac{\mathbf{r}}{r^3} + \mathbf{a}_t. \tag{6.30}$$

Also, as in the case of Cowell's method, the positions $\mathbf{r}_q$ and velocities $\mathbf{v}_q$ of the perturbing bodies can be determined at the beginning of each integration step by various analytical expressions, and their values for each of the six steps of the Runge-Kutta method may be found by numerical integration using their two-body accelerations. Thus, we have for each perturbing body during the interval of the integration step:

$$\frac{d\mathbf{r}_q}{d\tau} = \mathbf{v}_q \tag{6.31}$$

$$\frac{d\mathbf{v}_q}{d\tau} = -(1+m_q)\frac{\mathbf{r}_q}{r_q^3}. \tag{6.32}$$

The Runge-Kutta algorithm for the method of Encke integrates the functions of Equations 6.25 and 6.26 by means of auxiliary vectors having the following general forms:

$$\mathbf{F}_j = hf(\delta \mathbf{v})_j \tag{6.33}$$
$$\mathbf{G}_j = hg(m, \mathbf{r}_\mu, r_\mu, \mathbf{r}, r, \mathbf{a}_t)_j, \tag{6.34}$$

where h is the modified time interval and $j = 1$ to 6. The integration is begun by setting

$$\delta \mathbf{r} = \mathbf{0}$$
$$\delta \mathbf{v} = \mathbf{0},$$

because the perturbed and two-body orbits coincide at the beginning of the step. At the end of the integration step, the perturbation increments for the celestial body's orbital motion are obtained from

$$\delta \mathbf{r} = \frac{1}{90}(7\mathbf{F}_1 + 32\mathbf{F}_3 + 12\mathbf{F}_4 + 32\mathbf{F}_5 + 7\mathbf{F}_6) \qquad (6.35)$$

$$\delta \mathbf{v} = \frac{1}{90}(7\mathbf{G}_1 + 32\mathbf{G}_3 + 12\mathbf{G}_4 + 32\mathbf{G}_5 + 7\mathbf{G}_6). \qquad (6.36)$$

Finally, the body's new position and velocity are given by Equations 6.23 and 6.24, namely,

$$\mathbf{r} = \mathbf{r}_\mu + \delta \mathbf{r} \qquad (6.37)$$
$$\mathbf{v} = \mathbf{v}_\mu + \delta \mathbf{v}. \qquad (6.38)$$

These elements become the starting point for the next integration step.

6.5 A Perturbed Ephemeris

Once a special perturbation method has been used to determine a celestial body's position and velocity at some given epoch, its corresponding right ascension and declination can be found by the process described in Section 5.5. The justification for applying this much effort to an orbit computation is, of course, dependent on the accuracy required of the result. If the purpose of the ephemeris is merely to find an approximate position with an uncertainty of one or two arcminutes, then the equations of two-body motion are probably sufficient. On the other hand, if it is necessary to predict the position of the body with an uncertainty of only an arcsecond or less over a period of a year or more, then perturbations must be taken into account.

Consideration should also be given to the fact that a special perturbation method can only yield accurate results when the initial values of the vector elements are accurately known. If the elements are not sufficiently determined, one can attempt to improve these parameters by subjecting them to a least-squares procedure using a set of reliable observations. Two methods for doing this are described in Chapter 12.

6.6 Computer Programs

6.6.1 Program ATTRACT

Program ATTRACT computes solar and planetary attractions to provide an overview of their relative magnitudes. ATTRACT is only a demonstration program and not a prerequisite for any other program. The vector elements for the *perturbing planets* are taken directly from Section E of *The Astronomical Almanac*. That table gives the velocities in terms of astronomical units per day. The program converts these data to modified time units.

Program Algorithm

Define	Line Number
FNVS(X,Y,Z): vector squaring function	1070
FNMG(X,Y,Z): vector magnitude function	1080
FNDP(X1,Y1,Z1,X2,Y2,Z2): dot product function	1090

Given	Line Number
name of celestial body for information only	1160
equinox for information only	1160
k: gravitational constant	1160
m: mass of celestial body	1170
epoch time of celestial body's elements	1180
position elements of celestial body	1190
velocity elements of celestial body in modified time units	1200
number of perturbing bodies	1240
epoch time of perturber elements	1240
For each perturber:	1260-1390

name for information only	1280
m_q: mass of each perturber	1290
position elements in astronomical units	1300
velocity elements in astronomical units per day	1310
velocity elements converted to modified time	1360
new epoch at the end of the interval of computation	1430
number of computation steps	1450
Compute	Line Number
step-size in days	1610
For the celestial body and each perturber:	1630-1950
elements for the closed elliptic f and g routine	1650-1710
For each step:	1750-1930
t, $\mathbf{r}$, $\mathbf{v}$, $\mathbf{r}_q$, $\mathbf{v}_q$, and $\mathbf{p}_q$ by means of closed elliptic f and g expressions	1770-1910
combined mass for celestial body	2030
For each step:	2050-2190
r: magnitude of $\mathbf{r}$	2070
r^3	2080
$\ddot{\mathbf{r}}_\mu$: two-body attraction by Equation 6.21	2110
magnitude of $\ddot{\mathbf{r}}_\mu$	2140
For each perturber:	2230-2490

6.6. COMPUTER PROGRAMS

    For each step: 2300-2460

        r_q: magnitude of $\mathbf{r}_q$ 2320

        p_q: magnitude of $\mathbf{p}_q$ 2330

        r_q^3 2340

        p_q^3 2350

        $\mathbf{a}_q$: perturbing attraction by Equation 6.19   2380

        magnitude of $\mathbf{a}_q$ 2410

End.

Program Listing

```
1000 CLS
1010 PRINT"# ATTRACT # SOLAR AND PLANETARY ATTRACTIONS"
1020 PRINT"# AN OVERVIEW"
1030 PRINT
1040 DEFDBL A-Z
1050 DEFINT I,K,N,Q
1060 REM
1070 DEF FNVS(X,Y,Z)=X*X+Y*Y+Z*Z
1080 DEF FNMG(X,Y,Z)=SQR(X*X+Y*Y+Z*Z)
1090 DEF FNDP(X1,Y1,Z1,X2,Y2,Z2)=X1*X2+Y1*Y2+Z1*Z2
1100 REM
1110 DIM T(20),R(9,20,3),V(9,20,3),P(9,20,3)
1120 REM
1130 G$="###.#######"
1140 J$="####.#####"
1150 REM
1160 READ N$(0), E$, K#
1170 READ M(0)
1180 READ J(0)
1190 READ RE(0,1),RE(0,2),RE(0,3)
1200 READ VE(0,1),VE(0,2),VE(0,3)
1210 REM
1220 REM
1230 REM
1240 READ NQ, TQ
1250 REM
1260 FOR Q=1 TO NQ
1270    REM
1280    READ N$(Q)
1290    READ M(Q)
1300    READ RE(Q,1),RE(Q,2),RE(Q,3)
1310    READ VE(Q,1),VE(Q,2),VE(Q,3)
1320    REM
1330    J(Q)=TQ
1340    REM
1350    FOR K=1 TO 3
1360       VE(Q,K)=VE(Q,K)/K#
1370    NEXT K
1380    REM
1390 NEXT Q
```

6.6. COMPUTER PROGRAMS

```
1400 REM
1410 REM
1420 REM
1430 INPUT">>>NEW EPOCH";TN
1440 PRINT
1450 INPUT">>>NUMBER OF STEPS";NS
1460 REM
1470 REM
1480 CLS
1490 PRINT"SOLAR AND PLANETARY ATTRACTIONS"
1500 PRINT N$(0)
1510 PRINT E$
1520 REM
1530 PRINT TAB(4)"T(I)";
1540 PRINT TAB(17)"A(X)";TAB(28)"A(Y)";TAB(39)"A(Z)";
1550 PRINT TAB(50)"A"
1560 REM
1570 REM
1580 REM *** ORBITS ***
1590 REM
1600 REM
1610 DT=(TN-J(0))/NS
1620 REM
1630 FOR Q=0 TO NQ
1640    REM
1650    M=1+M(Q)
1660    RV=FNDP(RE(Q,1),RE(Q,2),RE(Q,3),
        VE(Q,1),VE(Q,2),VE(Q,3))
1670    V2=FNVS(VE(Q,1),VE(Q,2),VE(Q,3))
1680    R0=FNMG(RE(Q,1),RE(Q,2),RE(Q,3))
1690    D0=RV/SQR(M)
1700    AI=2/R0-V2/M
1710    T0=J(Q)
1720    REM
1730    REM
1740    REM
1750    FOR I=0 TO NS
1760       REM
1770       T(I)=J(0)+I*DT
1780       REM
1790       GOSUB 16010 REM ELLIPSE/SUB
1800       REM
```

```
1810    F=1-CC/R0
1820    G=(R0*SS+D0*CC)/SQR(M)
1830    R=R0+C0*CC+D0*SS
1840    FP=-SQR(M)*SS/(R*R0)
1850    GP=1-CC/R
1860    REM
1870    FOR K=1 TO 3
1880       R(Q,I,K)=F*RE(Q,K)+G*VE(Q,K)
1890       V(Q,I,K)=FP*RE(Q,K)+GP*VE(Q,K)
1900       P(Q,I,K)=R(Q,I,K)-R(0,I,K)
1910    NEXT K
1920    REM
1930  NEXT I
1940　 REM
1950 NEXT Q
1960 REM
1970 REM
1980 REM *** ATTRACTIONS ***
1990 REM
2000 REM
2010 PRINT"SUN"
2020 REM
2030 M=1+M(0)
2040 REM
2050 FOR I=0 TO NS
2060    REM
2070    R=FNMG(R(0,I,1),R(0,I,2),R(0,I,3))
2080    R3=R*R*R
2090    REM
2100    FOR K=1 TO 3
2110       AT(K)=-M*R(0,I,K)/R3
2120    NEXT K
2130    REM
2140    A=FNMG(AT(1),AT(2),AT(3))
2150    REM
2160    PRINT USING J$;T(I);
2170    PRINT USING G$;AT(1),AT(2),AT(3);A
2180    REM
2190 NEXT I
2200 REM
2210 LINE INPUT"";L$
2220 REM
```

6.6. COMPUTER PROGRAMS

```
2230 FOR Q=1 TO NQ
2240   REM
2250   CLS
2260   PRINT N$(Q)
2270   REM
2280   M=M(Q)
2290   REM
2300   FOR I=0 TO NS
2310     REM
2320     R=FNMG(R(Q,I,1),R(Q,I,2),R(Q,I,3))
2330     P=FNMG(P(Q,I,1),P(Q,I,2),P(Q,I,3))
2340     R3=R*R*R
2350     P3=P*P*P
2360     REM
2370     FOR K=1 TO 3
2380       AT(K)=M(Q)*(P(Q,I,K)/P3-R(Q,I,K)/R3)
2390     NEXT K
2400     REM
2410     A=FNMG(AT(1),AT(2),AT(3))
2420     REM
2430     PRINT USING J$;T(I);
2440     PRINT USING G$;AT(1),AT(2),AT(3),A
2450     REM
2460   NEXT I
2470   REM
2480   LINE INPUT"";L$
2490 NEXT Q
2500 REM
2510 PRINT"ATTRACT"
2520 END
16000 STOP
16010 REM # ELLIPSE/SUB # CLOSED ELLIPTIC F&G
16020 REM
16030 CO=1-RO*AI
16040 SO=DO*SQR(AI)
16050 N#=K#*SQR(M)*SQR(AI*AI*AI)
16060 WW=N#*(T(I)-T0)
16070 GG=WW
16080   FG=GG-CO*SIN(GG)-SO*COS(GG)+SO-WW
16090   IF ABS(FG)<.0000001# THEN 16130
16100   DF=1-CO*COS(GG)+SO*SIN(GG)
16110   GG=GG-FG/DF
```

```
16120    GOTO 16080
16130 CC=(1-COS(GG))/AI
16140 SS=SIN(GG)/SQR(AI)
16150 RETURN
20000 REM # CELESTIAL BODY #
20010 REM
30000 DATA   "MARS", "J2000.0", 0.017202099#
30010 DATA   0.000000323#
30020 DATA   6280.5#
30030 DATA  -0.8888462#, 1.2418782#, 0.5936583#
30040 DATA  -0.6523424#,-0.3449870#,-0.1405741#
30050 REM
40000 REM # PERTURBERS #
40010 REM
40020 DATA   6, 6280.5#
40030 REM
41000 DATA   "MERCURY"
41010 DATA   0.000000166#
41020 DATA   0.1693419#,-0.3559908#,-0.2077172#
41030 DATA   0.02036314#, 0.01152157#, 0.00404131#
42000 DATA   "VENUS"
42010 DATA   0.000002448#
42020 DATA   0.6457331#, 0.3130627#, 0.0999533#
42030 DATA  -0.00920284#, 0.01616581#, 0.00785453#
43000 DATA   "EARTH-MOON"
43010 DATA   0.000003040#
43020 DATA   0.6632717#,-0.7045336#,-0.3054775#
43030 DATA   0.01273999#, 0.01025921#, 0.00444834#
44000 'DATA   "MARS"
44010 'DATA   0.000000323#
44020 'DATA  -0.8888462#, 1.2418782#, 0.5936583#
44030 'DATA  -0.01122166#,-0.00593450#,-0.00241817#
45000 DATA   "JUPITER"
45010 DATA   0.000954791#
45020 DATA   3.406998#,-3.434686#,-1.555282#
45030 DATA   0.005504353#, 0.005026155#, 0.002020327#
46000 DATA   "SATURN"
46010 DATA   0.000285878#
46020 DATA  -5.344798#,-7.833691#,-3.005450#
46030 DATA   0.004409801#,-0.002715426#,-0.001310925#
47000 DATA   "URANUS"
47010 DATA   0.000043554#
```

6.6. COMPUTER PROGRAMS

```
47020 DATA -4.33594#,-17.05279#,-7.40711#
47030 DATA  0.003807513#,-0.000963315#,-0.000475940#
48000 DATA  "NEPTUNE"
48010 DATA  0.000051776#
48020 DATA  1.41455#,-27.95496#,-11.47747#
48030 DATA  0.0031224404#, 0.000184341#,-0.000002187#
49000 DATA   "PLUTO"
49010 DATA  0.000000008#
49020 DATA -23.55598#,-18.12624#, 1.44015#
49030 DATA  0.002025708#,-0.002522970#,-0.001395285#
```

6.6.2 Program COWELL

Program COWELL computes astrometric positions for perturbed heliocentric orbits by direct numerical integration using a fifth-order Runge-Kutta special perturbation routine. The motions of the perturbing planets are computed by closed f and g expressions using their two-body accelerations. The vector elements for the perturbing planets are taken directly from Section E of *The Astronomical Almanac*. That table gives the velocities in terms of astronomical units per day. The program converts these data to modified time units.

Program Algorithm

Define	Line Number
FNVS(X,Y,Z): vector squaring function	1090
FNMG(X,Y,Z): vector magnitude function	1100
FNDP(X1,Y1,Z1,X2,Y2,Z2): dot product function	1110
FNASN(X): inverse sine function	1120
FNACN(X): inverse cosine function	1130
FNAK(M,PK,P,RK,R): attraction function for Equation 6.6	1150
FNF(VK): velocity function for Equation 6.8	1160
FNG(M,RK,R,AK): acceleration function for Equation 6.5	1170

Given	Line Number
angle to radian conversion factor	1220
$1/c$: light-time constant	1320
name of the object for information only	1340
equinox for information only	1340
k: gravitational constant	1340

6.6. COMPUTER PROGRAMS

m: mass of celestial body	1350
epoch time of celestial body's elements	1360
position elements of celestial body	1370
velocity elements of celestial body in modified time units	1380
new epoch at the end of the interval of computation	1400
position of the Sun at the new epoch	1420
number of perturbing bodies	1460
string variable for brief comment	1460
epoch time of perturber elements	1460
For each perturber:	1480-1530
name for information only	1490
m_q: mass	1500
position elements in astronomical units	1510
velocity elements in astronomical units per day	1520
the number of computation steps to reach the new epoch	1660
Compute	Line Number
For each perturber:	1700-1880
velocity elements in modified time units	1750
orbit elements for closed f and g expressions	1780-1860
step-size in days	2070
For each computation step:	2090-2530

t: time at each step	2130
initial position of celestial body for Runge-Kutta routine	2160
initial velocity of celestial body for Runge-Kutta routine	2170
For each perturber:	2220-2330
By closed elliptic f and g expressions:	2260
initial position for Runge-Kutta routine	2290
initial velocity for Runge-Kutta routine	2300
By Runge-Kutta special perturbation subroutine:	2400
r: perturbed position by Equation 6.16	14210 & 2430
v: perturbed velocity by Equation 6.17	14220 & 2440
r: magnitude of the radius vector	2470
Right ascension and declination	2620-3100

End.

6.6. COMPUTER PROGRAMS

Program Listing

```
1000 CLS
1010 PRINT"# COWELL # RA AND DEC"
1020 PRINT"# FROM PERTURBED ORBITAL MOTION"
1030 PRINT"# BY THE METHOD OF COWELL"
1040 PRINT"# USING RUNGE-KUTTA FIVE"
1050 PRINT
1060 DEFDBL A-Z
1070 DEFINT I,J,K,N,Q
1080 REM
1090 DEF FNVS(X,Y,Z)=X*X+Y*Y+Z*Z
1100 DEF FNMG(X,Y,Z)=SQR(X*X+Y*Y+Z*Z)
1110 DEF FNDP(X1,Y1,Z1,X2,Y2,Z2)=X1*X2+Y1*Y2+Z1*Z2
1120 DEF FNASN(X)=ATN(X/SQR(-X*X+1))
1130 DEF FNACN(X)=-ATN(X/SQR(-X*X+1))+1.5707963263#
1140 REM
1150 DEF FNAK(M,PK,P,RK,R)=M*(PK/(P*P*P)-RK/(R*R*R))
1160 DEF FNF(VK)=VK
1170 DEF FNG(M,RK,R,AK)=-(1+M)*RK/(R*R*R)+AK
1180 REM
1190 DIM T(200)
1200 DIM R(200,3),V(200,3)
1210 REM
1220 Q1#=.017453293#
1230 G$="####.#######"
1240 J$="####.#####"
1250 A$="#####.######"
1260 D$="######.#####"
1270 AM$="###"
1280 AC$="####.##"
1290 DM$="######"
1300 DC$="####.#"
1310 REM
1320 READ AB
1330 REM
1340 READ N$(0), E$, K#
1350 READ M(0)
1360 READ T(0)
1370 READ R(0,1),R(0,2),R(0,3)
1380 READ V(0,1),V(0,2),V(0,3)
1390 REM
```

```
1400 READ TF
1410 REM
1420 READ RR(1),RR(2),RR(3)
1430 REM
1440 REM
1450 REM
1460 READ NQ, C$, TQ
1470 REM
1480 FOR Q=1 TO NQ
1490   READ N$(Q)
1500   READ M(Q)
1510   READ RE(Q,1),RE(Q,2),RE(Q,3)
1520   READ VE(Q,1),VE(Q,2),VE(Q,3)
1530 NEXT Q
1540 REM
1550 REM
1560 REM
1570 PRINT N$(0);C$
1580 PRINT"PERTURBER EPOCH:";TAB(19) TQ
1590 PRINT"INITIAL EPOCH:";TAB(20);
1600 PRINT USING J$;T(0)
1610 PRINT"FINAL EPOCH:";TAB(20);
1620 PRINT USING J$;TF
1630 PRINT"ELAPSED TIME:";TAB(20);
1640 PRINT USING J$;(TF-T(0))
1650 PRINT
1660 INPUT">>>NUMBER OF STEPS";NS
1670 REM
1680 REM
1690 REM
1700 FOR Q=1 TO NQ
1710   REM
1720   REM * PERTURBER ELEMENTS *
1730   REM
1740   FOR K=1 TO 3
1750     VE(Q,K)=VE(Q,K)/K#
1760   NEXT K
1770   REM
1780   M=1+M(Q)
1790   RV=FNDP(RE(Q,1),RE(Q,2),RE(Q,3),
       VE(Q,1),VE(Q,2),VE(Q,3))
1800   V2=FNVS(VE(Q,1),VE(Q,2),VE(Q,3))
```

6.6. COMPUTER PROGRAMS

```
1810    RO(Q)=FNMG(RE(Q,1),RE(Q,2),RE(Q,3))
1820    DO(Q)=RV/SQR(M)
1830    AI(Q)=2/RO(Q)-V2/M
1840    CO(Q)=1-RO(Q)*AI(Q)
1850    SO(Q)=DO(Q)*SQR(AI(Q))
1860    N#(Q)=K#*SQR(M)*SQR(AI(Q)*AI(Q)*AI(Q))
1870    REM
1880 NEXT Q
1890 REM
1900 REM
1910 CLS
1920 PRINT"RA AND DEC"
1930 PRINT"FROM PERTURBED ORBITAL MOTION"
1940 PRINT"BY THE METHOD OF COWELL"
1950 PRINT"USING RUNGE-KUTTA FIVE"
1960 PRINT N$(0)
1970 PRINT E$
1980 PRINT"NQ";NQ;C$;TQ
1990 REM
2000 PRINT TAB(4)"T(I)";
2010 PRINT TAB(18)"X";TAB(30)"Y";TAB(42)"Z";
2020 PRINT TAB(54)"R"
2030 REM
2040 REM
2050 REM
2060 TN=TF
2070 DT=(TN-T(0))/NS
2080 REM
2090 FOR I=0 TO NS
2100    REM
2110    IF I=0 THEN 2470
2120    REM
2130    T(I)=T(I-1)+DT
2140    REM
2150    FOR K=1 TO 3
2160       RQ(0,K)=R(I-1,K)
2170       VQ(0,K)=V(I-1,K)
2180    NEXT K
2190    REM
2200    REM
2210    REM
2220    FOR Q=1 TO NQ
```

```
2230      REM
2240      REM * PERTURBER ORBITS *
2250      REM
2260      GOSUB 12010 REM PORB/SUB
2270      REM
2280      FOR K=1 TO 3
2290         RQ(Q,K)=F*RE(Q,K)+G*VE(Q,K)
2300         VQ(Q,K)=FP*RE(Q,K)+GP*VE(Q,K)
2310      NEXT K
2320      REM
2330   NEXT Q
2340   REM
2350   REM
2360   REM
2370      REM
2380      REM * NUMERICAL INTEGRATION *
2390      REM
2400      GOSUB 13010 REM CWL5/SUB
2410      REM
2420      FOR K=1 TO 3
2430         R(I,K)=RK(0,K)
2440         V(I,K)=VK(0,K)
2450      NEXT K
2460      REM
2470      R=FNMG(R(I,1),R(I,2),R(I,3))
2480      REM
2490      PRINT USING J$;T(I);
2500      PRINT USING G$;R(I,1),R(I,2),R(I,3),R;
2510      REM
2520      LINE INPUT"";L$
2530   NEXT I
2540   REM
2550   I=I-1
2560   REM
2570   PRINT TAB(11);
2580   PRINT USING G$;V(I,1),V(I,2),V(I,3)
2590   REM
2600   REM
2610   REM
2620      REM
2630      REM * LIGHT-TIME *
2640      REM
```

6.6. COMPUTER PROGRAMS

```
2650     FOR K=1 TO 3
2660       P(K)=R(I,K)+RR(K)
2670     NEXT K
2680     REM
2690     P=FNMG(P(1),P(2),P(3))
2700     REM
2710     AP=AB*P
2720     IF T(I)-(TF-AP)<.00001 THEN 2890
2730     T(I)=TF-AP
2740     DT=T(I)-T(I-1)
2750     REM
2760     GOSUB 13010 REM CWL5/SUB
2770     REM
2780     FOR K=1 TO 3
2790       R(I,K)=RK(0,K)
2800     NEXT K
2810     REM
2820     GOTO 2650
2830 REM
2840 REM
2850 REM
2860     REM
2870     REM * RA AND DEC *
2880     REM
2890     FOR K=1 TO 3
2900       LL(K)=P(K)/P
2910     NEXT K
2920     CD=SQR(1-LL(3)*LL(3))
2930     CX=LL(1)/CD
2940     SX=LL(2)/CD
2950     REM
2960     GOSUB 16010 REM ARC/SUB
2970     REM
2980     A=X/(15*Q1#)
2990     D=FNASN(LL(3))/Q1#
3000     REM
3010     AH=FIX(A)
3020     AM=FIX((A-AH)*60)
3030     AC=(A-AH-AM/60)*3600
3040     REM
3050     DD=FIX(D)
3060     DM=FIX((D-DD)*60)
```

```
3070    DC=(D-DD-DM/60)*3600
3080    REM
3090    REM
3100    REM
3110 PRINT
3120 PRINT TAB(4)"T(I)";
3130 PRINT TAB(18)"A(I)";TAB(30)"D(I)";TAB(42)"P(I)"
3140 REM
3150 PRINT USING J$;TF;
3160 PRINT USING A$;A;
3170 PRINT USING D$;D;
3180 PRINT USING G$;P
3190 REM
3200 PRINT TAB(13);
3210 PRINT USING AM$;AM;
3220 PRINT USING AC$;AC;
3230 PRINT USING DM$;DM;
3240 PRINT USING DC$;DC
3250 PRINT
3260 PRINT"COWELL"
3270 END
12000 STOP
12010 REM # PORB/SUB # COMPUTES PERTURBER ORBITS
12020 REM # BY CLOSED ELLIPTIC F&G EXPRESSIONS
12030 WQ=N#(Q)*(T(I-1)-TQ)
12040 GG=WQ
12050    FG=GG-C0(Q)*SIN(GG)-S0(Q)*COS(GG)+S0(Q)-WQ
12060    IF ABS(FG)<.0000001# THEN 12100
12070    DF=1-C0(Q)*COS(GG)+S0(Q)*SIN(GG)
12080    GG=GG-FG/DF
12090    GOTO 12050
12100 CC=(1-COS(GG))/AI(Q)
12110 SS=SIN(GG)/SQR(AI(Q))
12120 F=1-CC/R0(Q)
12130 G=(R0(Q)*SS+D0(Q)*CC)/SQR(1+M(Q))
12140 R=R0(Q)+C0(Q)*CC+D0(Q)*SS
12150 FP=-SQR(1+M(Q))*SS/(R*R0(Q))
12160 GP=1-CC/R
12170 RETURN
13000 STOP
13010 REM # CWL5/SUB # RK5 NUMERICAL INTEGRATION
13020 REM # FOR THE METHOD OF COWELL
```

6.6. COMPUTER PROGRAMS

```
13030 REM
13040 H=K#*DT
13050 REM
13060 REM ********** STEP 1 **********
13070 REM
13080 FOR Q=0 TO NQ
13090   FOR K=1 TO 3
13100     RK(Q,K)=RQ(Q,K)
13110     VK(Q,K)=VQ(Q,K)
13120     PK(Q,K)=RK(Q,K)-RK(0,K)
13130   NEXT K
13140 NEXT Q
13150 REM
13160 GOSUB 15010 REM SUBSUB
13170 REM
13180 FOR K=1 TO 3
13190   FOR Q=0 TO NQ
13200     F1(Q,K)=FX(Q,K)
13210     G1(Q,K)=GX(Q,K)
13220   NEXT Q
13230 NEXT K
13240 REM
13250 REM ********** STEP 2 **********
13260 REM
13270 FOR Q=0 TO NQ
13280   FOR K=1 TO 3
13290     RK(Q,K)=RQ(Q,K)+F1(Q,K)/4
13300     VK(Q,K)=VQ(Q,K)+G1(Q,K)/4
13310     PK(Q,K)=RK(Q,K)-RK(0,K)
13320   NEXT K
13330 NEXT Q
13340 REM
13350 GOSUB 15010 REM SUBSUB
13360 REM
13370 FOR K=1 TO 3
13380   FOR Q=0 TO NQ
13390     F2(Q,K)=FX(Q,K)
13400     G2(Q,K)=GX(Q,K)
13410   NEXT Q
13420 NEXT K
13430 REM
13440 REM ********** STEP 3 **********
```

```
13450 REM
13460 FOR Q=0 TO NQ
13470   FOR K=1 TO 3
13480     RK(Q,K)=RQ(Q,K)+(F1(Q,K)+F2(Q,K))/8
13490     VK(Q,K)=VQ(Q,K)+(G1(Q,K)+G2(Q,K))/8
13500     PK(Q,K)=RK(Q,K)-RK(0,K)
13510   NEXT K
13520 NEXT Q
13530 REM
13540 GOSUB 15010 REM SUBSUB
13550 REM
13560 FOR K=1 TO 3
13570   FOR Q=0 TO NQ
13580     F3(Q,K)=FX(Q,K)
13590     G3(Q,K)=GX(Q,K)
13600   NEXT Q
13610 NEXT K
13620 REM
13630 REM ********** STEP 4 **********
13640 REM
13650 FOR Q=0 TO NQ
13660   FOR K=1 TO 3
13670     RK(Q,K)=RQ(Q,K)-(F2(Q,K)-2*F3(Q,K))/2
13680     VK(Q,K)=VQ(Q,K)-(G2(Q,K)-2*G3(Q,K))/2
13690     PK(Q,K)=RK(Q,K)-RK(0,K)
13700   NEXT K
13710 NEXT Q
13720 REM
13730 GOSUB 15010 REM SUBSUB
13740 REM
13750 FOR K=1 TO 3
13760   FOR Q=0 TO NQ
13770     F4(Q,K)=FX(Q,K)
13780     G4(Q,K)=GX(Q,K)
13790   NEXT Q
13800 NEXT K
13810 REM
13820 REM ********** STEP 5 **********
13830 REM
13840 FOR Q=0 TO NQ
13850   FOR K=1 TO 3
13860     RK(Q,K)=RQ(Q,K)+(3*F1(Q,K)+9*F4(Q,K))/16
```

6.6. COMPUTER PROGRAMS

```
13870     VK(Q,K)=VQ(Q,K)+(3*G1(Q,K)+9*G4(Q,K))/16
13880     PK(Q,K)=RK(Q,K)-RK(0,K)
13890   NEXT K
13900 NEXT Q
13910 REM
13920 GOSUB 15010 REM SUBSUB
13930 REM
13940 FOR K=1 TO 3
13950   FOR Q=0 TO NQ
13960     F5(Q,K)=FX(Q,K)
13970     G5(Q,K)=GX(Q,K)
13980   NEXT Q
13990 NEXT K
14000 REM
14010 REM ********** STEP 6 **********
14020 REM
14030 FOR Q=0 TO NQ
14040   FOR K=1 TO 3
14050     RK(Q,K)=RQ(Q,K)-(3*F1(Q,K)-2*F2(Q,K)-12*F3(Q,K)+
          12*F4(Q,K)-8*F5(Q,K))/7
14060     VK(Q,K)=VQ(Q,K)-(3*G1(Q,K)-2*G2(Q,K)-12*G3(Q,K)+
          12*G4(Q,K)-8*G5(Q,K))/7
14070     PK(Q,K)=RK(Q,K)-RK(0,K)
14080   NEXT K
14090 NEXT Q
14100 REM
14110 GOSUB 15010 REM SUBSUB
14120 REM
14130 FOR K=1 TO 3
14140   F6(0,K)=FX(0,K)
14150   G6(0,K)=GX(0,K)
14160 NEXT K
14170 REM
14180 REM ********** RESULT **********
14190 REM
14200 FOR K=1 TO 3
14210   RK(0,K)=RQ(0,K)+(7*F1(0,K)+32*F3(0,K)+12*F4(0,K)+
        32*F5(0,K)+7*F6(0,K))/90
14220   VK(0,K)=VQ(0,K)+(7*G1(0,K)+32*G3(0,K)+12*G4(0,K)+
        32*G5(0,K)+7*G6(0,K))/90
14230 NEXT K
14240 RETURN
```

```
15000 STOP
15010 REM # CWL5 SUBSUBROUTINE #
15020 REM
15030 FOR Q=1 TO NQ
15040   RK(Q,0)=FNMG(RK(Q,1),RK(Q,2),RK(Q,3))
15050   PK(Q,0)=FNMG(PK(Q,1),PK(Q,2),PK(Q,3))
15060   FOR K=1 TO 3
15070     AK(Q,K)=FNAK(M(Q),PK(Q,K),PK(Q,0),RK(Q,K),RK(Q,0))
15080   NEXT K
15090 NEXT Q
15100 REM
15110 FOR K=1 TO 3
15120   AK(0,K)=0
15130   FOR Q=1 TO NQ
15140     AK(0,K)=AK(0,K)+AK(Q,K)
15150   NEXT Q
15160 NEXT K
15170 REM
15180 RK(0,0)=FNMG(RK(0,1),RK(0,2),RK(0,3))
15190 REM
15200 FOR K=1 TO 3
15210   FX(0,K)=H*FNF(VK(0,K))
15220   GX(0,K)=H*FNG(M(0),RK(0,K),RK(0,0),AK(0,K))
15230   FOR Q=1 TO NQ
15240     FX(Q,K)=H*FNF(VK(Q,K))
15250     GX(Q,K)=H*FNG(M(Q),RK(Q,K),RK(Q,0),0)
15260   NEXT Q
15270 NEXT K
15280 RETURN
16000 STOP
16010 REM # ARC/SUB # COMPUTES X FROM SIN(X) AND COS(X)
16020 IF ABS(SX)<=.707107 THEN X=FNASN(ABS(SX))
16030 IF ABS(CX)<=.707107 THEN X=FNACN(ABS(CX))
16040 IF CX>=0 AND SX>=0 THEN X=X
16050 IF CX<0 AND SX>=0 THEN X=180*Q1#-X
16060 IF CX<0 AND SX<0 THEN X=180*Q1#+X
16070 IF CX>=0 AND SX<0 THEN X=360*Q1#-X
16080 RETURN
20000 DATA 0.005775519#
20010 REM
20020 REM # CELESTIAL BODY #
20030 REM
```

6.6. COMPUTER PROGRAMS

```
30000 DATA  "MARS", "J2000.0", 0.017202099#
30010 DATA  0.000000323#
30020 DATA  6280.5#
30030 DATA -0.8888462#, 1.2418782#, 0.5936583#
30040 DATA -0.6523424#,-0.3449870#,-0.1405741#
30050 REM
39000 REM # SUN #
39010 REM
39020 DATA 7000.5#,-0.5197797#,+0.8008713#,+0.3472473#
39030 REM
40000 REM # PERTURBERS #
40010 REM
40020 'DATA  0, "(TWO-BODY)", 0
40030 DATA  5, "(MVEJS)", 6280.5#
40040 REM
41000 DATA  "MERCURY"
41010 DATA  0.000000166#
41020 DATA  0.1693419#,-0.3559908#,-0.2077172#
41030 DATA  0.02036314#, 0.01152157#, 0.00404131#
42000 DATA  "VENUS"
42010 DATA  0.000002448#
42020 DATA  0.6457331#, 0.3130627#, 0.0999533#
42030 DATA -0.00920284#, 0.01616581#, 0.00785453#
43000 DATA  "EARTH-MOON"
43010 DATA  0.000003040#
43020 DATA  0.6632717#,-0.7045336#,-0.3054775#
43030 DATA  0.01273999#, 0.01025921#, 0.00444834#
44000 'DATA  "MARS"
44010 'DATA  0.000000323#
44020 'DATA -0.8888462#, 1.2418782#, 0.5936583#
44030 'DATA -0.01122166#,-0.00593450#,-0.00241817#
45000 DATA  "JUPITER"
45010 DATA  0.000954791#
45020 DATA  3.406998#,-3.434686#,-1.555282#
45030 DATA  0.005504353#, 0.005026155#, 0.002020327#
46000 DATA  "SATURN"
46010 DATA  0.000285878#
46020 DATA -5.344798#,-7.833691#,-3.005450#
46030 DATA  0.004409801#,-0.002715426#,-0.001310925#
47000 'DATA  "URANUS"
47010 'DATA  0.000043554#
47020 'DATA -4.33594#,-17.05279#,-7.40711#
```

```
47030 'DATA   0.003807513#,-0.000963315#,-0.000475940#
48000 'DATA   "NEPTUNE"
48010 'DATA   0.000051776#
48020 'DATA   1.41455#,-27.95496#,-11.47747#
48030 'DATA   0.0031224404#, 0.000184341#,-0.000002187#
49000 'DATA    "PLUTO"
49010 'DATA   0.000000008#
49020 'DATA  -23.55598#,-18.12624#,  1.44015#
49030 'DATA   0.002025708#,-0.002522970#,-0.001395285#
```

6.6. COMPUTER PROGRAMS

6.6.3 Program ENCKE

Program ENCKE computes astrometric positions for perturbed heliocentric orbits using the universal formulation and a fifth-order Runge-Kutta special perturbation routine. The motions of the perturbing planets are computed by closed f and g expressions using their two-body accelerations. The vector elements for the perturbing planets are taken directly from Section E of *The Astronomical Almanac*. That table gives the velocities in terms of astronomical units per day. The program converts these data to modified time units.

Program Algorithm

Define	Line Number
FNVS(X,Y,Z): vector squaring function	1090
FNMG(X,Y,Z): vector magnitude function	1100
FNDP(X1,Y1,Z1,X2,Y2,Z2): dot product function	1110
FNASN(X): inverse sine function	1120
FNACN(X): inverse cosine function	1130
FNAK(M,PK,P,RK,R): attraction function for Equation 6.19	1150
FNV(V): velocity function for Equations 6.27, 6.29, and 6.31	1160
FNA(M,RK,R,AK): acceleration function for Equations 6.28, 6.30, and 6.32	1170
FNF(VK): velocity function for Equation 6.25	1180
FNG(M,RU,R1,RP,R2,AK): acceleration function for Equation 6.26	1190

Given	Line Number
angle to radian conversion factor	1240
$1/c$: light-time constant	1340
name of the object for information only	1360

288   CHAPTER 6. SPECIAL PERTURBATIONS

    equinox for information only   1360

    k: gravitational constant   1360

    m: mass of celestial body   1370

    epoch time of celestial body's elements   1380

    position elements of celestial body   1390

    velocity elements of celestial body in modified time units   1400

    new epoch at the end of the interval of computation   1420

    position of the Sun at the new epoch   1440

    number of perturbing bodies   1480

    string variable for brief comment   1480

    epoch time of perturber elements   1480

    For each perturber: 1500-1550

        name for information only   1510

        m_q: mass   1520

        position elements in astronomical units   1530

        velocity elements in astronomical units per day   1540

        the number of computation steps to reach the new epoch 1680

Compute   Line Number

    coefficients for the universal variables   1720-1750

    For each perturber: 1790-1970

6.6. COMPUTER PROGRAMS

velocity elements in modified time units	1840
orbit elements for closed f and g expressions	1870-1950
step-size in days	2160
For each computation step:	2180-2710
t: time at each step	2220
For the celestial body:	2260-2360
position at beginning of step by universal variables	2320
velocity at beginning of step by universal variables	2330
position at beginning of step for Runge-Kutta routine	2340
velocity at beginning of step for Runge-Kutta routine	2350
For each perturber:	2400-2510
position at beginning of step by closed f and g expressions	2470
velocity at beginning of step by closed f and g expressions	2480
By Runge-Kutta special perturbation subroutine:	2580
$\delta\mathbf{r}$: perturbation of position by Equation 6.35	14770
$\delta\mathbf{v}$: perturbation of velocity by Equation 6.36	14780
$\mathbf{r}$: perturbed position by Equation 6.37	2610
$\mathbf{v}$: perturbed velocity by Equation 6.38	2620
r: magnitude of the radius vector	2650
Right ascension and declination	2800-3280

End.

Program Listing

```
1000 CLS
1010 PRINT"# ENCKE # RA AND DEC"
1020 PRINT"# FROM PERTURBED ORBITAL MOTION"
1030 PRINT"# BY UNIVERSAL VARIABLES"
1040 PRINT"# AND RUNGE-KUTTA FIVE"
1050 PRINT
1060 DEFDBL A-Z
1070 DEFINT I,J,K,N,Q
1080 REM
1090 DEF FNVS(X,Y,Z)=X*X+Y*Y+Z*Z
1100 DEF FNMG(X,Y,Z)=SQR(X*X+Y*Y+Z*Z)
1110 DEF FNDP(X1,Y1,Z1,X2,Y2,Z2)=X1*X2+Y1*Y2+Z1*Z2
1120 DEF FNASN(X)=ATN(X/SQR(-X*X+1))
1130 DEF FNACN(X)=-ATN(X/SQR(-X*X+1))+1.5707963263#
1140 REM
1150 DEF FNAK(M,PK,P,RK,R)=M*(PK/(P*P*P)-RK/(R*R*R))
1160 DEF FNV(VK)=VK
1170 DEF FNA(M,RK,R,AK)=-(1+M)*RK/(R*R*R)+AK
1180 DEF FNF(VK)=VK
1190 DEF FNG(M,RU,R1,RP,R2,AK)=(1+M)*(RU/(R1*R1*R1)-
     RP/(R2*R2*R2))+AK
1200 REM
1210 DIM T(200),B(19)
1220 DIM R(200,3),V(200,3)
1230 REM
1240 Q1#=.017453293#
1250 G$="####.#######"
1260 J$="####.#####"
1270 A$="#####.######"
1280 D$="######.#####"
1290 AM$="###"
1300 AC$="####.##"
1310 DM$="######"
1320 DC$="####.#"
1330 REM
1340 READ AB
1350 REM
1360 READ N$(0), E$, K#
1370 READ M(0)
1380 READ T(0)
```

6.6. COMPUTER PROGRAMS

```
1390 READ R(0,1),R(0,2),R(0,3)
1400 READ V(0,1),V(0,2),V(0,3)
1410 REM
1420 READ TF
1430 REM
1440 READ RR(1),RR(2),RR(3)
1450 REM
1460 REM
1470 REM
1480 READ NQ, C$, TQ
1490 REM
1500 FOR Q=1 TO NQ
1510   READ N$(Q)
1520   READ M(Q)
1530   READ RE(Q,1),RE(Q,2),RE(Q,3)
1540   READ VE(Q,1),VE(Q,2),VE(Q,3)
1550 NEXT Q
1560 REM
1570 REM
1580 REM
1590 PRINT N$(0);C$
1600 PRINT"PERTURBER EPOCH:";TAB(19) TQ
1610 PRINT"INITIAL EPOCH:";TAB(20);
1620 PRINT USING J$;T(0)
1630 PRINT"FINAL EPOCH:";TAB(20);
1640 PRINT USING J$;TF
1650 PRINT"ELAPSED TIME:";TAB(20);
1660 PRINT USING J$;(TF-T(0))
1670 PRINT
1680 INPUT">>>NUMBER OF STEPS";NS
1690 REM
1700 REM
1710 REM
1720 B(1)=1
1730 FOR J=2 TO 19
1740   B(J)=B(J-1)/J
1750 NEXT J
1760 REM
1770 REM
1780 REM
1790 FOR Q=1 TO NQ
1800   REM
```

```
1810    REM * PERTURBER ELEMENTS *
1820    REM
1830    FOR K=1 TO 3
1840       VE(Q,K)=VE(Q,K)/K#
1850    NEXT K
1860    REM
1870    M=1+M(Q)
1880    RV=FNDP(RE(Q,1),RE(Q,2),RE(Q,3),
        VE(Q,1),VE(Q,2),VE(Q,3))
1890    V2=FNVS(VE(Q,1),VE(Q,2),VE(Q,3))
1900    R0(Q)=FNMG(RE(Q,1),RE(Q,2),RE(Q,3))
1910    D0(Q)=RV/SQR(M)
1920    AI(Q)=2/R0(Q)-V2/M
1930    C0(Q)=1-R0(Q)*AI(Q)
1940    S0(Q)=D0(Q)*SQR(AI(Q))
1950    N#(Q)=K#*SQR(M)*SQR(AI(Q)*AI(Q)*AI(Q))
1960    REM
1970 NEXT Q
1980 REM
1990 REM
2000 CLS
2010 PRINT"RA AND DEC"
2020 PRINT"FROM PERTURBED ORBITAL MOTION"
2030 PRINT"BY UNIVERSAL VARIABLES"
2040 PRINT"AND RUNGE-KUTTA FIVE"
2050 PRINT N$(0)
2060 PRINT E$
2070 PRINT"NQ";NQ;C$;TQ
2080 REM
2080 PRINT TAB(4)"T(I)";
2100 PRINT TAB(18)"X";TAB(30)"Y";TAB(42)"Z";
2110 PRINT TAB(54)"R"
2120 REM
2130 REM
2140 REM
2150 TN=TF
2160 DT=(TN-T(0))/NS
2170 REM
2180 FOR I=0 TO NS
2190    REM
2200    IF I=0 THEN 2650
2210    REM
```

6.6. COMPUTER PROGRAMS

```
2220    T(I)=T(I-1)+DT
2230    REM
2240    REM
2250    REM
2260      REM
2270      REM * REFERENCE ORBIT *
2280      REM
2290      GOSUB 11010 REM RORB/SUB
2300      REM
2310      FOR K=1 TO 3
2320        R(I,K)=F*R(I-1,K)+G*V(I-1,K)
2330        V(I,K)=FP*R(I-1,K)+GP*V(I-1,K)
2340        RQ(0,K)=R(I-1,K)
2350        VQ(0,K)=V(I-1,K)
2360      NEXT K
2370    REM
2380    REM
2390    REM
2400    FOR Q=1 TO NQ
2410      REM
2420      REM * PERTURBER ORBITS *
2430      REM
2440      GOSUB 12010 REM PORB/SUB
2450      REM
2460      FOR K=1 TO 3
2470        RQ(Q,K)=F*RE(Q,K)+G*VE(Q,K)
2480        VQ(Q,K)=FP*RE(Q,K)+GP*VE(Q,K)
2490      NEXT K
2500      REM
2510    NEXT Q
2520    REM
2530    REM
2540    REM
2550      REM
2560      REM * NUMERICAL INTEGRATION *
2570      REM
2580      GOSUB 13010 REM ENK5/SUB
2590      REM
2600      FOR K=1 TO 3
2610        R(I,K)=R(I,K)+DR(0,K)
2620        V(I,K)=V(I,K)+DV(0,K)
2630      NEXT K
```

```
2640      REM
2650      R=FNMG(R(I,1),R(I,2),R(I,3))
2660      REM
2670      PRINT USING J$;T(I);
2680      PRINT USING G$;R(I,1),R(I,2),R(I,3),R;
2690      REM
2700      LINE INPUT"";L$
2710 NEXT I
2720 REM
2730 I=I-1
2740 REM
2750 PRINT TAB(11);
2760 PRINT USING G$;V(I,1),V(I,2),V(I,3)
2770 REM
2780 REM
2790 REM
2800      REM
2810      REM * LIGHT-TIME *
2820      REM
2830      FOR K=1 TO 3
2840         P(K)=R(I,K)+RR(K)
2850      NEXT K
2860      REM
2870      P=FNMG(P(1),P(2),P(3))
2880      REM
2890      AP=AB*P
2900      IF T(I)-(TF-AP)<.00001 THEN 3070
2910      T(I)=TF-AP
2920      DT=T(I)-T(I-1)
2930      REM
2940      GOSUB 11010 REM RORB/SUB
2950      REM
2960      FOR K=1 TO 3
2970         R(I,K)=F*R(I-1,K)+G*V(I-1,K)
2980      NEXT K
2990      REM
3000      GOTO 2830
3010 REM
3020 REM
3030 REM
3040      REM
3050      REM * RA AND DEC *
```

6.6. COMPUTER PROGRAMS

```
3060    REM
3070    FOR K=1 TO 3
3080       LL(K)=P(K)/P
3090    NEXT K
3100    CD=SQR(1-LL(3)*LL(3))
3110    CX=LL(1)/CD
3120    SX=LL(2)/CD
3130    REM
3140    GOSUB 16010 REM ARC/SUM
3150    REM
3160    A=X/(15*Q1#)
3170    D=FNASN(LL(3))/Q1#
3180    REM
3190    AH=FIX(A)
3200    AM=FIX((A-AH)*60)
3210    AC=(A-AH-AM/60)*3600
3220    REM
3230    DD=FIX(D)
3240    DM=FIX((D-DD)*60)
3250    DC=(D-DD-DM/60)*3600
3260    REM
3270    REM
3280    REM
3290 PRINT
3300 PRINT TAB(4)"T(I)";
3310 PRINT TAB(18)"A(I)";TAB(30)"D(I)";TAB(42)"P(I)"
3320 REM
3330 PRINT USING J$;TF;
3340 PRINT USING A$;A;
3350 PRINT USING D$;D;
3360 PRINT USING G$;P
3370 REM
3380 PRINT TAB(13);
3390 PRINT USING AM$;AM;
3400 PRINT USING AC$;AC;
3410 PRINT USING DM$;DM;
3420 PRINT USING DC$;DC
3430 PRINT
3440 PRINT"ENCKE"
3450 END
11000 STOP
11010 REM # RORB/SUB # COMPUTES REFERENCE ORBIT
```

296   CHAPTER 6. SPECIAL PERTURBATIONS

```
11020 REM # UNIVERSAL VARIABLES
11030 M=1+M(0)
11040 R0=FNMG(R(I-1,1),R(I-1,2),R(I-1,3))
11050 D0=FNDP(R(I-1,1),R(I-1,2),R(I-1,3),
      V(I-1,1),V(I-1,2),V(I-1,3))/SQR(M)
11060 AI=2/R0-FNVS(V(I-1,1),V(I-1,2),V(I-1,3))/M
11070 C0=1-R0*AI
11080 WW=K#*SQR(M)*DT
11090 XX=WW/R0
11100   X2=XX*XX
11110   XA=X2*AI
11120   X3=X2*XX
11130   CC=X2*(B(2)-XA*(B(4)-XA*(B(6)-XA*(B(8)-XA*(B(10)-
        XA*(B(12)-XA*(B(14)-XA*(B(16)-XA*(B(18)))))))))
11140   UU=X3*(B(3)-XA*(B(5)-XA*(B(7)-XA*(B(9)-XA*(B(11)-
        XA*(B(13)-XA*(B(15)-XA*(B(17)-XA*(B(19)))))))))
11150   SS=XX-UU*AI
11160   FX=R0*XX+C0*UU+D0*CC-WW
11170   IF ABS(FX)<.00000005# THEN 11210
11180   DF=R0+C0*CC+D0*SS
11190   XX=XX-FX/DF
11200   GOTO 11100
11210 F=1-CC/R0
11220 G=(R0*SS+D0*CC)/SQR(M)
11230 R=R0+C0*CC+D0*SS
11240 FP=-SQR(M)*SS/(R*R0)
11250 GP=1-CC/R
11260 RETURN
12000 STOP
12010 REM # PORB/SUB # COMPUTES PERTURBER ORBIT
12020 REM # CLOSED ELLIPTIC F&G EXPRESSIONS
12030 WQ=N#(Q)*(T(I-1)-TQ)
12040 GG=WQ
12050   FG=GG-C0(Q)*SIN(GG)-S0(Q)*COS(GG)+S0(Q)-WQ
12060   IF ABS(FG)<.0000001# THEN 12100
12070   DF=1-C0(Q)*COS(GG)+S0(Q)*SIN(GG)
12080   GG=GG-FG/DF
12090   GOTO 12050
12100 CC=(1-COS(GG))/AI(Q)
12110 SS=SIN(GG)/SQR(AI(Q))
12120 F=1-CC/R0(Q)
12130 G=(R0(Q)*SS+D0(Q)*CC)/SQR(1+M(Q))
```

6.6. COMPUTER PROGRAMS

```
12140 R=R0(Q)+C0(Q)*CC+D0(Q)*SS
12150 FP=-SQR(1+M(Q))*SS/(R*R0(Q))
12160 GP=1-CC/R
12170 RETURN
13000 STOP
13010 REM # ENK5/SUB # RK5 NUMERICAL INTEGRATION
13020 REM # FOR THE METHOD OF ENCKE
13030 REM
13040 H=K#*DT
13050 REM
13060 REM ********** STEP 1 **********
13070 REM
13080 FOR K=1 TO 3
13090    DV(0,K)=0
13100    RP(0,K)=RQ(0,K)
13110    VP(0,K)=VQ(0,K)
13120 NEXT K
13130 REM
13140 FOR Q=0 TO NQ
13150    FOR K=1 TO 3
13160       RK(Q,K)=RQ(Q,K)
13170       VK(Q,K)=VQ(Q,K)
13180       PK(Q,K)=RK(Q,K)-RP(0,K)
13190    NEXT K
13200 NEXT Q
13210 REM
13220 GOSUB 15010 REM SUBSUB
13230 REM
13240 FOR K=1 TO 3
13250    L1(K)=LX(K)
13260    S1(K)=SX(K)
13270    FOR Q=0 TO NQ
13280       U1(Q,K)=UX(Q,K)
13290       W1(Q,K)=WX(Q,K)
13300    NEXT Q
13310    F1(K)=FX(K)
13320    G1(K)=GX(K)
13330 NEXT K
13340 REM
13350 REM ********** STEP 2 **********
13360 REM
13370 FOR K=1 TO 3
```

```
13380     DV(0,K)=G1(K)/4
13390     RP(0,K)=RQ(0,K)+L1(K)/4
13400     VP(0,K)=VQ(0,K)+S1(K)/4
13410 NEXT K
13420 REM
13430 FOR Q=0 TO NQ
13440    FOR K=1 TO 3
13450       RK(Q,K)=RQ(Q,K)+U1(Q,K)/4
13460       VK(Q,K)=VQ(Q,K)+W1(Q,K)/4
13470       PK(Q,K)=RK(Q,K)-RP(0,K)
13480    NEXT K
13490 NEXT Q
13500 REM
13510 GOSUB 15010 REM SUBSUB
13520 REM
13530 FOR K=1 TO 3
13540    L2(K)=LX(K)
13550    S2(K)=SX(K)
13560    FOR Q=0 TO NQ
13570       U2(Q,K)=UX(Q,K)
13580       W2(Q,K)=WX(Q,K)
13590    NEXT Q
13600    F2(K)=FX(K)
13610    G2(K)=GX(K)
13620 NEXT K
13630 REM
13640 REM ********** STEP 3 **********
13650 REM
13660 FOR K=1 TO 3
13670    DV(0,K)=(G1(K)+G2(K))/8
13680    RP(0,K)=RQ(0,K)+(L1(K)+L2(K))/8
13690    VP(0,K)=VQ(0,K)+(S1(K)+S2(K))/8
13700 NEXT K
13710 REM
13720 FOR Q=0 TO NQ
13730    FOR K=1 TO 3
13740       RK(Q,K)=RQ(Q,K)+(U1(Q,K)+U2(Q,K))/8
13750       VK(Q,K)=VQ(Q,K)+(W1(Q,K)+W2(Q,K))/8
13760       PK(Q,K)=RK(Q,K)-RP(0,K)
13770    NEXT K
13780 NEXT Q
13790 REM
```

6.6. COMPUTER PROGRAMS

```
13800 GOSUB 15010 REM SUBSUB
13810 REM
13820 FOR K=1 TO 3
13830    L3(K)=LX(K)
13840    S3(K)=SX(K)
13850    FOR Q=0 TO NQ
13860       U3(Q,K)=UX(Q,K)
13870       W3(Q,K)=WX(Q,K)
13880    NEXT Q
13890    F3(K)=FX(K)
13900    G3(K)=GX(K)
13910 NEXT K
13920 REM
13930 REM ********** STEP 4 **********
13940 REM
13950 FOR K=1 TO 3
13960    DV(0,K)=-(G2(K)-2*G3(K))/2
13970    RP(0,K)=RQ(0,K)-(L2(K)-2*L3(K))/2
13980    VP(0,K)=VQ(0,K)-(S2(K)-2*S3(K))/2
13990 NEXT K
14000 REM
14010 FOR Q=0 TO NQ
14020    FOR K-1 TO 3
14030       RK(Q,K)=RQ(Q,K)-(U2(Q,K)-2*U3(Q,K))/2
14040       VK(Q,K)=VQ(Q,K)-(W2(Q,K)-2*W3(Q,K))/2
14050       PK(Q,K)=RK(Q,K)-RP(0,K)
14060    NEXT K
14070 NEXT Q
14080 REM
14090 GOSUB 15010 REM SUBSUB
14100 REM
14110 FOR K=1 TO 3
14120    L4(K)=LX(K)
14130    S4(K)=SX(K)
14140    FOR Q=0 TO NQ
14150       U4(Q,K)=UX(Q,K)
14160       W4(Q,K)=WX(Q,K)
14170    NEXT Q
14180    F4(K)=FX(K)
14190    G4(K)=GX(K)
14200 NEXT K
14210 REM
```

```
14220 REM ********** STEP 5 **********
14230 REM
14240 FOR K=1 TO 3
14250    DV(0,K)=(3*G1(K)+9*G4(K))/16
14260    RP(0,K)=RQ(0,K)+(3*L1(K)+9*L4(K))/16
14270    VP(0,K)=VQ(0,K)+(3*S1(K)+9*S4(K))/16
14280 NEXT K
14290 REM
14300 FOR Q=0 TO NQ
14300    FOR K=1 TO 3
14300       RK(Q,K)=RQ(Q,K)+(3*U1(Q,K)+9*U4(Q,K))/16
14300       VK(Q,K)=VQ(Q,K)+(3*W1(Q,K)+9*W4(Q,K))/16
14300       PK(Q,K)=RK(Q,K)-RP(0,K)
14300    NEXT K
14300 NEXT Q
14300 REM
14300 GOSUB 15010 REM SUBSUB
14300 REM
14400 FOR K=1 TO 3
14410    L5(K)=LX(K)
14420    S5(K)=SX(K)
14430    FOR Q=0 TO NQ
14440       U5(Q,K)=UX(Q,K)
14450       W5(Q,K)=WX(Q,K)
14460    NEXT Q
14470    F5(K)=FX(K)
14480    G5(K)=GX(K)
14490 NEXT K
14500 REM
14510 REM ********** STEP 6 **********
14520 REM
14530 FOR K=1 TO 3
14540    DV(0,K)=-(3*G1(K)-2*G2(K)-12*G3(K)+12*G4(K)-8*G5(K))/7
14550    RP(0,K)=RQ(0,K)-(3*L1(K)-2*L2(K)-12*L3(K)+12*L4(K)-
       8*L5(K))/7
14560    VP(0,K)=VQ(0,K)-(3*S1(K)-2*S2(K)-12*S3(K)+12*S4(K)-
       8*S5(K))/7
14570 NEXT K
14580 REM
14590 FOR Q=0 TO NQ
14600    FOR K=1 TO 3
14610       RK(Q,K)=RQ(Q,K)-(3*U1(Q,K)-2*U2(Q,K)-12*U3(Q,K)+
```

6.6. COMPUTER PROGRAMS

```
              12*U4(Q,K)-8*U5(Q,K))/7
14620    VK(Q,K)=VQ(Q,K)-(3*W1(Q,K)-2*W2(Q,K)-12*W3(Q,K)+
              12*W4(Q,K)-8*W5(Q,K))/7
14630    PK(Q,K)=RK(Q,K)-RP(0,K)
14640   NEXT K
14650 NEXT Q
14660 REM
14670 GOSUB 15010 REM SUBSUB
14680 REM
14690 FOR K=1 TO 3
14700    F6(K)=FX(K)
14710    G6(K)=GX(K)
14720 NEXT K
14730 REM
14740 REM ********** RESULT **********
14750 REM
14760 FOR K=1 TO 3
14770    DR(0,K)=(7*F1(K)+32*F3(K)+12*F4(K)+32*F5(K)+
              7*F6(K))/90
14780    DV(0,K)=(7*G1(K)+32*G3(K)+12*G4(K)+32*G5(K)+
              7*G6(K))/90
14790 NEXT K
14800 RETURN
15000 STOP
15010 REM # ENK5 SUBSUBROUTINE #
15020 REM
15030 FOR Q=1 TO NQ
15040    RK(Q,0)=FNMG(RK(Q,1),RK(Q,2),RK(Q,3))
15050    PK(Q,0)=FNMG(PK(Q,1),PK(Q,2),PK(Q,3))
15060    FOR K=1 TO 3
15070       AK(Q,K)=FNAK(M(Q),PK(Q,K),PK(Q,0),RK(Q,K),RK(Q,0))
15080    NEXT K
15090 NEXT Q
15100 REM
15110 FOR K=1 TO 3
15120    AK(0,K)=0
15130    FOR Q=1 TO NQ
15140       AK(0,K)=AK(0,K)+AK(Q,K)
15150    NEXT Q
15160 NEXT K
15170 REM
15180 RK(0,0)=FNMG(RK(0,1),RK(0,2),RK(0,3))
```

```
15190 RP(0,0)=FNMG(RP(0,1),RP(0,2),RP(0,3))
15200 REM
15210 FOR K=1 TO 3
15220    LX(K)=H*FNV(VP(0,K))
15230    SX(K)=H*FNA(M(0),RP(0,K),RP(0,0),AK(0,K))
15240    FOR Q=0 TO NQ
15250       UX(Q,K)=H*FNV(VK(Q,K))
15260       WX(Q,K)=H*FNA(M(Q),RK(Q,K),RK(Q,0),0)
15270    NEXT Q
15280    FX(K)=H*FNF(DV(0,K))
15290    GX(K)=H*FNG(M(0),RK(0,K),RK(0,0),
         RP(0,K),RP(0,0),AK(0,K))
15300 NEXT K
15310 RETURN
16000 STOP
16010 REM # ARC/SUB # COMPUTES X FROM SIN(X) AND COS(X)
16020 IF ABS(SX)<=.707107 THEN X=FNASN(ABS(SX))
16030 IF ABS(CX)<=.707107 THEN X=FNACN(ABS(CX))
16040 IF CX>=0 AND SX>=0 THEN X=X
16050 IF CX<0 AND SX>=0 THEN X=180*Q1#-X
16060 IF CX<0 AND SX<0 THEN X=180*Q1#+X
16070 IF CX>=0 AND SX<0 THEN X=360*Q1#-X
16080 RETURN
20000 DATA 0.005775519#
20010 REM
20020 REM # CELESTIAL BODY #
20030 REM
30000 DATA   "URANUS","J2000.0",0.017202099#
30010 DATA   0.000043554#
30020 DATA   6280.5#
30030 DATA   -4.33594#, -17.05279#,-7.40711#
30040 DATA   +0.2213400#,-0.0559999#,-0.0276676#
30050 REM
39000 REM # SUN #
39010 REM
39020 DATA 7760.5#,-0.8672422#,+0.4774827#,+0.2070265#
39030 REM
40000 REM # PERTURBERS #
40010 REM
40020 'DATA   0, "(TWO-BODY)",  0
40030 DATA   7, "(MVEMJSN)", 6280.5#
40040 REM
```

6.6. COMPUTER PROGRAMS

```
41000 DATA   "MERCURY"
41010 DATA   0.000000166#
41020 DATA   0.1693419#,-0.3559908#,-0.2077172#
41030 DATA   0.02036314#,+0.01152157#, 0.00404131#
42000 DATA   "VENUS"
42010 DATA   0.000002448#
42020 DATA   0.6457331#, 0.3130627#, 0.0999533#
42030 DATA  -0.00920284#, 0.01616581#, 0.00785453#
43000 DATA   "EARTH-MOON"
43010 DATA   0.000003040#
43020 DATA   0.6632717#,-0.7045336#,-0.3054775#
43030 DATA   0.01273999#, 0.01025921#, 0.00444834#
44000 DATA   "MARS"
44010 DATA   0.000000323#
44020 DATA  -0.8888462#, 1.2418782#, 0.5936583#
44030 DATA  -0.01122166#,-0.00593450#, 0.00241817#
45000 DATA   "JUPITER"
45010 DATA   0.000954791#
45020 DATA   3.406998#,-3.434686#,-1.555282#
45030 DATA   0.005504353#, 0.005026155#, 0.002020327#
46000 DATA   "SATURN"
46010 DATA   0.000285878#
46020 DATA  -5.344798#,-7.833691#,-3.005450#
46030 DATA   0.004409801#,-0.002715426#,-0.001310925#
47000 'DATA   "URANUS"
47010 'DATA   0.000043554#
47020 'DATA  -4.33594#,-17.05279#,-7.40711#
47030 'DATA   0.003807513#,-0.000963315#,-0.000475940#
48000 DATA   "NEPTUNE"
48010 DATA   0.000051776#
48020 DATA   1.41455#,-27.95496#,-11.47747#
48030 DATA   0.0031224404#, 0.000184341#,-0.000002187#
49000 'DATA    "PLUTO"
49010 'DATA   0.000000008#
49020 'DATA  -23.55598#,-18.12624#, 1.44015#
49030 'DATA   0.002025708#,-0.002522970#,-0.001395285#
```

6.7 Numerical Examples

6.7.1 Solar and Planetary Attractions

Problem

The mass and vector orbital elements for the planet Mars are listed below for the epoch 1985 August 3 [9].

- $m = 0.000000323$
- $t_0 = 2446280.5$
- $\mathbf{r}_0 = \{-0.8888462, +1.2418782, +0.5936583\}$
- $\dot{\mathbf{r}}_0 = \{-0.6523424, -0.3449870, -0.1405741\}$

Compute the attractions of the Sun, Mercury, Venus, Earth-Moon, Jupiter, Saturn, and Uranus at eight dates between this epoch and JD 2447760.5. The vector orbital elements of the major planets for the epoch JD 2446280.5 are given in Section E of *The Astronomical Almanac 1985*.

Solution

Use the given information to write data lines 30010 to 43030 and 45000 to 47030 as shown at the end of program ATTRACT. Run the program and answer the prompts as follows:

>>>NEW EPOCH? 7760.5#

>>>NUMBER OF STEPS? 8

6.7. NUMERICAL EXAMPLES

Results

```
SOLAR AND PLANETARY ATTRACTIONS
MARS
J2000.0
     T(I)         A(X)         A(Y)         A(Z)          A
SUN
  6280.50000   0.2020569  -0.2823098  -0.1349533   0.3724758
  6465.50000   0.3497524   0.1528139   0.0606225   0.3864633
  6650.50000  -0.2964878   0.3758590   0.1804166   0.5115908
  6835.50000  -0.2385092  -0.3490189  -0.1536249   0.4497794
  7020.50000   0.3007499  -0.1789061  -0.0901976   0.3613773
  7205.50000   0.2631973   0.2969605   0.1290796   0.4172768
  7390.50000  -0.4882828   0.1670106   0.0898177   0.5238128
  7575.50000  -0.0317749  -0.3730676  -0.1702509   0.4113082
  7760.50000   0.3562532  -0.0562426  -0.0354392   0.3624024
MERCURY
  6280.50000  -0.0000003   0.0000006   0.0000004   0.0000008
  6465.50000  -0.0000007   0.0000005   0.0000004   0.0000010
  6650.50000  -0.0000014   0.0000000   0.0000001   0.0000014
  6835.50000  -0.0000011  -0.0000013  -0.0000006   0.0000018
  7020.50000   0.0000005  -0.0000014  -0.0000008   0.0000017
  7205.50000   0.0000012  -0.0000004  -0.0000003   0.0000013
  7390.50000   0.0000009   0.0000003   0.0000000   0.0000010
  7575.50000   0.0000006   0.0000005   0.0000002   0.0000008
  7760.50000   0.0000003   0.0000006   0.0000003   0.0000008
VENUS
  6280.50000  -0.0000036  -0.0000024  -0.0000008   0.0000044
  6465.50000  -0.0000031   0.0000025   0.0000013   0.0000042
  6650.50000  -0.0000014   0.0000051   0.0000023   0.0000057
  6835.50000   0.0000041   0.0000010   0.0000002   0.0000042
  7020.50000   0.0000056  -0.0000040  -0.0000020   0.0000072
  7205.50000  -0.0000014  -0.0000037  -0.0000016   0.0000042
  7390.50000  -0.0000073   0.0000020   0.0000011   0.0000077
  7575.50000  -0.0000024   0.0000030   0.0000015   0.0000042
  7760.50000   0.0000032   0.0000032   0.0000013   0.0000047
EARTH-MOON
  6280.50000  -0.0000017   0.0000017   0.0000007   0.0000025
  6465.50000   0.0000028  -0.0000010  -0.0000005   0.0000030
  6650.50000  -0.0000053   0.0000155   0.0000084   0.0000184
  6835.50000   0.0000014  -0.0000022  -0.0000010   0.0000028
  7020.50000  -0.0000019   0.0000015   0.0000007   0.0000025
```

7205.50000	0.0000026	-0.0000008	-0.0000003	0.0000028
7390.50000	-0.0000156	-0.0000012	0.0000008	0.0000157
7575.50000	0.0000017	-0.0000024	-0.0000011	0.0000032
7760.50000	-0.0000021	0.0000013	0.0000005	0.0000025

JUPITER

6280.50000	-0.0000112	0.0000102	0.0000045	0.0000158
6465.50000	-0.0000076	0.0000104	0.0000047	0.0000137
6650.50000	0.0000223	0.0000071	0.0000030	0.0000236
6835.50000	0.0000108	-0.0000130	-0.0000062	0.0000180
7020.50000	-0.0000115	-0.0000090	-0.0000037	0.0000150
7205.50000	-0.0000138	-0.0000076	-0.0000028	0.0000160
7390.50000	-0.0000081	0.0000076	0.0000036	0.0000116
7575.50000	0.0000218	0.0000286	0.0000109	0.0000375
7760.50000	0.0000108	-0.0000038	-0.0000020	0.0000116

SATURN

6280.50000	0.0000005	0.0000002	0.0000000	0.0000006
6465.50000	0.0000000	-0.0000009	-0.0000003	0.0000009
6650.50000	-0.0000005	-0.0000002	-0.0000001	0.0000006
6835.50000	0.0000001	0.0000006	0.0000002	0.0000007
7020.50000	0.0000004	0.0000003	0.0000001	0.0000005
7205.50000	0.0000003	-0.0000008	-0.0000003	0.0000009
7390.50000	-0.0000004	-0.0000002	-0.0000001	0.0000005
7575.50000	-0.0000001	0.0000006	0.0000003	0.0000007
7760.50000	0.0000003	0.0000004	0.0000002	0.0000005

URANUS

6280.50000	0.0000000	0.0000000	0.0000000	0.0000000
6465.50000	0.0000000	-0.0000000	-0.0000000	0.0000000
6650.50000	-0.0000000	-0.0000000	-0.0000000	0.0000000
6835.50000	-0.0000000	0.0000000	0.0000000	0.0000000
7020.50000	0.0000000	0.0000000	0.0000000	0.0000000
7205.50000	0.0000000	-0.0000000	-0.0000000	0.0000000
7390.50000	-0.0000000	-0.0000000	-0.0000000	0.0000000
7575.50000	-0.0000000	0.0000000	0.0000000	0.0000000
7760.50000	0.0000000	0.0000000	0.0000000	0.0000000

ATTRACT

6.7. NUMERICAL EXAMPLES

6.7.2 The Motion of Mars

Problem

Use the method of Cowell to compute the motion of Mars from 1985 August 3 to 1987 July 24, taking into account the attractions of Mercury, Venus, Earth-Moon, Jupiter, and Saturn. The orbital elements of Mars are listed below [9]:

- $m = 0.000000323$
- $t_0 = 2446280.5$
- $\mathbf{r}_0 = \{-0.8888462, +1.2418782, +0.5936583\}$
- $\dot{\mathbf{r}}_0 = \{-0.6523424, -0.3449870, -0.1405741\}$

Assume the following geocentric position of the Sun for 1987 July 24 [10]:

- $t = 2447000.5$
- $\mathbf{R} = \{-0.5197797, +0.8008713, +0.3472473\}$

Use 144 steps to cover the interval of 720 days, and compare the computed right ascension and declination with that reduced from data published in *The Astronomical Almanac 1987*.

Solution

Use the given information to write data lines 20000 to 39020, 40030 to 43030, and 45000 to 46030 as shown at the end of program COWELL. Run the program and answer the prompt as follows:

```
>>>NUMBER OF STEPS? 144
```

Results

RA AND DEC
FROM PERTURBED ORBITAL MOTION
BY THE METHOD OF COWELL
USING RUNGE-KUTTA FIVE
MARS
J2000.0
NQ 5 (MVEJS) 6280.5

T(I)	X	Y	Z	R
6280.50000	-0.8888462	1.2418782	0.5936583	1.6385174
6285.50000	-0.9441932	1.2111722	0.5810728	1.6419758
6290.50000	-0.9979627	1.1784424	0.5675164	1.6452145
6295.50000	-1.0500747	1.1437551	0.5530173	1.6482296
6300.50000	-1.1004523	1.1071783	0.5376046	1.6510172
6305.50000	-1.1490212	1.0687818	0.5213083	1.6535739
6310.50000	-1.1957105	1.0286367	0.5041592	1.6558966
6315.50000	-1.2404520	0.9868158	0.4861887	1.6579825
6320.50000	-1.2831803	0.9433933	0.4674290	1.6598291
6325.50000	-1.3238331	0.8984448	0.4479133	1.6614341
6330.50000	-1.3623507	0.8520472	0.4276752	1.6627958
6335.50000	-1.3986765	0.8042786	0.4067489	1.6639125
6340.50000	-1.4327564	0.7552187	0.3851695	1.6647828
6345.50000	-1.4645395	0.7049480	0.3629726	1.6654058
6350.50000	-1.4939772	0.6535484	0.3401945	1.6657808
6355.50000	-1.5210242	0.6011030	0.3168720	1.6659074
6360.50000	-1.5456377	0.5476959	0.2930425	1.6657852
6365.50000	-1.5677776	0.4934124	0.2687440	1.6654146
6370.50000	-1.5874070	0.4383388	0.2440153	1.6647959
6375.50000	-1.6044916	0.3825625	0.2188953	1.6639299
6380.50000	-1.6189999	0.3261721	0.1934240	1.6628175
6385.50000	-1.6309035	0.2692570	0.1676415	1.6614600
6390.50000	-1.6401767	0.2119077	0.1415886	1.6598590
6395.50000	-1.6467970	0.1542158	0.1153068	1.6580164
6400.50000	-1.6507446	0.0962737	0.0888379	1.6559344
6405.50000	-1.6520032	0.0381749	0.0622243	1.6536154
6410.50000	-1.6505594	-0.0199862	0.0355090	1.6510623
6415.50000	-1.6464028	-0.0781143	0.0087354	1.6482780
6420.50000	-1.6395268	-0.1361132	-0.0180525	1.6452661
6425.50000	-1.6299276	-0.1938859	-0.0448104	1.6420304
6430.50000	-1.6176053	-0.2513344	-0.0714934	1.6385747
6435.50000	-1.6025632	-0.3083602	-0.0980560	1.6349036

6.7. NUMERICAL EXAMPLES

```
6440.50000  -1.5848086  -0.3648638  -0.1244526   1.6310219
     .
     .
     .
6640.50000   0.6866022  -1.1072725  -0.5264467   1.4052121
     .
     .
     .
6800.50000   1.1351773   0.8278058   0.3489604   1.4476406
     .
     .
     .
6980.50000  -1.0297844   1.1577492   0.5588879   1.6471778
6985.50000  -1.0808502   1.1219136   0.5438336   1.6500553
6990.50000  -1.1301365   1.0842308   0.5278838   1.6527027
6995.50000  -1.1775711   1.0447711   0.5110689   1.6551160
7000.50000  -1.2230846   1.0036068   0.4934202   1.6572948
            -0.5177352  -0.4882215  -0.2099165

   T(I)        A(I)         D(I)        P(I)
7000.50000   8.933377    18.52700    2.6458497
             56   0.16    31   37.2
```

COWELL

Discussion of Results

The COWELL computation carries Mars a little more than once around the Sun. The result is summarized as follows:

- $t = 2447000.5$

- $\mathbf{r} = \{-1.2230846, +1.0036068, +0.4934202\}$

- $\dot{\mathbf{r}} = \{-0.5177352, -0.4882215, -0.2099165\}$

- $r = 1.0572948$

- $\alpha = 8^{\rm h}.933377 = 8^{\rm h}56^{\rm m}0^{\rm s}.16$

- $\delta = 18°.52700 = 18°31'37.2''$

- $p = 2.6458497$

For comparison, the vector orbital elements of Mars given in *The Astronomical Almanac 1987* for JD 2447000.5 can also be used in program COWELL or RADEC to produce an astrometric position on this same date. The computation requires only one step of zero length. When this is accomplished, the result is

- $t = 2447000.5$
- $\mathbf{r} = \{-1.2230815, +1.0036100, +0.4934217\}$
- $\dot{\mathbf{r}} = \{-0.5177368, -0.4882201, -0.2099157\}$
- $r = 1.6572949$
- $\alpha = 8^h.933370 = 8^h 56^m 0^s.13$
- $\delta = 18°.52703 = 18°31'37.3''$
- $p = 2.6458503$

Alternatively, one could use the procedures demonstrated in Sections 2.10.5 and 2.10.6, where programs ADAPP and ADCESS are employed to reduce the apparent position of Mars given in *The Astronomical Almanac 1987* for July 24 to an astrometric position on the same date.

As a matter of interest, running COWELL without taking the planetary attractions into account will produce the following two-body astrometric position for JD 2447000.5:

- $t = 2447000.5$
- $\mathbf{r} = \{-1.2227938, +1.0037363, +0.4934715\}$
- $\dot{\mathbf{r}} = \{-0.5179000, -0.4881732, -0.2098869\}$
- $r = 1.6571739$
- $\alpha = 8^h.932922 = 8^h 55^m 58^s.52$
- $\delta = 18°.07082 = 18°31'43.7''$
- $p = 2.6457627$

6.7. NUMERICAL EXAMPLES

6.7.3 The Motion of Uranus

Problem

Use the method of Encke to compute the motion of the planet Uranus from 1985 August 3 to 1989 August 22, taking into account the attractions of Mercury, Venus, Earth-Moon, Mars, Jupiter, Saturn, and Neptune. The orbital elements of Uranus for the initial epoch are as follows [9]:

- $m = 0.000043554$
- $t_0 = 2446280.5$
- $\mathbf{r}_0 = \{-4.33594, -17.05279, -7.40711\}$
- $\dot{\mathbf{r}}_0 = \{+0.2213400, -0.0559999, -0.0276676\}$

Assume the following geocentric position of the Sun for 1989 August 22 [11]:

- $t = 2447760.5$
- $\mathbf{R} = \{-0.8672422, +0.4774827, +0.2070265\}$

Use 74 steps to cover the interval of 1480 days, and compare the computed right ascension and declination with that reduced from data published in *The Astronomical Almanac 1989*.

Solution

Use the given information to write data lines 20000 to 39020, 40030 to 46030, and 48000 to 48030 as shown at the end of program ENCKE. Run the program and answer the prompt as follows:

```
>>>NUMBER OF STEPS? 74
```

Results

```
RA AND DEC
FROM PERTURBED ORBITAL MOTION
BY UNIVERSAL VARIABLES
AND RUNGE-KUTTA FIVE
URANUS
J2000.0
NQ 7 (MVEMJSN) 6280.5
     T(I)          X             Y             Z            R
  6280.50000   -4.3359400   -17.0527900   -7.4071100   19.0909220
  6300.50000   -4.2597548   -17.0719094   -7.4165649   19.0945289
  6320.50000   -4.1835005   -17.0907350   -7.4258922   19.0981354
  6340.50000   -4.1071782   -17.1092672   -7.4350919   19.1017415
  6360.50000   -4.0307890   -17.1275056   -7.4441639   19.1053470
  6380.50000   -3.9543342   -17.1454501   -7.4531082   19.1089518
  6400.50000   -3.8778152   -17.1631004   -7.4619245   19.1125557
  6420.50000   -3.8012334   -17.1804566   -7.4706129   19.1161587
  6440.50000   -3.7245902   -17.1975183   -7.4791732   19.1197605
  6460.50000   -3.6478873   -17.2142855   -7.4876053   19.1233611
  6480.50000   -3.5711264   -17.2307581   -7.4959092   19.1269606
  6500.50000   -3.4943089   -17.2469365   -7.5040849   19.1305595
  6520.50000   -3.4174360   -17.2628209   -7.5121326   19.1341577
  6540.50000   -3.3405090   -17.2784111   -7.5200522   19.1377552
  6560.50000   -3.2635292   -17.2937072   -7.5278438   19.1413522
  6580.50000   -3.1864978   -17.3087092   -7.5355072   19.1449484
  6600.50000   -3.1094160   -17.3234169   -7.5430424   19.1485438
  6620.50000   -3.0322852   -17.3378300   -7.5504493   19.1521379
  6640.50000   -2.9551069   -17.3519482   -7.5577277   19.1557305
  6660.50000   -2.8778831   -17.3657715   -7.5648776   19.1593216
  6680.50000   -2.8006155   -17.3793001   -7.5718989   19.1629114
  6700.50000   -2.7233056   -17.3925342   -7.5787919   19.1665001
  6720.50000   -2.6459549   -17.4054741   -7.5855565   19.1700880
  6740.50000   -2.5685650   -17.4181203   -7.5921930   19.1736756
  6760.50000   -2.4911371   -17.4304733   -7.5987015   19.1772633
  6780.50000   -2.4136720   -17.4425335   -7.6050823   19.1808515
  6800.50000   -2.3361706   -17.4543008   -7.6113353   19.1844399
  6820.50000   -2.2586339   -17.4657752   -7.6174604   19.1880284
  6840.50000   -2.1810631   -17.4769564   -7.6234576   19.1916166
  6860.50000   -2.1034595   -17.4878445   -7.6293269   19.1952045
  6880.50000   -2.0258243   -17.4984390   -7.6350680   19.1987915
  6900.50000   -1.9481592   -17.5087399   -7.6406808   19.2023775
```

6.7. NUMERICAL EXAMPLES

6920.50000	-1.8704660	-17.5187473	-7.6461654	19.2059625
6940.50000	-1.7927462	-17.5284615	-7.6515219	19.2095469
6960.50000	-1.7150011	-17.5378829	-7.6567505	19.2131308
6980.50000	-1.6372321	-17.5470117	-7.6618511	19.2167144
7000.50000	-1.5594403	-17.5558482	-7.6668240	19.2202979
7020.50000	-1.4816269	-17.5643927	-7.6716693	19.2238814
7040.50000	-1.4037930	-17.5726453	-7.6763870	19.2274651
7060.50000	-1.3259395	-17.5806060	-7.6809771	19.2310486
7080.50000	-1.2480679	-17.5882746	-7.6854395	19.2346318
7100.50000	-1.1701797	-17.5956513	-7.6897742	19.2382144
7120.50000	-1.0922763	-17.6027364	-7.6939814	19.2417968
7140.50000	-1.0143590	-17.6095300	-7.6980610	19.2453791
7160.50000	-0.9364291	-17.6160325	-7.7020132	19.2489613
7180.50000	-0.8584881	-17.6222443	-7.7058382	19.2525437
7200.50000	-0.7805369	-17.6281659	-7.7095361	19.2561267
7220.50000	-0.7025763	-17.6337976	-7.7131071	19.2597106
7240.50000	-0.6246072	-17.6391394	-7.7165512	19.2632950
7260.50000	-0.5466306	-17.6441912	-7.7198684	19.2668798
7280.50000	-0.4686475	-17.6489530	-7.7230585	19.2704646
7300.50000	-0.3906592	-17.6534247	-7.7261217	19.2740492
7320.50000	-0.3126670	-17.6576060	-7.7290577	19.2776333
7340.50000	-0.2346724	-17.6614970	-7.7318664	19.2812164
7360.50000	-0.1566772	-17.6650977	-7.7345480	19.2847986
7380.50000	-0.0786829	-17.6684087	-7.7371027	19.2883803
7400.50000	-0.0006908	-17.6714305	-7.7395305	19.2919617
7420.50000	0.0772979	-17.6741634	-7.7418317	19.2955432
7440.50000	0.1552821	-17.6766081	-7.7440065	19.2991249
7460.50000	0.2332608	-17.6787649	-7.7460551	19.3027073
7480.50000	0.3112333	-17.6806342	-7.7479777	19.3062905
7500.50000	0.3891988	-17.6822160	-7.7497742	19.3098741
7520.50000	0.4671564	-17.6835102	-7.7514446	19.3134580
7540.50000	0.5451047	-17.6845167	-7.7529888	19.3170418
7560.50000	0.6230425	-17.6852357	-7.7544069	19.3206255
7580.50000	0.7009687	-17.6856672	-7.7556989	19.3242088
7600.50000	0.7788819	-17.6858114	-7.7568648	19.3277917
7620.50000	0.8567807	-17.6856685	-7.7579047	19.3313742
7640.50000	0.9346639	-17.6852391	-7.7588187	19.3349566
7660.50000	1.0125305	-17.6845234	-7.7596071	19.3385390
7680.50000	1.0903798	-17.6835217	-7.7602699	19.3421213
7700.50000	1.1682106	-17.6822339	-7.7608071	19.3457033
7720.50000	1.2460219	-17.6806602	-7.7612188	19.3492850
7740.50000	1.3238125	-17.6788006	-7.7615049	19.3528661

```
7760.50000    1.4015814  -17.6766550   -7.7616655   19.3564463
              0.2260117    0.0066521   -0.0002843

   T(I)          A(I)         D(I)         P(I)
7760.50000    18.118538   -23.70303    18.7928053
                  7  6.74    -42 -10.9
```

ENCKE

Discussion of Results

The ENCKE computation follows Uranus for about four of the 84 years it takes the planet to complete one orbit around the Sun. The result may be summarized as follows:

- $t = 2447760.5$
- $\mathbf{r} = \{+1.4015814, -17.6766550, -7.7616655\}$
- $\dot{\mathbf{r}} = \{+0.2260117, +0.0066521, -0.0002843\}$
- $r = 19.3564463$
- $\alpha = 18^{\mathrm{h}}.118538 = 18^{\mathrm{h}}7^{\mathrm{m}}6^{\mathrm{s}}.74$
- $\delta = -23°.70303 = -23°42'10.9''$
- $p = 18.7928053$

For comparison, the vector orbital elements of Uranus given in *The Astronomical Almanac 1989* for JD 2447760.5 can be used in program ENCKE to produce an astrometric position on this same date. When this is done, the result is

- $t = 2447760.5$
- $\mathbf{r} = \{+1.4015700, -17.6766500, -7.7616600\}$
- $\dot{\mathbf{r}} = \{+0.2260107, +0.0066522, -0.0002840\}$
- $r = 19.3564387$
- $\alpha = 18^{\mathrm{h}}.118536 = 18^{\mathrm{h}}7^{\mathrm{m}}6^{\mathrm{s}}.73$
- $\delta = -23°.70303 = -23°42'10.9''$
- $p = 18.7927996$

6.7. NUMERICAL EXAMPLES

Running ENCKE without taking planetary attractions into account will produce the following two-body astrometric position for Uranus on JD 2447760.5:

- $t = 2447760.5$
- $\mathbf{r} = \{+1.4115429, -17.6795945, -7.7631999\}$
- $\dot{\mathbf{r}} = \{+0.2266822, +0.0066952, -0.0002849\}$
- $r = 19.3604697$
- $\alpha = 18^h.120728 = 18^h 7^m 14^s.62$
- $\delta = -23°.70333 = -23°42'12.0''$
- $p = 18.7963994$

References

[1] Herget, *The Computation of Orbits* Edwards Brothers, Inc., 1948.

[2] Dubyago, *The Determination of Orbits*, The Macmillan Company, New York, 1961.

[3] Brouwer and Clemence, *Celestial Mechanics*, Academic Press, New York, 1961.

[4] Roy, *Orbital Motion*, Adam Hilger Ltd., Bristol, 1978.

[5] Baker, *Astrodynamics: Applications and Advanced Topics*, Academic Press, 1967.

[6] Bate, Mueller, and White, *Fundamentals of Astrodynamics*, Dover Publications Inc., 1971.

[7] Danby, *Fundamentals of Celestial Mechanics*, Willmann-Bell, Inc., 1988.

[8] *Planetary and Lunar Coordinates for the Years 1984-2000*, U.S. Government Printing Office, 1983.

[9] *The Astronomical Almanac 1985*, U.S. Government Printing Office, 1984.

[10] *The Astronomical Almanac 1987*, U.S. Government Printing Office, 1986.

[11] *The Astronomical Almanac 1989*, U.S. Government Printing Office, 1988.

Chapter 7

Applied Numerical Methods

7.1 Introduction

This chapter forms a bridge between the application of the principles of celestial mechanics to problems where the orbital elements are known and the application of those principles to problems in which it is necessary to discover or improve the elements. The vehicle for this transition is a set of powerful numerical methods which apply arithmetic and logical operations to solve complex functions, handle large systems of simultaneous equations, and analyze data gleaned from observations.

Our interest in these numerical methods is utilitarian. Thus, the discussions are brief, leading directly to BASIC routines which can be used in various combinations in longer programs. For a thorough treatment of this subject, the reader should consult the references or similar texts. References 1 and 3 are especially recommended to those new to this field of applied mathematics.

7.2 Finding the Root of an Equation

The *root* of an equation is a value of the independent variable for which the function equals zero. In other words, given an equation of the general form

$$y = f(x), \qquad (7.1)$$

where $f(x)$ is any function of the independent variable x, then its roots are those values of x for which

$$f(x) = 0. \qquad (7.2)$$

Figure 7.1 illustrates a situation where Equation 7.2 is satisfied by three values of x: a, b, and c.

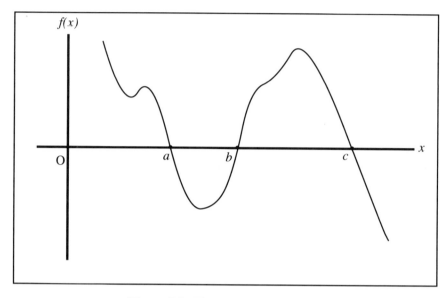

Figure 7.1: Three roots a, b, and c.

The first step in finding an accurate value for the root of an equation is to determine an approximate value. This can be done by computing $f(x)$ at a series of discrete values of x over a range which includes the root. Consider the situation depicted in Figure 7.2. The points x_l and x_h represent the respective low and high values of x which bound the range chosen to include the root x_r. From a table of computed values for $f(x)$, a value of x which causes $f(x)$ to change algebraic sign can be taken as an approximate root. For example, suppose we have computed the following table of discrete values of x and $f(x)$:

x	$f(x)$
1.0	-2.351
1.1	-2.144
1.2	-1.399
1.3	0.387
1.4	4.017
1.5	9.720

Then it would be reasonable to choose either 1.2 or 1.3 as an approximate value for the root.

A simple BASIC routine will generate and display a table of values for x and $f(x)$ from which an approximate root can be chosen. If FNF(X) is a user defined function in BASIC which represents $f(x)$, then

7.2. FINDING THE ROOT OF AN EQUATION

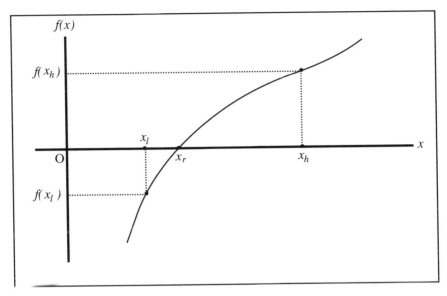

Figure 7.2: Points x_l and x_h bound the range which includes the root x_r.

```
1100 INPUT">>>CHOOSE XLOW AND XHIGH";XL,XH
1130 PRINT
1140 PRINT TAB(2)"N";TAB(13)"X";TAB(19)"F(X)"
1150 REM
1160 DX=(XH-XL)/10
1170 REM
1180 FOR N=0 TO 10
1190    X=XL+N*DX
1200    PRINT N;TAB(6);
1210    PRINT USING L$; X,FNF(X)
1220 NEXT N
```

The results of this routine can be used as the starting point for finding a much more accurate value of the root by either the *bisection method* or the *Newton-Raphson method*.

7.2.1 The Bisection Method

The bisection method determines a root by successively dividing in half the interval of the independent variable over which the function $f(x)$ changes sign. At each step, a new value of the root is chosen at the midpoint of the interval, and the process is repeated until the trial value of the root is sufficiently accurate for

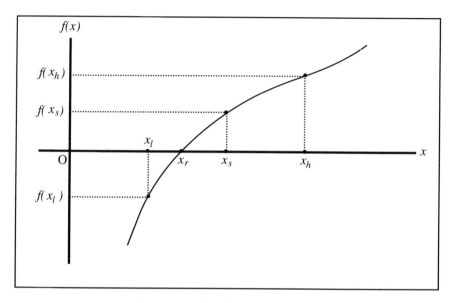

Figure 7.3: The bisection method.

the requirements of the problem. Compare the steps of the following algorithm with the depiction of Figure 7.3:

Step 1 Choose x_l and x_h as lower and higher values of x which bound a range in which the the function changes sign.

Step 2 Estimate the root by computing

$$x_s = \frac{x_l + x_h}{2}.$$

Step 3 If $|f(x_s)|$ is less than a given tolerance, go to Step 5, otherwise continue with Step 4.

Step 4 Use the following logical operations to determine on which side of x_s the true root x_r lies:

- If $f(x_l)f(x_s) < 0$, the root lies to the left of x_s. Therefore, let $x_h = x_s$ and go to Step 2.
- If $f(x_l)f(x_s) > 0$, the root lies to the right of x_s. Therefore, let $x_l = x_s$ and go to Step 2.

Step 5 Stop computation. Take x_s as the value of the root.

7.2. FINDING THE ROOT OF AN EQUATION

If FNF(X) represents the function $f(x)$, then the bisection algorithm can be implemented by the following BASIC routine:

```
1240 INPUT">>>BRACKET WITH XLOW AND XHIGH";XL,XH
1250 REM
1260    XS=(XL+XH)/2
1270    IF ABS(FNF(XS))<.0000001# GOTO 1320
1280    SN=FNF(XL)*FNF(XS)
1290    IF SN<0 THEN XH=XS
1300    IF SN>0 THEN XL=XS
1310    GOTO 1260
1320 PRINT
1350 PRINT"ROOT";TAB(10);
1360 PRINT USING G$;XS
```

7.2.2 The Newton-Raphson Method

The Newton-Raphson method is one of the most commonly used techniques for finding a root. Starting from an approximate value of the root, the method uses the first derivative of the function to extrapolate an improved value of the root. This procedure is repeated until a sufficiently accurate root is computed by successive refinements.

Consider the situation shown in Figure 7.4, where x_i represents an initial approximation of the root, and x_s is an improved value of the root obtained by extrapolating down to the x-axis by means of a tangent line from the point P_t. Now, the first derivative of a function evaluated at a given point is equivalent to the slope of the curve at that point. Therefore, in this case, we have

$$f'(x_i) = \text{slope of the curve at } P_t,$$

where $f'(x_i)$ represents the value of the first derivative of $f(x)$ evaluated at x_i. Consequently, since it is also true that

$$\text{slope of the curve at } P_t = \tan\varphi,$$

where φ is the interior angle at x_s, then it follows that

$$f'(x_i) = \tan\varphi.$$

From the geometry of Figure 7.4, we can rewrite this last equation to obtain

$$f'(x_i) = \frac{f(x_i) - 0}{x_i - x_s},$$

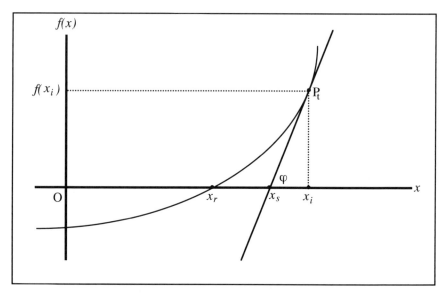

Figure 7.4: The Newton-Raphson method.

where the second term of the numerator is zero because the base of the right triangle containing φ lies on the x-axis. Thus,

$$f'(x_i) = \frac{f(x_i)}{x_i - x_s}. \tag{7.3}$$

Finally, rearranging this last equation yields

$$x_s = x_i - \frac{f(x_i)}{f'(x_i)}. \tag{7.4}$$

Equation 7.4 is the *Newton-Raphson formula* [1].

Given a function $f(x)$ and its first derivative $f'(x)$, the Newton-Raphson algorithm is as follows:

Step 1 Choose an initial approximation x_i for the root.

Step 2 If $|f(x_i)|$ is less than a given tolerance, go to Step 5, otherwise continue with Step 3.

Step 3 Estimate the root by using the Newton-Raphson formula

$$x_s = x_i - \frac{f(x_i)}{f'(x_i)}.$$

7.3. SOLVING A SYSTEM OF LINEAR EQUATIONS

Step 4 Let $x_i = x_s$ and go to Step 2.

Step 5 Stop computation. Take x_i as the value of the root.

This algorithm can be very easily implemented in BASIC. If FNF(X) and FNDF(X) are user defined functions representing $f(x)$ and $f'(x)$, respectively, then

```
1250 INPUT">>>APPROXIMATE SOLUTION X";X
1260 REM
1270    IF ABS(FNF(X))<.0000001# GOTO 1300
1280    X=X-FNF(X)/FNDF(X)
1290    GOTO 1270
1300 PRINT
1330 PRINT"ROOT";TAB(10);
1340 PRINT USING G$;X
```

7.3 Solving a System of Linear Equations

In order to handle many of the problems which arise in orbit computation, one must be able to solve large systems of simultaneous linear equations. Regardless of their size, such systems will be of the following general form:

$$\begin{aligned} a_{11}x_1 + a_{12}x_2 + \cdots + a_{1n}x_n &= b_1 \\ a_{21}x_1 + a_{22}x_2 + \cdots + a_{2n}x_n &= b_2 \\ a_{31}x_1 + a_{32}x_2 + \cdots + a_{3n}x_n &= b_3 \\ &\vdots \\ a_{n1}x_1 + a_{n2}x_2 + \cdots + a_{nn}x_n &= b_n, \end{aligned} \quad (7.5)$$

where the a_{ij} and b_i are constants and n equals the number of equations. Equations 7.5 can be solved by a technique called *Gauss elimination*. This procedure is accomplished in two phases: forward elimination of the unknowns and solution through back-substitution. We will first describe a simple version, known as *naive Gauss elimination*, which assumes that a division by zero cannot occur during forward elimination. We will then combine this with a process called *partial pivoting* to produce an algorithm which automatically switches the order of the equations during forward elimination to prevent a division by zero [1].

7.3.1 Naive Gauss Elimination

Although the Gauss elimination process is applicable to any number of simultaneous linear equations where there are an equal number of unknowns, we will use the following system of three equations to illustrate the method:

$$a_{11}x_1 + a_{12}x_2 + a_{13}x_3 = b_1 \qquad (7.6)$$
$$a_{21}x_1 + a_{22}x_2 + a_{23}x_3 = b_2 \qquad (7.7)$$
$$a_{31}x_1 + a_{32}x_2 + a_{33}x_3 = b_3. \qquad (7.8)$$

Forward elimination. Forward elimination is begun by dividing Equation 7.6 by the coefficient of the first unknown term to obtain

$$x_1 + a'_{12}x_2 + a'_{13}x_3 = b'_1, \qquad (7.9)$$

where

$$a'_{12} = \frac{a_{12}}{a_{11}}$$

$$a'_{13} = \frac{a_{13}}{a_{11}}$$

$$b'_1 = \frac{b_1}{a_{11}}.$$

Next, multiply Equation 7.9 by a_{21} and subtract the result form Equation 7.7 to eliminate $a_{21}x_1$ and obtain

$$a'_{22}x_2 + a'_{23}x_3 = b'_2,$$

where

$$a'_{22} = a_{22} - a_{21}a'_{12}$$
$$a'_{23} = a_{23} - a_{21}a'_{13}$$
$$b'_2 = b_2 - a_{21}b'_1.$$

Similarly, multiplying Equation 7.9 by a_{31} and subtracting the result from Equation 7.8 will eliminate the $a_{31}x_1$ term to produce

$$a'_{32}x_2 + a'_{33}x_3 = b'_3,$$

where

$$a'_{32} = a_{32} - a_{31}a'_{12}$$
$$a'_{33} = a_{33} - a_{31}a'_{13}$$
$$b'_3 = b_3 - a_{31}b'_1.$$

The original system of equations has now been transformed into the following:

$$a_{11}x_1 + a_{12}x_2 + a_{13}x_3 = b_1 \qquad (7.10)$$
$$a'_{22}x_2 + a'_{23}x_3 = b'_2 \qquad (7.11)$$
$$a'_{32}x_2 + a'_{33}x_3 = b'_3. \qquad (7.12)$$

7.3. SOLVING A SYSTEM OF LINEAR EQUATIONS

In the foregoing procedure, Equation 7.10 was the *pivot equation* and a_{11} the *pivot coefficient*.

If Equation 7.11 is now chosen as the pivot equation and a'_{22} as the pivot coefficient, repeating the above process to eliminate the x_2 term from Equation 7.12 will yield the following *upper triangular system* of equations:

$$a_{11}x_1 + a_{12}x_2 + a_{13}x_3 = b_1 \qquad (7.13)$$
$$a'_{22}x_2 + a'_{23}x_3 = b'_2 \qquad (7.14)$$
$$a''_{33}x_3 = b''_3, \qquad (7.15)$$

where the double prime indicates that those coefficients have been modified twice.

Back-substitution. Notice that Equation 7.15 can be solved immediately for x_3, that is

$$x_3 = \frac{b''_3}{a''_{33}}. \qquad (7.16)$$

This result can now be substituted into Equation 7.14 so that x_2 can be found. Finally, x_1 can be evaluated by substituting x_2 and x_3 into Equation 7.13.

7.3.2 Partial Pivoting

When the Gauss elimination is applied to a system of equations there will often occur situations where the pivot coefficient is zero or very nearly zero. Therefore, it is always best to compute in double-precision and include a routine which will determine the largest available coefficient before choosing the pivot equation. This is called *partial pivoting* because only the order of the equations is altered, leaving the order of the unknowns unchanged. The BASIC subroutine listed below includes partial pivoting along with naive Gauss elimination. The number of equations is equal to N, and, for convenience, the b_i of Equations 7.5 are represented in the BASIC implementation by A(I,N+1).

```
19000 STOP
19010 REM # GELIM/SUB # GAUSS ELIMINATION
19020 FOR I=1 TO (N-1)

19030     REM ***** PARTIAL PIVOTING *****
19040     JP=I
19050     PE!=ABS(A(I,I))
19060     FOR J=(I+1) TO N
19070       CE!=ABS(A(J,I))
19080       IF CE!-PE! < 0 GOTO 19130
19090       PE!=CE!
```

```
19100      JP=J
19110      NEXT J
19120      REM
19130      IF JP=I GOTO 19210
19140      REM
19150      FOR K=I TO (N+1)
19160         HE=A(I,K)
19170         A(I,K)=A(JP,K)
19180         A(JP,K)=HE
19190      NEXT K

19200      REM ***** FORWARD ELIMINATION *****
19210      FOR J=(I+1) TO N
19220         FOR K=(I+1) TO (N+1)
19230            A(J,K)=A(J,K)-A(J,I)*A(I,K)/A(I,I)
19240         NEXT K
19250         A(J,I)=0
19260      NEXT J
19270 NEXT I

19280 REM ******* BACK SUBSTITUTION *******
19290 XU(N)=A(N,N+1)/A(N,N)
19300 FOR I=(N-1) TO 1 STEP -1
19310    SS=0
19320    FOR K=(I+1) TO N
19330       SS=SS+A(I,K)*XU(K)
19340    NEXT K
19350    XU(I)=(A(I,N+1)-SS)/A(I,I)
19360 NEXT I
19370 RETURN
```

Subroutine GELIM/SUB is included in programs PGRESS and MGRESS at the end of this chapter.

7.4 Polynomial Interpolation

Suppose we have a set of precise numerical quantities f_i which are the values of a certain function $f(x)$ corresponding to discrete values x_i of the independent variable, where $i = 1$ to n. If we wish to find a particular value f of the function at a given point x, we must interpolate. As illustrated in Figure 7.5, this can be accomplished by creating a polynomial curve that passes through points surrounding (x, f) which can be evaluated at x to produce a good approximation

7.4. POLYNOMIAL INTERPOLATION

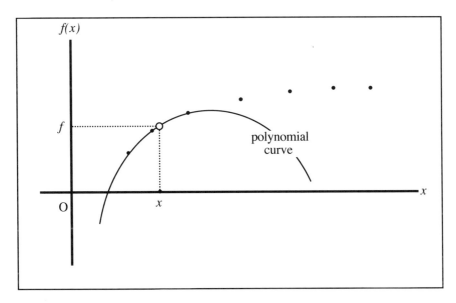

Figure 7.5: An interpolating polynomial through three points.

of the true value f. Although several methods exist to interpolate tabular data, the *Lagrange interpolating polynomial* is convenient for our application. It can be used to interpolate equally or unequally spaced data and is easily modified to accomplish numerical differentiation. Letting f represent the value of the function at x, the Lagrange interpolating polynomial of order $n - 1$ can be written

$$f = f_1 \frac{l_1}{g_1} + f_2 \frac{l_2}{g_2} + \cdots + f_n \frac{l_n}{g_n}, \qquad (7.17)$$

where

$$\begin{aligned}
g_1 &= (x_1 - x_2)(x_1 - x_3) \cdots (x_1 - x_n) \\
g_2 &= (x_2 - x_1)(x_2 - x_3) \cdots (x_2 - x_n) \\
g_3 &= (x_3 - x_1)(x_3 - x_2) \cdots (x_3 - x_n) \\
&\vdots \\
g_n &= (x_n - x_1)(x_n - x_2) \cdots (x_n - x_{n-1}),
\end{aligned} \qquad (7.18)$$

and

$$\begin{aligned}
l_1 &= (x - x_2)(x - x_3) \cdots (x - x_n) \\
l_2 &= (x - x_1)(x - x_3) \cdots (x - x_n) \\
l_3 &= (x - x_1)(x - x_2) \cdots (x - x_n)
\end{aligned}$$

$$\vdots$$
$$l_n = (x - x_1)(x - x_2) \cdots (x - x_{n-1}). \tag{7.19}$$

To illustrate the application of this polynomial, consider the situation where only three values of the tabular function are known. In this case, Equations 7.18 yield the g_i terms

$$\begin{aligned} g_1 &= (x_1 - x_2)(x_1 - x_3) \\ g_2 &= (x_2 - x_1)(x_2 - x_3) \\ g_3 &= (x_3 - x_1)(x_3 - x_2), \end{aligned}$$

and Equations 7.19 yield the l_i terms

$$\begin{aligned} l_1 &= (x - x_2)(x - x_3) \\ l_2 &= (x - x_1)(x - x_3) \\ l_3 &= (x - x_1)(x - x_2). \end{aligned}$$

Finally, Equation 7.17 produces the second order polynomial

$$f = f_1 \frac{l_1}{g_1} + f_2 \frac{l_2}{g_2} + f_3 \frac{l_3}{g_3}. \tag{7.20}$$

The following BASIC routine will generate the Lagrange interpolating polynomial [4]:

```
1090 INPUT">>>INTERPOLATION POINT X";X
1100 REM
1110 READ N$,NP
1120 REM
1130 FOR I=1 TO NP
1140    READ X(I),F(I)
1150 NEXT I
1160 REM
1170 FOR I=1 TO NP
1180    GX(I)=1
1190    LO(I)=1
1220    FOR J=1 TO NP
1230       IF J=I GOTO 1280
1240       GX(I)=GX(I)*(X(I)-X(J))
1270       LO(I)=LO(I)*(X-X(J))
1280    NEXT J
1290 NEXT I
```

7.5. POLYNOMIAL REGRESSION

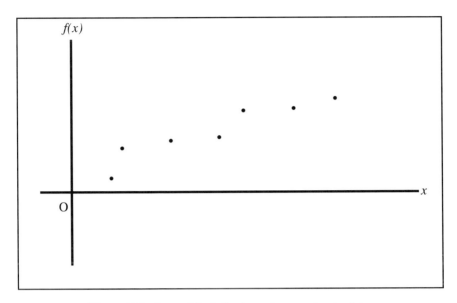

Figure 7.6: A graphical display of approximate data.

```
1300 REM
1310 F0=0
1340 FOR I=1 TO NP
1350   F0=F0+F(I)*LO(I)/GX(I)
1380 NEXT I
1390 REM
1490 PRINT"POINT X";:PRINT TAB(12);
1500 PRINT USING G$;X
1510 PRINT
1520 PRINT"F";:PRINT TAB(12);
1530 PRINT USING G$;FO
```

7.5 Polynomial Regression

Suppose that Figure 7.6 is a graphical depiction of a set of approximate numerical quantities f_i which correspond to discrete values x_i of the independent variable, where $i = 1$ to n. If an interpolating polynomial were used to represent this data, the result might be as shown in Figure 7.7. Although the curve passes exactly through each point, it oscillates between points to produce a poor fit to the overall trend of the data. Figure 7.8 shows the same data represented by a curve derived by the method of *least squares polynomial regression*. This

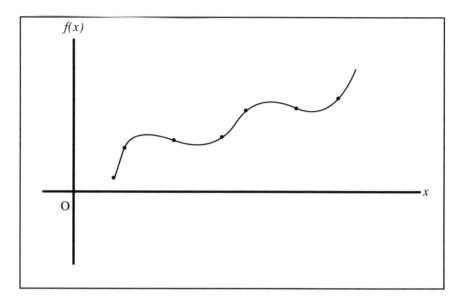

Figure 7.7: An exact fit by an interpolating polynomial.

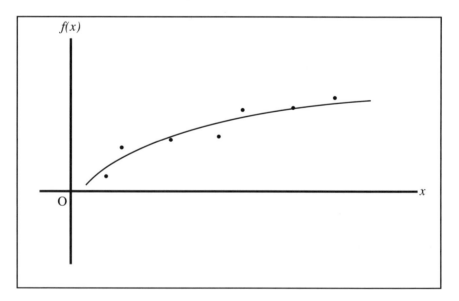

Figure 7.8: A least squares polynomial fit.

7.5. POLYNOMIAL REGRESSION

technique produces a power series polynomial

$$f(x) = c_0 + c_1 x + c_2 x^2 + \cdots + c_m x^m \tag{7.21}$$

that minimizes the sum of the squares of the discrepancies between the polynomial curve and each data point. These discrepancies are known as *residuals*. The polynomial order m needed to obtain an adequate fit to the data must be determined empirically. However, it should be less than $n-1$, otherwise the curve will pass exactly through all the points as was the case for the interpolating polynomial [1,5].

Although least squares regression can fit a high-order polynomial to a large set of data, we will use a second-order example to explain the procedure. Thus, given n sets of data one may write the following n *equations of condition*:

$$\begin{aligned} A + Bx_1 + Cx_1^2 &= f_1 \\ A + Bx_2 + Cx_2^2 &= f_2 \\ &\vdots \\ A + Bx_n + Cx_n^2 &= f_n, \end{aligned} \tag{7.22}$$

where the x_i and f_i are known quantities, and the coefficients A, B, and C must be determined by the method of least squares.

Now, for any adopted values of the coefficients, the residuals s_i can be represented as

$$\begin{aligned} s_1 &= f_1 - A - Bx_1 - Cx_1^2 \\ s_2 &= f_2 - A - Bx_2 - Cx_2^2 \\ &\vdots \\ s_n &= f_n - A - Bx_n - Cx_n^2. \end{aligned} \tag{7.23}$$

Squaring each equation and adding, the sum of the squares of the residuals is given by

$$S = \sum_{i=1}^{n} (f_i - A - Bx_i - Cx_i^2)^2. \tag{7.24}$$

Applying the principle of least squares, A, B, and C must be given values which minimize S. This will occur for values which cause the first derivative of S to be zero. Therefore, we take the partial derivative of S with respect to each of the coefficients to obtain

$$\frac{\partial S}{\partial A} = -2 \sum (f_i - A - Bx_i - Cx_i^2)$$

$$\frac{\partial S}{\partial B} = -2\sum x_i(f_i - A - Bx_i - Cx_i^2)$$

$$\frac{\partial S}{\partial C} = -2\sum x_i^2(f_i - A - Bx_i - Cx_i^2).$$

Setting these partial derivatives equal to zero, the result can be written

$$\sum(f_i - A - Bx_i - Cx_i^2) = 0$$
$$\sum(f_ix_i - Ax_i - Bx_i^2 - Cx_i^3) = 0$$
$$\sum(f_ix_i^2 - Ax_i^2 - Bx_i^3 - Cx_i^4) = 0,$$

which can again be rearranged to form the system of *normal equations*

$$An + B\sum x_i + C\sum x_i^2 = \sum f_i$$
$$A\sum x_i + B\sum x_i^2 + C\sum x_i^3 = \sum f_ix_i$$
$$A\sum x_i^2 + B\sum x_i^3 + C\sum x_i^4 = \sum f_ix_i^2, \qquad (7.25)$$

where all the summations are from $i = 1$ to n. Although these normal equations may appear to be complex, they are only simple linear equations where A, B, and C are the unknowns, and the summation terms are all numerical quantities which can be calculated from the given data. Notice that the number of equations now equals the number of unknowns. Finally, let the normal equations be written as follows:

$$a_{11}A + a_{12}B + a_{13}C = b_1$$
$$a_{21}A + a_{22}B + a_{23}C = b_2$$
$$a_{31}A + a_{32}B + a_{33}C = b_3, \qquad (7.26)$$

where the a_{ij} and b_i equate to the summation terms occupying the corresponding positions in Equations 7.25. Because Equations 7.26 have the same form as Equations 7.5, Gauss elimination can be used to solve for A, B, and C.

The BASIC routine shown below computes the coefficients a_{ij} and b_i for the normal equations given the number data points NP, the order of the polynomial M, and the zero point for the expansion X. Its output is designed for use in subroutine GELIM/SUB.

```
1390 REM * POWERS OF X(I) *
1400 REM
1410 FOR I=1 TO NP
1420    XX(I)=X(I)-X
1430    XP(I,0)=1
```

```
1440    FOR P=1 TO (M+M)
1450      XP(I,P)=XP(I,P-1)*XX(I)
1460    NEXT P
1470 NEXT I
1480 REM
1490 REM * NORMAL EQUATION COEFFICIENTS *
1500 REM
1510 FOR J=1 TO (M+1)
1520    FOR K=1 TO (M+1)
1530      A(J,K)=0
1540      FOR I=1 TO NP
1550        A(J,K)=A(J,K)+XP(I,J+K-2)
1560      NEXT I
1570    NEXT K
1580    REM
1590    A(J,M+2)=0
1600    FOR I=1 TO NP
1610      A(J,M+2)=A(J,M+2)+XP(I,J-1)*F(I)
1620    NEXT I
1630 NEXT J
```

This polynomial regression routine is part of program PGRESS, which is given at the end of this chapter along with a numerical example.

7.6 Multiple Linear Regression

The technique of least squares regression can easily be extended to the case where the f_i are values of a linear function of two or more variables, rather than a polynomial function of one variable. In other words,

$$f(x_1, x_2, \ldots, x_n) = c_1 x_1 + c_2 x_2 + \cdots + c_n x_n. \tag{7.27}$$

For example, assume that it is appropriate to describe a set of measured data by a linear function of three independent variables,

$$f(x_1, x_2, x_3) = Ax_1 + Bx_2 + Cx_3, \tag{7.28}$$

so that the following n *equations of condition* can be written using the given data [6]:

$$\begin{aligned} Ax_{11} + Bx_{12} + Cx_{13} &= f_1 \\ Ax_{21} + Bx_{22} + Cx_{23} &= f_2 \\ &\vdots \\ Ax_{n1} + Bx_{n2} + Cx_{n3} &= f_n, \end{aligned} \tag{7.29}$$

where the x_{ik} and f_i are known quantities, and the coefficients A, B, and C must be determined by the method of least squares.

Applying the principle of least squares regression to minimize the sum of the squares of the residuals

$$\begin{aligned} s_1 &= f_1 - Ax_{11} - Bx_{12} - Cx_{13} \\ s_2 &= f_2 - Ax_{21} - Bx_{22} - Cx_{23} \\ &\vdots \\ s_n &= f_n - Ax_{n1} - Bx_{n2} - Cx_{n3}, \end{aligned} \qquad (7.30)$$

we obtain

$$S = \sum_{i=1}^{n}(f_i - Ax_{i1} - Bx_{i2} - Cx_{i3})^2. \qquad (7.31)$$

Following the same process used for polynomial regression, we differentiate S with respect to A, B, and C and set the resulting expressions equal to zero. Thus, we can obtain the *normal equations*

$$\begin{aligned} A\sum x_{i1}x_{i1} + B\sum x_{i1}x_{i2} + C\sum x_{i1}x_{i3} &= \sum x_{i1}f_i \\ A\sum x_{i2}x_{i1} + B\sum x_{i2}x_{i2} + C\sum x_{i2}x_{i3} &= \sum x_{i2}f_i \\ A\sum x_{i3}x_{i1} + B\sum x_{i3}x_{i2} + C\sum x_{i3}x_{i3} &= \sum x_{i3}f_i, \end{aligned} \qquad (7.32)$$

where, again, the summations are from $i = 1$ to n and are numerical quantities which can be calculated from the given data. Finally, let the normal equations be written in the general form

$$\begin{aligned} a_{11}A + a_{12}B + a_{13}C &= b_1 \\ a_{21}A + a_{22}B + a_{23}C &= b_2 \\ a_{31}A + a_{32}B + a_{33}C &= b_3, \end{aligned} \qquad (7.33)$$

where the a_{ij} and b_i equate to the summation terms occupying the corresponding positions in Equations 7.32. Gauss elimination can now be used to solve for A, B, and C.

The BASIC routine below computes the quantities a_{ij} and b_i for the normal equations given the number data points NP and the number of linear variables M. Its output is compatible with subroutine GELIM/SUB.

```
1300 REM * NORMAL EQUATION COEFFICIENTS *
1310 REM
1320 FOR J=1 TO M
1330    FOR K=1 TO (M+1)
1340       A(J,K)=0
```

7.7. NUMERICAL DIFFERENTIATION

```
1350      FOR I=1 TO NP
1360        A(J,K)=A(J,K)+X(I,J)*X(I,K)
1370      NEXT I
1380    NEXT K
1390 NEXT J
```

A multiple linear regression program called MGRESS is given at the end of this chapter along with a numerical example.

7.7 Numerical Differentiation

The numerical methods of polynomial interpolation and polynomial regression can be easily extended to encompass numerical differentiation. Although the procedures are general, we will only consider the first and second derivatives. Furthermore, for convenience, differentiation with respect to the independent variable will be indicated by the same notation as that used for modified time.

7.7.1 The Interpolating Polynomial

Recalling the Lagrange interpolating polynomial discussed in Section 7.4, we can describe a set of precise numerical data by an equation of the form

$$f = f_1 \frac{l_1}{g_1} + f_2 \frac{l_2}{g_2} + \cdots, \tag{7.34}$$

where the f_i are given, and the g_i and l_i are defined by Equations 7.18 and 7.19, respectively. Therefore, the f_i and g_i are constants, and only the l_i contain the independent variable x. Thus, differentiating Equation 7.34 twice with respect to x, the result is

$$\dot{f} = f_1 \frac{\dot{l}_1}{g_1} + f_2 \frac{\dot{l}_2}{g_2} + \cdots \tag{7.35}$$

$$\ddot{f} = f_1 \frac{\ddot{l}_1}{g_1} + f_2 \frac{\ddot{l}_2}{g_2} + \cdots \tag{7.36}$$

The problem is to find expressions for the $\dot{l}_i$ and $\ddot{l}_i$ which will be valid for any number of tabular values x_i and f_i. The solution can be found by going directly to the BASIC routine of Section 7.4 to derive *recurrence equations* for these derivatives. From line 1270 we have

$$\text{L0(I)} = \text{L0(I)} * (\text{X} - \text{X(J)}), \tag{7.37}$$

where, in the context of the FOR-NEXT loop, the L0(I) on the right side of the equal sign is the value of l_i from the previous step. Letting L1(I) represent $\dot{l}_i$, differentiating Equation 7.37 yields

$$\text{L1(I)} = \text{L1(I)} * (\text{X} - \text{X(J)}) + \text{L0(I)}. \tag{7.38}$$

Differentiating a second time, we obtain

$$\text{L2(I)} = \text{L2(I)} * (\text{X} - \text{X(J)}) + 2 * \text{L1(I)}, \tag{7.39}$$

where L2(I) represents $\ddot{l}_i$.

In the same fashion, we can use line 1350 of the BASIC routine to write recurrence equations for $\dot{f}$ and $\ddot{f}$. Thus,

$$\text{F0} = \text{F0} + \text{F(I)} * \text{L0(I)}/\text{GX(I)}, \tag{7.40}$$

where F0 on the right side of the equation represents the value of f from the previous step. Differentiating Equation 7.40 once produces

$$\text{F1} = \text{F1} + \text{F(I)} * \text{L1(I)}/\text{GX(I)}, \tag{7.41}$$

where F1 represents $\dot{f}$ and the value of L1(I) is obtained from Equation 7.38. Differentiating Equation 7.40 a second time yields

$$\text{F2} = \text{F2} + \text{F(I)} * \text{L2(I)}/\text{GX(I)}, \tag{7.42}$$

where F2 represents $\ddot{f}$ and L2(I) is determined by Equation 7.39.

The BASIC routine below is similar to the one used in program PTERP, which is given at the end of this chapter. It will generate the Lagrange interpolating polynomial along with its first and second derivatives:

```
1090 INPUT">>>INTERPOLATION POINT X";X
1100 REM
1110 READ N$,NP
1120 REM
1130 FOR I=1 TO NP
1140   READ X(I),F(I)
1150 NEXT I
1160 REM
1170 FOR I=1 TO NP
1180   GX(I)=1
1190   L0(I)=1
1200   L1(I)=0
1210   L2(I)=0
```

7.7. NUMERICAL DIFFERENTIATION

```
1220    FOR J=1 TO NP
1230      IF J=I GOTO 1280
1240      GX(I)=GX(I)*(X(I)-X(J))
1250      L2(I)=L2(I)*(X-X(J))+2*L1(I)
1260      L1(I)=L1(I)*(X-X(J))+L0(I)
1270      L0(I)=L0(I)*(X-X(J))
1280    NEXT J
1290 NEXT I
1300 REM
1310 F0=0
1320 F1=0
1330 F2=0
1340 FOR I=1 TO NP
1350    F0=F0+F(I)*L0(I)/GX(I)
1360    F1=F1+F(I)*L1(I)/GX(I)
1370    F2=F2+F(I)*L2(I)/GX(I)
1380 NEXT I
1390 REM
1490 PRINT"POINT X";:PRINT TAB(12);
1500 PRINT USING G$;X
1510 PRINT
1520 PRINT"F";:PRINT TAB(12);
1530 PRINT USING G$;F0
1540 PRINT"F'";:PRINT TAB(12);
1550 PRINT USING G$;F1
1560 PRINT"F''";:PRINT TAB(12);
1570 PRINT USING G$;F2
```

7.7.2 The Regression Polynomial

Numerical differentiation of the least squares regression polynomial is very straightforward. Given the power series expansion

$$f(x) = c_0 + c_1 x + c_2 x^2 + c_3 x^3 + \cdots, \qquad (7.43)$$

we differentiate twice with respect to x and obtain

$$\dot{f}(x) = c_1 + 2c_2 x + 3c_3 x^2 + \cdots \qquad (7.44)$$
$$\ddot{f}(x) = 2c_2 + 6c_3 x + \cdots \qquad (7.45)$$

The above equations allow us to compute $f(x)$ and its first and second derivatives at any point on the regression curve by simply choosing that point as the origin of the series expansion. In that case, x will be zero by definition, and Equations 7.43, 7.44, and 7.45 reduce to

$$f = c_0 \tag{7.46}$$
$$\dot{f} = c_1 \tag{7.47}$$
$$\ddot{f} = 2c_2, \tag{7.48}$$

where f, $\dot{f}$, and $\ddot{f}$ represent $f(x)$ and its first and second derivatives at $x = 0$. Numerical differentiation of the regression polynomial is included as part of program PGRESS.

7.8 Computer Programs

7.8.1 Program PTERP

Program PTERP computes polynomial interpolation and numerical differentiation by the method of Lagrange. The interpolation algorithm is discussed in Section 7.4 and the differentiation algorithm in Section 7.7.1.

Program Listing

```
1000 CLS
1010 PRINT"# PTERP # POLYNOMIAL INTERPOLATION"
1020 PRINT"# BY THE METHOD OF LAGRANGE"
1030 PRINT
1040 DEFDBL A-Z
1050 DEFINT I,J,N
1060 REM
1070 G$="####.#######"
1080 REM
1090 INPUT">>>INTERPOLATION POINT X";X
1100 CLS
1110 READ N$,NP
1120 REM
1130 FOR I=1 TO NP
1140    READ X(I),F(I)
1150 NEXT I
1160 REM
1170 FOR I=1 TO NP
1180    GX(I)=1
1190    L0(I)=1
1200    L1(I)=0
1210    L2(I)=0
1220    FOR J=1 TO NP
1230       IF J=I GOTO 1280
1240       GX(I)=GX(I)*(X(I)-X(J))
1250       L2(I)=L2(I)*(X-X(J))+2*L1(I)
1260       L1(I)=L1(I)*(X-X(J))+L0(I)
1270       L0(I)=L0(I)*(X-X(J))
1280    NEXT J
1290 NEXT I
1300 REM
1310 F0=0
```

```
1320 F1=0
1330 F2=0
1340 FOR I=1 TO NP
1350   F0=F0+F(I)*L0(I)/GX(I)
1360   F1=F1+F(I)*L1(I)/GX(I)
1370   F2=F2+F(I)*L2(I)/GX(I)
1380 NEXT I
1390 REM
1400 PRINT"POLYNOMIAL INTERPOLATION"
1410 PRINT N$
1420 PRINT
1430 PRINT"INITIAL DATA"
1440 PRINT
1450 PRINT TAB(6)"X(I)";TAB(18)"F(I)"
1460 FOR I=1 TO NP
1470   PRINT USING G$;X(I),F(I)
1480 NEXT I
1490 REM
1500 LINE INPUT"";L$
1510 CLS
1520 PRINT"FUNCTION AND DERIVATIVES"
1530 PRINT
1540 PRINT"POINT X";:PRINT TAB(12);
1550 PRINT USING G$;X
1560 PRINT
1570 PRINT"F";:PRINT TAB(12);
1580 PRINT USING G$;F0
1590 PRINT"F'";:PRINT TAB(12);
1600 PRINT USING G$;F1
1610 PRINT"F''";:PRINT TAB(12);
1620 PRINT USING G$;F2
1630 END
30000 DATA"EXAMPLE",4
30010 DATA 15, 0.9238643#
30020 DATA 17, 0.9099594#
30030 DATA 19, 0.8949607#
30040 DATA 21, 0.8788878#
```

7.8. COMPUTER PROGRAMS

7.8.2 Program PGRESS

Program PGRESS computes polynomial regression and numerical differentiation by the method of least squares. The polynomial regression algorithm is discussed in Section 7.5 and the differentiation algorithm in Section 7.7.2.

Program Listing

```
1000 CLS
1010 PRINT"# PGRESS # POLYNOMIAL REGRESSION"
1020 PRINT"# BY THE METHOD OF LEAST SQUARES"
1030 PRINT
1040 DEFDBL A-Z
1050 DEFINT I,J,K,M,N,P
1060 REM
1070 G$="####.#######"
1080 S$="#######.#######"
1090 REM
1100 READ N$,NP
1110 REM
1120 DIM X(NP),F(NP)
1130 REM
1140 FOR I=1 TO NP
1150    READ X(I),F(I)
1160 NEXT I
1170 REM
1180 PRINT"NUMBER OF DATA POINTS =";NP
1190 PRINT
1200 INPUT">>>ORDER OF POLYNOMIAL";M
1210 REM
1220 IF M>(NP-1) THEN 1230 ELSE 1270
1230 CLS
1240 PRINT">>> ORDER TO HIGH * TRY AGAIN <<<"
1250 PRINT
1260 GOTO 1180
1270 PRINT
1280 INPUT">>>EXPANSION POINT X";X
1290 CLS
1300 DIM XX(NP),XP(NP,M+M),A(M+1,M+2),XU(M+1),C(M)
1310 REM
1320 PRINT"POLYNOMIAL REGRESSION"
1330 PRINT N$
```

```
1340 PRINT
1350 PRINT"INITIAL DATA"
1360 PRINT
1370 PRINT TAB(6)"X(I)";TAB(18)"F(I)"
1380 FOR I=1 TO NP
1390    PRINT USING G$;X(I),F(I)
1400 NEXT I
1410 REM
1420 REM * POWERS OF X(I) *
1430 REM
1440 FOR I=1 TO NP
1450    XX(I)=X(I)-X
1460    XP(I,0)=1
1470    FOR P=1 TO (M+M)
1480       XP(I,P)=XP(I,P-1)*XX(I)
1490    NEXT P
1500 NEXT I
1510 REM
1520 REM * NORMAL EQUATION COEFFICIENTS *
1530 REM
1540 FOR J=1 TO (M+1)
1550    FOR K=1 TO (M+1)
1560       A(J,K)=0
1570       FOR I=1 TO NP
1580          A(J,K)=A(J,K)+XP(I,J+K-2)
1590       NEXT I
1600    NEXT K
1610    REM
1620    A(J,M+2)=0
1630    FOR I=1 TO NP
1640       A(J,M+2)=A(J,M+2)+XP(I,J-1)*F(I)
1650    NEXT I
1660 NEXT J
1670 REM
1680 N=(M+1)
1690 REM
1700 REM * GAUSS ELIMINATION *
1710 REM
1720 GOSUB 19010 REM GELIM/SUB
1730 REM
1740 FOR P=0 TO M
1750    C(P)=XU(P+1)
```

7.8. COMPUTER PROGRAMS

```
1760 NEXT P
1770 REM
1780 LINE INPUT"";L$
1790 CLS
1800 PRINT"POLYNOMIAL COEFFICIENTS"
1810 PRINT
1820 PRINT"POINT X";:PRINT TAB(10);
1830 PRINT USING S$;X
1840 PRINT
1850 FOR P=0 TO M
1860   PRINT"C(";P;")";:PRINT TAB(10);
1870   PRINT USING S$;C(P)
1880 NEXT P
1890 REM
1900 LINE INPUT"";L$
1910 CLS
1920 PRINT"FUNCTION AND DERIVATIVES"
1930 PRINT
1940 PRINT"POINT X";:PRINT TAB(10);
1950 PRINT USING S$;X
1960 PRINT
1970 PRINT"F";:PRINT TAB(10);
1980 PRINT USING S$;C(0)
1990 PRINT"F'";:PRINT TAB(10);
2000 PRINT USING S$;C(1)
2010 PRINT"F''";:PRINT TAB(10);
2020 PRINT USING S$;C(2)*2
2030 END
19000 STOP
19010 REM # GELIM/SUB # GAUSS ELIMINATION
19020 FOR I=1 TO (N-1)
19030    REM
19040    JP=I
19050    PE!=ABS(A(I,I))
19060    FOR J=(I+1) TO N
19070       CE!=ABS(A(J,I))
19080       IF CE!-PE! < 0 GOTO 19130
19090       PE!=CE!
19100       JP=J
19110    NEXT J
19120    REM
19130    IF JP=I GOTO 19210
```

```
19140    REM
19150    FOR K=I TO (N+1)
19160      HE=A(I,K)
19170      A(I,K)=A(JP,K)
19180      A(JP,K)=HE
19190    NEXT K
19200    REM
19210    FOR J=(I+1) TO N
19220      FOR K=(I+1) TO (N+1)
19230        A(J,K)=A(J,K)-A(J,I)*A(I,K)/A(I,I)
19240      NEXT K
19250      A(J,I)=0
19260    NEXT J
19270 NEXT I
19280 REM
19290 XU(N)=A(N,N+1)/A(N,N)
19300 FOR I=(N-1) TO 1 STEP -1
19310    SS=0
19320    FOR K=(I+1) TO N
19330      SS=SS+A(I,K)*XU(K)
19340    NEXT K
19350    XU(I)=(A(I,N+1)-SS)/A(I,I)
19360 NEXT I
19370 RETURN
30000 DATA"EXAMPLE",11
30010 DATA 10, 0.954#
30020 DATA 12, 0.943#
30030 DATA 14, 0.930#
30040 DATA 16, 0.917#
30050 DATA 18, 0.903#
30060 DATA 20, 0.887#
30070 DATA 22, 0.870#
30080 DATA 24, 0.853#
30090 DATA 26, 0.834#
30100 DATA 28, 0.814#
30110 DATA 30, 0.794#
```

7.8. COMPUTER PROGRAMS

7.8.3 Program MGRESS

Program MGRESS computes multiple linear regression by the method of least squares. The multiple linear regression algorithm is discussed in Section 7.6.

Program Listing

```
1000 CLS
1010 PRINT"# MGRESS # MULTIPLE LINEAR REGRESSION"
1020 PRINT"# BY THE METHOD OF LEAST SQUARES"
1030 REM
1040 DEFDBL A-Z
1050 DEFINT I,J,K,M,N
1060 REM
1070 G$="####.##"
1080 S$="#######.#######"
1090 REM
1100 READ N$,NP,M
1110 REM
1120 DIM X(NP,M+1),A(M,M+1),XU(M),C(M)
1130 REM
1140 FOR I=1 TO NP
1150   FOR K=1 TO (M+1)
1160     READ X(I,K)
1170   NEXT K
1180 NEXT I
1190 REM
1200 LINE INPUT"";L$
1210 CLS
1220 PRINT"MULTIPLE LINEAR REGRESSION"
1230 PRINT N$
1240 PRINT
1250 PRINT"INITIAL DATA"
1260 PRINT
1270 FOR I=1 TO NP
1280   FOR K=1 TO (M+1)
1290     PRINT USING G$;X(I,K);
1300   NEXT K
1310   PRINT
1320 NEXT I
1330 REM
1340 REM * NORMAL EQUATION COEFFICIENTS *
```

```
1350 REM
1360 FOR J=1 TO M
1370   FOR K=1 TO (M+1)
1380     A(J,K)=0
1390     FOR I=1 TO NP
1400       A(J,K)=A(J,K)+X(I,J)*X(I,K)
1410     NEXT I
1420   NEXT K
1430 NEXT J
1440 REM
1450 N=M
1460 REM
1470 REM * GAUSS ELIMINATION *
1480 REM
1490 GOSUB 19010 REM GELIM/SUB
1500 REM
1510 FOR K=1 TO M
1520   C(K)=XU(K)
1530 NEXT K
1540 REM
1550 LINE INPUT"";L$
1560 CLS
1570 PRINT"MULTIPLE LINEAR COEFFICIENTS"
1580 PRINT
1590 FOR K=1 TO M
1600   PRINT"C(";K;")";:PRINT TAB(10);
1610   PRINT USING S$;C(K)
1620 NEXT K
1630 END
19000 STOP
19010 REM # GELIM/SUB # GAUSS ELIMINATION
19020 FOR I=1 TO (N-1)
19030   REM
19040   JP=I
19050   PE!=ABS(A(I,I))
19060   FOR J=(I+1) TO N
19070     CE!=ABS(A(J,I))
19080     IF CE!-PE! < 0 GOTO 19130
19090     PE!=CE!
19100     JP=J
19110   NEXT J
19120   REM
```

7.8. COMPUTER PROGRAMS

```
19130    IF JP=I GOTO 19210
19140    REM
19150    FOR K=I TO (N+1)
19160      HE=A(I,K)
19170      A(I,K)=A(JP,K)
19180      A(JP,K)=HE
19190    NEXT K
19200    REM
19210    FOR J=(I+1) TO N
19220      FOR K=(I+1) TO (N+1)
19230        A(J,K)=A(J,K)-A(J,I)*A(I,K)/A(I,I)
19240      NEXT K
19250      A(J,I)=0
19260    NEXT J
19270 NEXT I
19280 REM
19290 XU(N)=A(N,N+1)/A(N,N)
19300 FOR I=(N-1) TO 1 STEP -1
19310    SS=0
19320    FOR K=(I+1) TO N
19330      SS=SS+A(I,K)*XU(K)
19340    NEXT K
19350    XU(I)=(A(I,N+1)-SS)/A(I,I)
19360 NEXT I
19370 RETURN
30000 DATA"EXAMPLE",7,4
30010 DATA 0.4#, 0.3#,-0.2#, 0.1#, 0.8#
30020 DATA 0.1#, 0.4#, 0.3#,-0.2#, 1.0#
30030 DATA 0.4#,-0.1#, 0.2#, 0.3#, 2.0#
30040 DATA 0.3#, 0.4#,-0.1#, 0.2#, 1.6#
30050 DATA 0.2#,-0.3#, 0.4#, 0.1#, 1.2#
30060 DATA 0.3#, 0.2#, 0.1#,-0.4#,-0.6#
30070 DATA 0.4#,-0.2#, 0.1#,-0.3#,-0.9#
```

7.9 Numerical Examples

7.9.1 Polynomial Interpolation and Differentiation

Problem

Given the four sets of data listed below, interpolate the value of the function at $x = 18$ and compute its first and second derivatives.

x_i	f_i
15	0.9238643
17	0.9099594
19	0.8949607
21	0.8788878

Solution

Use the given data to write lines 30000 to 30040 as shown at the end of program PTERP. Run the program and answer the prompt as follows:

```
>>>INTERPOLATION POINT X? 18
```

Results

```
POLYNOMIAL INTERPOLATION
EXAMPLE

INITIAL DATA

    X(I)          F(I)
  15.0000000    0.9238643
  17.0000000    0.9099594
  19.0000000    0.8949607
  21.0000000    0.8788878

FUNCTION AND DERIVATIVES

POINT X      18.0000000

F            0.9025956
F'          -0.0074998
F''         -0.0002710
```

7.9. NUMERICAL EXAMPLES

7.9.2 Polynomial Regression and Differentiation

Problem

Given the eleven sets of data listed below, use a third-order polynomial regression to determine the first and second derivatives of the function at $x = 18$.

x_i	f_i
10	0.954
12	0.943
14	0.930
16	0.917
18	0.903
20	0.887
22	0.870
24	0.853
26	0.834
28	0.814
30	0.794

Solution

Use the data given above to write lines 30000 to 30110 as shown at the end of program PGRESS. Run the program and answer the prompts as follows:

```
>>>ORDER OF POLYNOMIAL? 3

>>>EXPANSION POINT X? 18
```

Results

POLYNOMIAL REGRESSION
EXAMPLE

INITIAL DATA

X(I)	F(I)
10.0000000	0.9540000
12.0000000	0.9430000
14.0000000	0.9300000
16.0000000	0.9170000
18.0000000	0.9030000
20.0000000	0.8870000
22.0000000	0.8700000
24.0000000	0.8530000
26.0000000	0.8340000
28.0000000	0.8140000
30.0000000	0.7940000

POLYNOMIAL COEFFICIENTS

POINT X   18.0000000

C(0)	0.9025478
C(1)	-0.0075282
C(2)	-0.0001337
C(3)	0.0000005

FUNCTION AND DERIVATIVES

POINT X   18.0000000

F	0.9025478
F'	-0.0075282
F''	-0.0002675

7.9. NUMERICAL EXAMPLES

7.9.3 Multiple Linear Regression

Problem

Listed below are seven linear equations of condition containing four unknown constant coefficients C_i. Use multiple linear regression to determine the values of the C_i which best represent the data.

$$0.40C_1 + 0.30C_2 - 0.20C_3 + 0.10C_4 = +0.80$$
$$0.10C_1 + 0.40C_2 + 0.30C_3 - 0.20C_4 = +1.00$$
$$0.40C_1 - 0.10C_2 + 0.20C_3 + 0.30C_4 = +2.00$$
$$0.30C_1 + 0.40C_2 - 0.10C_3 + 0.20C_4 = +1.60$$
$$0.20C_1 - 0.30C_2 + 0.40C_3 + 0.10C_4 = +1.20$$
$$0.30C_1 + 0.20C_2 + 0.10C_3 - 0.40C_4 = -0.60$$
$$0.40C_1 - 0.20C_2 + 0.10C_3 - 0.10C_4 = -0.90$$

Solution

Use the given data to write lines 30000 to 30070 as shown at the end of program MGRESS. Run the program.

Results

```
MULTIPLE LINEAR REGRESSION
EXAMPLE

INITIAL DATA

   0.40    0.30   -0.20    0.10    0.80
   0.10    0.40    0.30   -0.20    1.00
   0.40   -0.10    0.20    0.30    2.00
   0.30    0.40   -0.10    0.20    1.60
   0.20   -0.30    0.40    0.10    1.20
   0.30    0.20    0.10   -0.40   -0.60
   0.40   -0.20    0.10   -0.30   -0.90

MULTIPLE LINEAR COEFFICIENTS

C( 1 )          1.0000000
C( 2 )          2.0000000
C( 3 )          3.0000000
C( 4 )          4.0000000
```

References

[1] Chapra and Canale, *Numerical Methods for Engineers with Personal Computer Applications*, McGraw-Hill Book Co., 1985.

[2] Carnahan, Luther, and Wilkes, *Applied Numerical Methods*, John Wiley and Sons, 1969.

[3] Danby, *Computing Applications to Differential Equations*, Reston Publishing Co., Inc., 1985.

[4] Sinnott, "Astronomical Computing," *Sky & Telescope*, April 1984, p. 359.

[5] Bevington, *Data Reduction and Error Analysis for the Physical Sciences*, McGraw-Hill Book Co., 1969.

[6] Van de Kamp, *Principles of Astrometry*, W. H. Freeman and Company, 1967.

Chapter 8

Preliminary Orbit Data

8.1 Introduction

Consider again the situation illustrated in Figure 8.1, where the apparent motion of a celestial body is seen against the background of fixed stars by an observer on the surface of the Earth. Up to this point, we have only dealt with the problem of predicting such behavior when the elements of the orbit are well known. Since this is often not the case, methods for determining the orbital elements from observations of a celestial body's apparent motion are of great practical interest. Initial orbit determination requires the application of dynamic and geometric constraints to find a set of elements which will satisfactorily represent the observations when the known topocentric motion of the dynamical center is taken into account. This problem is not straightforward because the observer cannot directly measure the relatively simple motion of the body about the dynamical center. Instead, a combination of at least two independent motions is seen. Furthermore, an optical observer is hampered by the fact that only a projection of the body's motion against the celestial sphere can be recorded. Since each measurement yields only two spherical coordinates and gives no range information, at least three observations are required to provide six independent quantities which can be used to discover the six unknown orbital elements.

This chapter introduces the process by which a set of *preliminary elements* can be determined for geocentric and heliocentric orbits. Three classical schemes will be developed in subsequent chapters: the *method of Laplace*, the *method of Gauss*, and the *method of Olbers*. Finally, in Chapter 12, a method will be given for improving the preliminary elements by means of multiple linear least squares regression. The discussion of each of these procedures is far from exhaustive. For more thorough treatment of these topics, the interested reader may consult References 1 through 8 or similar texts.

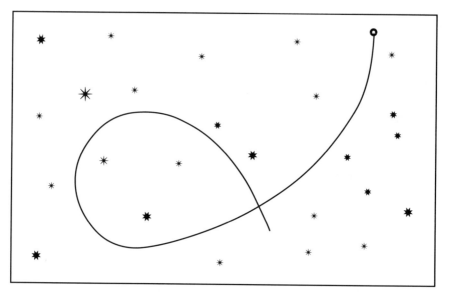

Figure 8.1: The apparent path of a celestial body.

8.2 Principal Constraints

The path of a celestial body circling the Earth or Sun can be approximated by a preliminary orbit which ignores all influences that perturb ideal two-body motion. Preliminary orbit determination treats the orbiting and central bodies as spherically symmetric masses accelerated by mutual Newtonian gravitational attraction. Thus,

$$\ddot{\mathbf{r}} = -\frac{\mu \mathbf{r}}{r^3}, \tag{8.1}$$

where the combined mass μ is assumed to be unity when the mass of the orbiting body is unknown. Equation 8.1 is the general dynamic constraint.

The geometry of the orbit determination problem is illustrated in Figure 8.2, where celestial body B is orbiting a dynamical center C which has its own motion relative to the location of the observer at O. Since the observer, dynamical center, and celestial body always form the corners of the fundamental vector triangle, we have, by Equation 2.60,

$$\mathbf{r} = p\mathbf{L} - \mathbf{R}, \tag{8.2}$$

which is the general geometric constraint.

8.3. THE TOPOCENTRIC VECTOR L

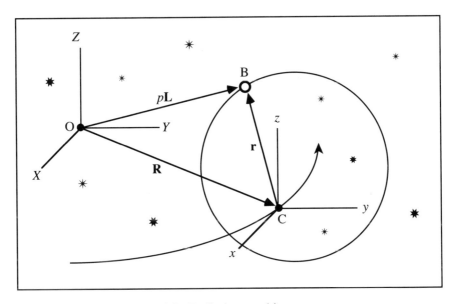

Figure 8.2: Preliminary orbit geometry.

8.3 The Topocentric Vector L

The changing apparent direction of an orbiting body as it moves across the celestial sphere can be recorded by measuring its position with respect to nearby comparison stars at a series of known times. Several methods for accomplishing this are described in References 9 through 16. Regardless of the method of measurement, we shall assume that the data have been reduced to right ascension and declination for the standard mean equinox J2000.0. Thus, if the position measurements were made at various times t_i, the initial observational data consist of the set

$$
\begin{array}{ccc}
t_1 & \alpha_1 & \delta_1 \\
t_2 & \alpha_2 & \delta_2 \\
t_3 & \alpha_3 & \delta_3 \\
\vdots & \vdots & \vdots \\
t_n & \alpha_n & \delta_n
\end{array}
$$

where $n \geq 3$. Therefore, according to Equation 2.56, the topocentric unit vector in the direction of the celestial body at a time t_i is given by

$$\mathbf{L}_i = \{L_{ix}, L_{iy}, L_{iz}\}, \qquad (8.3)$$

where

$$L_{ix} = \cos \delta_i \cos \alpha_i$$

$$L_{iy} = \cos \delta_i \sin \alpha_i$$
$$L_{iz} = \sin \delta_i . \tag{8.4}$$

Although the vectors $\mathbf{L}_i$ are sufficient to describe the apparent motion of an orbiting body for the methods of Gauss and Olbers, Laplacian orbit determination methods require the vectors

$$\mathbf{L}_t = \{L_{tx}, L_{ty}, L_{tz}\}$$
$$\dot{\mathbf{L}}_t = \{\dot{L}_{tx}, \dot{L}_{ty}, \dot{L}_{tz}\}$$
$$\ddot{\mathbf{L}}_t = \{\ddot{L}_{tx}, \ddot{L}_{ty}, \ddot{L}_{tz}\}, \tag{8.5}$$

where the derivatives are measured with respect to modified time and t represents a conveniently chosen epoch when the modified time is zero by definition.

Assume that the initial angular position data have been subjected to polynomial regression in order to obtain the quantities $\{t, \alpha_t, \dot{\alpha}_t, \ddot{\alpha}_t, \delta_t, \dot{\delta}_t, \ddot{\delta}_t\}$. Then, the components of $\mathbf{L}_t$ are

$$L_{tx} = \cos \delta_t \cos \alpha_t$$
$$L_{ty} = \cos \delta_t \sin \alpha_t$$
$$L_{tz} = \sin \delta_t , \tag{8.6}$$

and expressions for the components of $\dot{\mathbf{L}}_t$ and $\ddot{\mathbf{L}}_t$, in terms of the derivatives of α_t and δ_t, can be found by differentiating Equations 8.6 twice with respect to modified time. In the case of $\dot{\mathbf{L}}_t$, we obtain, after some simplification,

$$\dot{L}_{tx} = -\mathcal{X}\dot{\delta}_t - L_{ty}\dot{\alpha}_t$$
$$\dot{L}_{ty} = -\mathcal{Y}\dot{\delta}_t + L_{tx}\dot{\alpha}_t$$
$$\dot{L}_{tz} = +\mathcal{Z}\dot{\delta}_t , \tag{8.7}$$

where

$$\mathcal{X} = \sin \delta_t \cos \alpha_t$$
$$\mathcal{Y} = \sin \delta_t \sin \alpha_t$$
$$\mathcal{Z} = \cos \delta_t . \tag{8.8}$$

Differentiating Equations 8.7, we find the components of $\ddot{\mathbf{L}}_t$ to be

$$\ddot{L}_{tx} = -\dot{\mathcal{X}}\dot{\delta}_t - \mathcal{X}\ddot{\delta}_t - \dot{L}_{ty}\dot{\alpha}_t - L_{ty}\ddot{\alpha}_t$$
$$\ddot{L}_{ty} = -\dot{\mathcal{Y}}\dot{\delta}_t - \mathcal{Y}\ddot{\delta}_t + \dot{L}_{tx}\dot{\alpha}_t + L_{tx}\ddot{\alpha}_t$$
$$\ddot{L}_{tz} = +\dot{\mathcal{Z}}\dot{\delta}_t + \mathcal{Z}\ddot{\delta}_t , \tag{8.9}$$

where

$$\begin{aligned} \dot{\mathcal{X}} &= +L_{tx}\dot{\delta}_t - \mathcal{Y}\dot{\alpha}_t \\ \dot{\mathcal{Y}} &= +L_{ty}\dot{\delta}_t + \mathcal{X}\dot{\alpha}_t \\ \dot{\mathcal{Z}} &= -L_{tz}\dot{\delta}_t . \end{aligned} \quad (8.10)$$

Thus, $\mathbf{L}_t$, $\dot{\mathbf{L}}_t$, and $\ddot{\mathbf{L}}_t$ can all be determined by using Equations 8.6, 8.7, and 8.9 to compute the rectangular components.

The position angles α_t and δ_t, along with their first and second derivatives, may be obtained from the initial angular data using program ADGRESS listed in Section 8.5.1.

8.4 The Topocentric Vector R

We may assume that the necessary data are available in various published sources to determine the topocentric position and motion of the orbit's dynamical center with respect to the equinox J2000.0. Three vectors $\mathbf{R}_i$ are sufficient for the methods of Gauss and Olbers. However, the method of Laplace requires the determination of $\mathbf{R}_t$, $\dot{\mathbf{R}}_t$, and $\ddot{\mathbf{R}}_t$, where t represents the epoch used for the polynomial regression of the angular data described in Section 8.3. Although the methods used to determine $\mathbf{L}$ and its derivatives are essentially the same for geocentric and heliocentric orbits, two different approaches must be used when dealing with the vector $\mathbf{R}$.

8.4.1 Vector R for Geocentric Orbits

Figure 8.3 illustrates the motion of an Earth satellite B, where O is the location of the observer and C is the geocentric center of force. The diurnal rotation of the Earth will cause the geocenter to move with respect to the observer. If we assume that all observations are made from the same point on the Earth's surface, then, for each angular position of the satellite measured at a universal time t_i, there exists a common set

$$f \quad \phi \quad \Lambda \quad H$$

where f is the flattening, ϕ is the geodetic latitude, Λ is the east longitude, and H is the local terrain elevation. Furthermore, for each universal time t_i, the local mean sidereal time θ_i for the mean equinox of date is given by

$$\theta_i = \theta_g + \Lambda, \quad (8.11)$$

where, according to Section 2.4.3,

$$\theta_g = \theta_0 + 360°.98564724 \frac{\mathrm{UT}}{24}$$

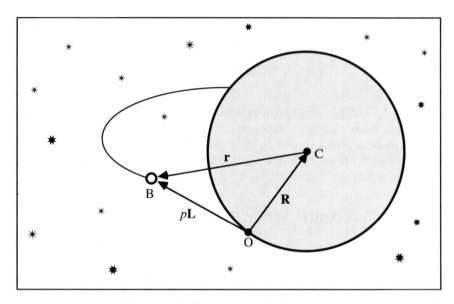

Figure 8.3: Geocentric orbital motion.

and

$$\theta_0 = 100°.4606184 + 36000°.77004\, J + 0°.000387933\, J^2$$
$$J = \frac{J_0 - 2451545.0}{36525}.$$

Accordingly, if we assume an inertial frame of reference centered on the observer and let

$$\mathbf{g}_i = \{g_{ix}, g_{iy}, g_{iz}\}$$

represent a *topocentric* vector to the geocenter at time t_i measured with respect to the *mean equinox of date*, then its rectangular components can be found by changing the algebraic signs of the right sides of Equations 2.28 through 2.30. Thus, we can write

$$\begin{aligned} g_{ix} &= -G_c \cos\phi \cos\theta_i \\ g_{iy} &= -G_c \cos\phi \sin\theta_i \\ g_{iz} &= -G_s \sin\phi, \end{aligned} \qquad (8.12)$$

where

$$\begin{aligned} F &= \sqrt{1 - (2f - f^2)\sin^2\phi} \\ G_c &= \frac{1}{F} + H \end{aligned}$$

8.4. THE TOPOCENTRIC VECTOR R

$$G_s = \frac{(1-f)^2}{F} + H. \tag{8.13}$$

Because the above equations yield $\mathbf{g}_i$ defined with respect to the mean equinox of date, rather than the standard equinox, the procedures of Section 2.8.5 must be used to reduce this vector to the equinox J2000.0 before it can be used along with $\mathbf{L}_i$ in an orbit computation. Thus, letting

$$\mathbf{R}_i = \{R_{ix}, R_{iy}, R_{iz}\} \tag{8.14}$$

represent the topocentric position of the geocenter measured with respect to the standard equinox at universal time t_i, its rectangular components are computed as follows:

$$\begin{aligned}
R_{ix} &= P_{11}g_{ix} + P_{21}g_{iy} + P_{31}g_{iz} \\
R_{iy} &= P_{12}g_{ix} + P_{22}g_{iy} + P_{32}g_{iz} \\
R_{iz} &= P_{13}g_{ix} + P_{23}g_{iy} + P_{33}g_{iz},
\end{aligned} \tag{8.15}$$

where the P-coefficients are the elements of the precession matrix given by Equations 2.81.

In order to employ the method of Laplace to determine a geocentric orbit, we must express the motion of the geocenter by using $\mathbf{R}_t$, $\dot{\mathbf{R}}_t$, and $\ddot{\mathbf{R}}_t$. The components of these vectors can be obtained by differentiating the topocentric vector

$$\mathbf{g}_t = \{g_{tx}, g_{ty}, g_{tz}\}$$

twice with respect to modified time and then reducing the results to the standard equinox J2000.0 by using the precession matrix. Thus, at the epoch t, which was used for the polynomial regression of the angular data, we have

$$\begin{aligned}
g_{tx} &= -G_c \cos\phi \cos\theta_t \\
g_{ty} &= -G_c \cos\phi \sin\theta_t \\
g_{tz} &= -G_s \sin\phi
\end{aligned} \tag{8.16}$$

with respect to the mean equinox of date. Expressions for the components of $\dot{\mathbf{g}}_t$ and $\ddot{\mathbf{g}}_t$ are derived by differentiating Equations 8.16 twice with respect to modified time. This procedure is relatively simple because the latitude ϕ and the Earth's rate of rotation $\dot{\theta}_t$ are constants. First, we obtain

$$\begin{aligned}
\dot{g}_{tx} &= +G_c \cos\phi \sin\theta_t \dot{\theta}_t \\
\dot{g}_{ty} &= -G_c \cos\phi \cos\theta_t \dot{\theta}_t \\
\dot{g}_{tz} &= 0.
\end{aligned}$$

If we make use of Equations 8.16, the above expressions can be simplified to yield

$$\dot{g}_{tx} = -g_{ty}\dot{\theta}_t$$
$$\dot{g}_{ty} = +g_{tx}\dot{\theta}_t$$
$$\dot{g}_{tz} = 0. \qquad (8.17)$$

Differentiating Equations 8.17 and remembering that $\dot{\theta}_t$ is a constant, we have

$$\ddot{g}_{tx} = -\dot{g}_{ty}\dot{\theta}_t$$
$$\ddot{g}_{ty} = +\dot{g}_{tx}\dot{\theta}_t$$
$$\ddot{g}_{tz} = 0.$$

Substituting Equations 8.17 for the corresponding terms in the above expressions, the result is

$$\ddot{g}_{tx} = -g_{tx}\dot{\theta}_t^2$$
$$\ddot{g}_{ty} = -g_{ty}\dot{\theta}_t^2$$
$$\ddot{g}_{tz} = 0. \qquad (8.18)$$

The numerical value of $\dot{\theta}_t$ can be found from Equation 8.11, where the expression for θ_g can be substituted to obtain

$$\theta = \theta_0 + 360°.98564724\frac{\text{UT}}{24} + \Lambda. \qquad (8.19)$$

Differentiating Equation 8.19 with respect to universal time while treating θ_0 as a constant, we get

$$\frac{d\theta}{dt} = 360.98564724 \text{ degrees/day},$$

which is also equivalent to

$$\frac{d\theta}{dt} = 0.0043752695 \text{ radians/minute}.$$

Thus, since this derivative has the same value at any time t, we have

$$\dot{\theta}_t = \frac{1}{k_e}\frac{d\theta}{dt}$$
$$\dot{\theta}_t = 0.0043752695/k_e, \qquad (8.20)$$

where $k_e = 0.07436680$. Therefore, the components of $\mathbf{g}_t$, $\dot{\mathbf{g}}_t$, and $\ddot{\mathbf{g}}_t$ can be computed from Equations 8.16, 8.17, and 8.18. Finally, the corresponding vectors referred to the standard equinox J2000.0, represented by

$$\mathbf{R}_t = \{R_{tx}, R_{ty}, R_{tz}\}$$
$$\dot{\mathbf{R}}_t = \{\dot{R}_{tx}, \dot{R}_{ty}, \dot{R}_{tz}\}$$
$$\ddot{\mathbf{R}}_t = \{\ddot{R}_{tx}, \ddot{R}_{ty}, \ddot{R}_{tz}\}, \qquad (8.21)$$

8.4. THE TOPOCENTRIC VECTOR R

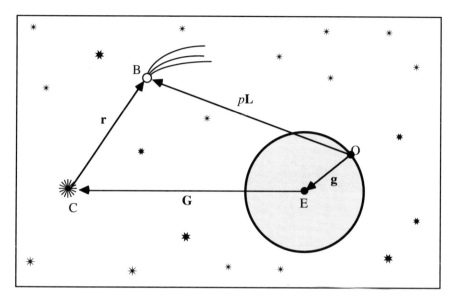

Figure 8.4: Heliocentric orbital motion.

are found by employing the precession matrix in the same manner as that used to obtain the $\mathbf{R}_i$ in Equations 8.15.

Program GEO in Section 8.5.2 may be used to compute the topocentric position and motion of the center of the Earth for any given universal time.

8.4.2 Vector R for Heliocentric Orbits

Figure 8.4 depicts the motion of a celestial body B through interplanetary space. The observer is positioned at O, E is the center of the Earth, and C is the heliocentric center of force within the Sun. Comparing this situation to that shown in Figure 8.5, we see that the topocentric vector $\mathbf{g}$ and the geocentric vector $\mathbf{G}$ can be combined to produce the topocentric vector

$$\mathbf{R} = \mathbf{g} + \mathbf{G}. \tag{8.22}$$

Therefore, to find $\mathbf{R}_i$ we must first compute $\mathbf{g}_i$ and $\mathbf{G}_i$.

The vector $\mathbf{g}_i$ defines the position of the geocenter with respect to the observer at the TDT instant t_i. As in Section 8.4.1, the local mean sidereal time θ_i, the latitude ϕ, and the east longitude Λ are used to compute the rectangular components of $\mathbf{g}_i$. Accordingly, if

$$A_e = 4.263523 \times 10^{-5} \tag{8.23}$$

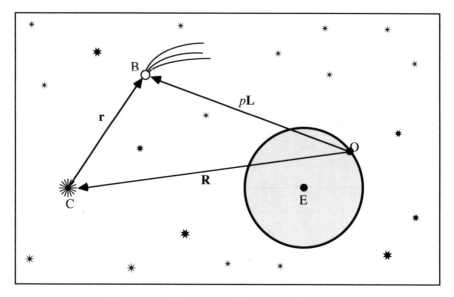

Figure 8.5: The vector **R** to the heliocenter.

is the radius of the Earth in the heliocentric system of units, then

$$\begin{aligned} g_{ix} &= -A_e \cos\phi \cos\theta_i \\ g_{iy} &= -A_e \cos\phi \sin\theta_i \\ g_{iz} &= -A_e \sin\phi. \end{aligned} \quad (8.24)$$

Because $\mathbf{g}_i$ is small when compared to $\mathbf{G}_i$, it is possible to ignore the difference between terrestrial dynamical time and universal time when computing the sidereal time by Equation 8.11. Furthermore, we may also ignore the fact that $\mathbf{g}_i$ is defined with respect to the mean equinox of date and simply assume this same value for the equinox J2000.0. In fact, it is often convenient to ignore $\mathbf{g}_i$ altogether when computing a preliminary heliocentric orbit; however, we shall include it to demonstrate how it may be taken into account.

The vector $\mathbf{G}_i$ expresses the geocentric position of the Sun at the TDT instant t_i. The daily geocentric rectangular solar coordinates for the mean equinox J2000.0 are published in Section C of *The Astronomical Almanac*. Therefore, the tabular values of Sun's coordinates can be interpolated by the procedures of Section 7.4 to yield the components of $\mathbf{G}_i$, and these can be used in Equation 8.22, along with the components of $\mathbf{g}_i$, to compute the rectangular components

$$\begin{aligned} R_{ix} &= g_{ix} + G_{ix} \\ R_{iy} &= g_{iy} + G_{iy} \end{aligned}$$

8.4. THE TOPOCENTRIC VECTOR R

$$R_{iz} = g_{iz} + G_{iz}. \tag{8.25}$$

Thus, with respect to the standard equinox J2000.0, the topocentric vector to the heliocenter at the TDT instant t_i is

$$\mathbf{R}_i = \{R_{ix}, R_{iy}, R_{iz}\}. \tag{8.26}$$

The method of Laplace requires the topocentric position, velocity, and acceleration of the Sun to be determined for a given terrestrial dynamical time t. We will assume that it is possible to ignore the short period fluctuations in the diurnal parallax **g**, so that the components of $\dot{\mathbf{R}}_t$ and $\ddot{\mathbf{R}}_t$ can be obtained directly from the interpolating polynomial used to calculate $\mathbf{G}_t$. When this is done, we have

$$\begin{aligned} \mathbf{R}_t &= \{R_{tx}, R_{ty}, R_{tz}\} \\ \dot{\mathbf{R}}_t &\approx \{\dot{G}_{tx}, \dot{G}_{ty}, \dot{G}_{tz}\} \\ \ddot{\mathbf{R}}_t &\approx \{\ddot{G}_{tx}, \ddot{G}_{ty}, \ddot{G}_{tz}\}. \end{aligned} \tag{8.27}$$

Program HELO in Section 8.5.3 may be used to determine the topocentric position and motion of the Sun for any given terrestrial dynamical time by interpolating the solar coordinates published in *The Astronomical Almanac*. Other methods may also be used which are based on general theories of solar motion [8,17].

8.5 Computer Programs

The computer programs presented in this section reduce the initial data for geocentric and heliocentric orbits to forms which are compatible with the orbit computation programs of the following chapters. All programs assume that position data are defined for the standard mean equinox J2000.0.

8.5.1 Program ADGRESS

Program ADGRESS performs polynomial regression of angular data by the method of least squares. Its primary purpose is to compute position and motion in right ascension and declination for use in the preliminary orbit method of Laplace.

Program Algorithm

Given	Line Number
name of the object for information only	1090
equinox for information only	1090
k: gravitational constant	1090
n: number of angular positions	1100
t_i, α_i, δ_i: $i = 1$ to n	1150

Choose	Line Number
m: order of the regression polynomial	1180
t: epoch time for $\tau = 0$	1260

Compute	Line Number
For $i = 1$ to n:	1430
$\quad \tau_i$: modified time intervals	1440
$\quad$ powers of τ_i	1470

8.5. COMPUTER PROGRAMS

For α and δ:	1510
normal equation coefficients	1550-1690
Gauss elimination	1730
polynomial coefficients	1760
α_t, δ_t: angular position at t	2000
$\dot{\alpha}_t, \dot{\delta}_t$: angular velocity at t	2020
$\ddot{\alpha}_t, \ddot{\delta}_t$: angular acceleration at t	2040

End.

Program Listing

```
1000 CLS
1010 PRINT"# ADGRESS # POLYNOMIAL REGRESSION"
1020 PRINT"# OF RA AND DEC DATA"
1030 PRINT
1040 DEFDBL A-Z
1050 DEFINT I,J,K,M,N,P
1060 REM
1070 G$="#####.#######"
1080 REM
1090 READ N$, E$, K#
1100 READ NP
1110 REM
1120 DIM T(NP),F(NP,2)
1130 REM
1140 FOR I=1 TO NP
1150    READ T(I),F(I,1),F(I,2)
1160 NEXT I
1170 REM
1180 INPUT">>>ORDER OF POLYNOMIAL";M
1190 REM
1200 IF M>(NP-1) THEN 1210 ELSE 1250
1210 CLS
1220 PRINT">>> ORDER TO HIGH * TRY AGAIN <<<"
1230 PRINT
1240 GOTO 1180
1250 PRINT
1260 INPUT">>>EPOCH TIME FOR TT=0";T
1270 PRINT
1280 DIM TT(NP),TP(NP,M+M),A(M+1,M+2),XU(M+1),C(M,2)
1290 CLS
1300 PRINT"POLYNOMIAL REGRESSION OF RA AND DEC"
1310 PRINT N$
1320 PRINT E$
1330 PRINT
1340 PRINT"INITIAL DATA *************************"
1350 PRINT
1360 PRINT TAB(7)"T(I)";TAB(20)"A(I)";TAB(33)"D(I)"
1370 FOR I=1 TO NP
1380    PRINT USING G$;T(I),F(I,1),F(I,2)
1390 NEXT I
```

8.5. COMPUTER PROGRAMS

```
1400 REM
1410 REM * POWERS OF TT(I) *
1420 REM
1430 FOR I=1 TO NP
1440   TT(I)=K#*(T(I)-T)
1450   TP(I,0)=1
1460   FOR P=1 TO (M+M)
1470     TP(I,P)=TP(I,P-1)*TT(I)
1480   NEXT P
1490 NEXT I
1500 REM
1510 FOR KK=1 TO 2
1520 REM
1530 REM * NORMAL EQUATION COEFFICIENTS *
1540 REM
1550 FOR J=1 TO (M+1)
1560   FOR K=1 TO (M+1)
1570     A(J,K)=0
1580     FOR I=1 TO NP
1590       A(J,K)=A(J,K)+TP(I,J+K-2)
1600     NEXT I
1610   NEXT K
1620   REM
1630   A(J,M+2)=0
1640   FOR I=1 TO NP
1650     A(J,M+2)=A(J,M+2)+TP(I,J-1)*F(I,KK)
1660   NEXT I
1670 NEXT J
1680 REM
1690 N=(M+1)
1700 REM
1710 REM * GAUSS ELIMINATION *
1720 REM
1730 GOSUB 19010 REM GELIM/SUB
1740 REM
1750 FOR P=0 TO M
1760   C(P,KK)=XU(P+1)
1770 NEXT P
1780 REM
1790 NEXT KK
1800 REM
1810 LINE INPUT"";L$
```

```
1820 CLS
1830 PRINT"POLYNOMIAL COEFFICIENTS **************"
1840 PRINT
1850 PRINT"EPOCH";:PRINT TAB(14);
1860 PRINT USING G$;T
1870 PRINT
1880 FOR P=0 TO M
1890    PRINT"C(";P;")";:PRINT TAB(14);
1900    PRINT USING G$;C(P,1),C(P,2)
1910 NEXT P
1920 LINE INPUT"";L$
1930 CLS
1940 PRINT"POSITION AND MOTION *****************"
1950 PRINT
1960 PRINT"EPOCH";:PRINT TAB(14);
1970 PRINT USING G$;T
1980 PRINT
1990 PRINT"F";:PRINT TAB(14);
2000 PRINT USING G$;C(0,1),C(0,2)
2010 PRINT"F'";:PRINT TAB(14);
2020 PRINT USING G$;C(1,1),C(1,2)
2030 PRINT"F''";:PRINT TAB(14);
2040 PRINT USING G$;C(2,1)*2,C(2,2)*2
2050 PRINT
2060 PRINT"ADGRESS"
2070 END
19000 STOP
19010 REM # GELIM/SUB # GAUSS ELIMINATION
19020 FOR I=1 TO (N-1)
19030    REM
19040    JP=I
19050    PE!=ABS(A(I,I))
19060    FOR J=(I+1) TO N
19070       CE!=ABS(A(J,I))
19080       IF CE!-PE! < 0 GOTO 19130
19090       PE!=CE!
19100       JP=J
19110    NEXT J
19120    REM
19130    IF JP=I GOTO 19210
19140    REM
19150    FOR K=I TO (N+1)
```

8.5. COMPUTER PROGRAMS

```
19160     HE=A(I,K)
19170     A(I,K)=A(JP,K)
19180     A(JP,K)=HE
19190   NEXT K
19200   REM
19210   FOR J=(I+1) TO N
19220     FOR K=(I+1) TO (N+1)
19230       A(J,K)=A(J,K)-A(J,I)*A(I,K)/A(I,I)
19240     NEXT K
19250     A(J,I)=0
19260   NEXT J
19270 NEXT I
19280 REM
19290 XU(N)=A(N,N+1)/A(N,N)
19300 FOR I=(N-1) TO 1 STEP -1
19310   SS=0
19320   FOR K=(I+1) TO N
19330     SS=SS+A(I,K)*XU(K)
19340   NEXT K
19350   XU(I)=(A(I,N+1)-SS)/A(I,I)
19360 NEXT I
19370 RETURN
30000 DATA "GEOS", "J2000.0", 0.07436680#
30010 DATA 9
40000 REM # GEOSOBS #
40010 DATA  95.08167#, 2.09624#, 2.84935#
40020 DATA  96.08167#, 2.73443#, 2.39724#
40030 DATA  97.08167#, 3.25454#, 1.78343#
40040 DATA  98.08167#, 3.68088#, 1.08683#
40050 DATA  99.08167#, 4.03544#, 0.35237#
40060 DATA 100.08167#, 4.33550#,-0.39552#
40070 DATA 101.08167#, 4.59395#,-1.14379#
40080 DATA 102.08167#, 4.82023#,-1.88538#
40090 DATA 103.08167#, 5.02129#,-2.61638#
40100 REM
50000 DATA "REBEK-JEWEL", "J2000.0", 0.017202099#
50010 DATA 10
60000 REM # RJOBS #
60010 DATA 6370.57744#, 5.41652#, 21.85272#
60020 DATA 6371.57632#, 5.35088#, 21.92988#
60030 DATA 6372.57518#, 5.28090#, 22.00440#
60040 DATA 6373.57402#, 5.20632#, 22.07521#
```

```
60050 DATA 6374.57284#, 5.12686#, 22.14104#
60060 DATA 6375.57163#, 5.04222#, 22.20041#
60070 DATA 6376.57040#, 4.95212#, 22.25161#
60080 DATA 6377.56916#, 4.85627#, 22.29266#
60090 DATA 6378.56789#, 4.75436#, 22.32127#
60100 DATA 6379.56659#, 4.64614#, 22.33487#
```

8.5. COMPUTER PROGRAMS

8.5.2 Program GEO

Program GEO computes the topocentric position, velocity, and acceleration of the geocenter for use in geocentric orbit computations.

Program Algorithm

Given	Line Number
angle to radian conversion factor	1070
A_e: Earth's equatorial radius in meters	1080
f: flattening factor	1090
Earth's rotation rate in radians/minute	1100
west longitude (°, ′, ″)	1150
latitude (°, ′, ″)	1150
terrain elevation in meters	1150
name of the object for information only	1170
equinox for information only	1170
k_e: gravitational constant	1170
year, month, day	1180
hour, minute, second	1190

Compute	Line Number
$\dot{\theta}_t$: Earth's rotation rate with respect to modified time	1210
west longitude in degrees	1220
ϕ: geodetic latitude in radians	1230
H: terrain elevation in Earth radii (er)	1240

UT: universal time in hours for epoch t — 1250

J_0: Julian day at 0^h UT — 1270

ST: sidereal time in degrees for epoch t — 1270

JD: Julian date for epoch t — 1290

θ_t: sidereal time in radians for epoch t — 1300

F: Equations 8.13 — 1320

G_c: Equations 8.13 — 1330

G_s: Equations 8.13 — 1340

$\mathbf{g}_t$: Equations 8.16 — 1360-1380

$\dot{\mathbf{g}}_t$: Equations 8.17 — 1400-1420

$\ddot{\mathbf{g}}_t$: Equations 8.18 — 1440-1460

elements of the precession matrix $\mathbf{P}$ — 1480

$\mathbf{R}_t$: Equations 8.15 — 1510

$\dot{\mathbf{R}}_t$: Equations 8.15 — 1520

$\ddot{\mathbf{R}}_t$: Equations 8.15 — 1530

End.

8.5. COMPUTER PROGRAMS

Program Listing

```
1000 CLS
1010 PRINT"# GEO # RR-VECTOR TO GEOCENTER"
1020 PRINT"# ALL TIMES ARE ASSUMED TO BE UT"
1030 REM
1040 DEFDBL A-Z
1050 DEFINT I,K
1060 REM
1070 Q1=.0174532925#
1080 AE=6378140#
1090 F=1/298.257#
1100 RM=.0043752695#
1110 G$="####.#######"
1120 S$="####.#####"
1130 J$="#######.#####"
1140 REM
1150 READ W1,W2,W3, B1,B2,B3, HT
1160 REM
1170 READ N$, E$, K#
1180 READ IY,IM,ID
1190 READ HR,MN,SC
1200 REM
1210 W=RM/K#
1220 WL=W1+W2/60+W3/3600
1230 BB=(B1+B2/60+B3/3600)*Q1
1240 HH=HT/AE
1250 UT=HR+MN/60+SC/3600
1260 REM
1270 GOSUB 14010 REM JOST/SUB
1280 REM
1290 JD=J0+UT/24
1300 SS=ST*Q1
1310 REM
1320 FF=SQR(1-(2*F-F*F)*SIN(BB)*SIN(BB))
1330 GC=1/FF+HH
1340 GS=(1-F)*(1-F)/FF+HH
1350 REM
1360 G(1)=-GC*COS(BB)*COS(SS)
1370 G(2)=-GC*COS(BB)*SIN(SS)
1380 G(3)=-GS*SIN(BB)
1390 REM
```

```
1400 V(1)=-G(2)*W
1410 V(2)=+G(1)*W
1420 V(3)= 0
1430 REM
1440 A(1)=-G(1)*W*W
1450 A(2)=-G(2)*W*W
1460 A(3)= 0
1470 REM
1480 GOSUB 12010 REM PMATRX/SUB
1490 REM
1500 FOR K=1 TO 3
1510    RR(K)=PP(1,K)*G(1)+PP(2,K)*G(2)+PP(3,K)*G(3)
1520    RV(K)=PP(1,K)*V(1)+PP(2,K)*V(2)+PP(3,K)*V(3)
1530    RA(K)=PP(1,K)*A(1)+PP(2,K)*A(2)+PP(3,K)*A(3)
1540 NEXT K
1550 LINE INPUT"";L$
1560 CLS
1570 PRINT"TOPOCENTRIC RR-VECTOR TO THE GEOCENTER"
1580 PRINT N$
1590 PRINT E$
1600 PRINT"DATE(UT)";TAB(11);IY;"/";IM;"/";ID
1610 PRINT"WL";TAB(12);:PRINT USING S$;WL
1620 PRINT"BB";TAB(12);:PRINT USING S$;BB/Q1
1630 PRINT"SS";TAB(12);:PRINT USING S$;SS/Q1
1640 PRINT"JO";TAB(9);:PRINT USING J$;JO
1650 PRINT"UT(MIN)";TAB(12);:PRINT USING S$;UT*60
1660 PRINT
1670 PRINT"RR(K)";TAB(12);
1680 PRINT USING G$;RR(1),RR(2),RR(3)
1690 REM
1700 PRINT"RR(K)'";TAB(12);
1710 PRINT USING G$;RV(1),RV(2),RV(3)
1720 REM
1730 PRINT"RR(K)''";TAB(12);
1740 PRINT USING G$;RA(1),RA(2),RA(3)
1750 PRINT
1760 PRINT"GEO"
1770 END
12000 STOP
12010 REM# PMATRX/SUB # PRECESSION ROTATION MATRIX
12020 TT=(JD-2451545#)/36525#
12030 PP(1,1)=1+(-29724*TT-13*TT*TT)*TT*.00000001#
```

8.5. COMPUTER PROGRAMS

```
12040 PP(1,2)=(-2236172!-677*TT+222*TT*TT)*TT*.00000001#
12050 PP(1,3)=(-971717!+207*TT+96*TT*TT)*TT*.00000001#
12060 PP(2,1)=-PP(1,2)
12070 PP(2,2)=1+(-25002*TT-15*TT*TT)*TT*.00000001#
12080 PP(2,3)=(-10865*TT)*TT*.00000001#
12090 PP(3,1)=-PP(1,3)
12100 PP(3,2)=PP(2,3)
12110 PP(3,3)=1+(-4721*TT)*TT*.00000001#
12120 RETURN
14000 STOP
14010 REM # JOST/SUB # CALCULATES JO AND LMST
14020 IY=IY+1900
14030 J0=367*IY-INT((7*(IY+INT((IM+9)/12)))/4)+
      INT(275*IM/9)+ID+1721013.5#
14040 J=(J0-2451545#)/36525#
14050 EL=360-WL
14060 S0=100.4606184#+36000.77004#*J+.000387933#*J*J
14070 S0=S0-360*INT(S0/360)
14080 SG=S0+360.98564724#*(UT/24)
14090 ST=SG+EL
14100 ST=ST-360*INT(ST/360)
14110 RETURN
20000 DATA 77,35,41, 37,31,33, 100
30000 DATA "GEOS", "J2000.0", 0.07436680#
30010 DATA 85,11,01
30020 DATA 1,39,04.9#
```

8.5.3 Program HELO

Program HELO computes the topocentric position, velocity, and acceleration of the Sun for use in heliocentric orbit computations.

Program Algorithm

Given	Line Number
angle to radian conversion factor	1070
A_e: Earth's equatorial radius in astronomical units (au)	1080
west longitude (°, ′, ″)	1140
latitude (°, ′, ″)	1140
name of the object for information only	1160
equinox for information only	1160
k: gravitational constant	1160
n: number of tabular $\mathbf{G}_p$ for interpolation	1170
$t_p, G_{px}, G_{py}, G_{pz}$: $p = 1$ to n	1200

Choose	Line Number
t: JD epoch for the interpolation	1230

Compute	Line Number
interpolating polynomial	1250
$\mathbf{G}_t$: Sun's geocentric position at t	1280
$\dot{\mathbf{G}}_t$: Sun's geocentric velocity at t	1290
$\ddot{\mathbf{G}}_t$: Sun's geocentric acceleration at t	1300
west longitude	1330

8.5. COMPUTER PROGRAMS

ϕ: geodetic latitude in radians	1340
J_0: Julian day corresponding to t	1360
fraction of the day on J_0	1370
ST: sidereal time in degrees	1390
θ_t: sidereal time in radians	1410
$\mathbf{g}_t$: Equations 8.24	1430-1450
$\mathbf{R}_t$: Equations 8.25	1480
$\dot{\mathbf{R}}_t$: Equations 8.27	1750
$\ddot{\mathbf{R}}_t$: Equations 8.27	1780

End.

Program Listing

```
1000 CLS
1010 PRINT"# HELO # RR-VECTOR TO HELIOCENTER"
1020 PRINT"# ALL TIMES ARE ASSUMED TO BE JD(TDT)"
1030 PRINT
1040 DEFDBL A-Z
1050 DEFINT K,P
1060 REM
1070 Q1=.0174532925#
1080 AE=.00004263523#
1090 G$="####.#######"
1100 S$="####.#####"
1110 JO$="#######.#"
1120 JD$="#######.#####"
1130 REM
1140 READ W1,W2,W3, B1,B2,B3
1150 REM
1160 READ N$, E$, K#
1170 READ NP
1180 REM
1190 FOR P=1 TO NP
1200    READ T(P),F(P,1),F(P,2),F(P,3)
1210 NEXT P
1220 REM
1230 INPUT">>>INTERPOLATION JD";T
1240 REM
1250 GOSUB 13010 REM INTERP/SUB
1260 REM
1270 FOR K=1 TO 3
1280    GG(K)=F0(K)
1290    GV(K)=F1(K)
1300    GA(K)=F2(K)
1310 NEXT K
1320 REM
1330 WL=W1+W2/60+W3/3600
1340 BB=(B1+B2/60+B3/3600)*Q1
1350 JD=T
1360 JO=INT(JD)+.5#
1370 DY=JD-JO
1380 REM
1390 GOSUB 14020 REM ST/SUB
```

8.5. COMPUTER PROGRAMS

```
1400 REM
1410 SS=ST*Q1
1420 REM
1430 G(1)=-AE*COS(BB)*COS(SS)
1440 G(2)=-AE*COS(BB)*SIN(SS)
1450 G(3)=-AE*SIN(BB)
1460 REM
1470 FOR K=1 TO 3
1480    RR(K)=G(K)+GG(K)
1490 NEXT K
1500 CLS
1510 PRINT"TOPOCENTRIC RR-VECTOR TO THE HELIOCENTER"
1520 PRINT N$
1530 PRINT E$
1540 PRINT"JD(TDT)";TAB(9);:PRINT USING JD$;JD
1550 PRINT"WL";TAB(12);:PRINT USING S$;WL
1560 PRINT"BB";TAB(12);:PRINT USING S$;BB/Q1
1570 PRINT"SS";TAB(12);:PRINT USING S$;SS/Q1
1580 REM
1590 FOR P=1 TO NP
1600    PRINT USING J0$;T(P);:PRINT TAB(12);
1610    PRINT USING G$;F(P,1),F(P,2),F(P,3)
1620 NEXT P
1630 LINE INPUT"";L$
1640 CLS
1650 PRINT"GG(K)";TAB(12);
1660 PRINT USING G$;F0(1),F0(2),F0(3)
1670 REM
1680 PRINT"G(K)";TAB(12);
1690 PRINT USING G$;G(1),G(2),G(3)
1700 PRINT
1710 PRINT"RR(K)";TAB(12);
1720 PRINT USING G$;RR(1),RR(2),RR(3)
1730 REM
1740 PRINT"RR(K)'";TAB(12);
1750 PRINT USING G$;GV(1),GV(2),GV(3)
1760 REM
1770 PRINT"RR(K)''";TAB(12);
1780 PRINT USING G$;GA(1),GA(2),GA(3)
1790 PRINT
1800 PRINT"ORDER";(NP-1)
1810 PRINT
```

```
1820 PRINT"HELO"
1830 END
13000 STOP
13010 REM # INTERP/SUB # LAGRANGE INTERPOLATION
13020 FOR P=1 TO NP
13030    GT(P)=1
13040    L0(P)=1
13050    L1(P)=0
13060    L2(P)=0
13070    FOR PJ=1 TO NP
13080      IF PJ=P GOTO 13130
13090      GT(P)=GT(P)*(T(P)-T(PJ))
13100      L2(P)=L2(P)*(T-T(PJ))+2*L1(P)
13110      L1(P)=L1(P)*(T-T(PJ))+L0(P)
13120      L0(P)=L0(P)*(T-T(PJ))
13130    NEXT PJ
13140 NEXT P
13150 FOR K=1 TO 3
13160    F0(K)=0
13170    F1(K)=0
13180    F2(K)=0
13190    FOR P=1 TO NP
13200      F0(K)=F0(K)+F(P,K)*L0(P)/GT(P)
13210      F1(K)=F1(K)+F(P,K)*L1(P)/(GT(P)*K#)
13220      F2(K)=F2(K)+F(P,K)*L2(P)/(GT(P)*K#*K#)
13230    NEXT P
13240 NEXT K
13250 RETURN
14000 STOP
14010 REM # ST/SUB # CALCULATES ST
14020 J=(J0-2451545#)/36525#
14030 EL=360-WL
14040 S0=100.4606184#+36000.77004#*J+.000387933#*J*J
14050 S0=S0-360*INT(S0/360)
14060 SG=S0+360.98564724#*DY
14070 ST=SG+EL
14080 ST=ST-360*INT(ST/360)
14090 RETURN
20000 DATA 77,35,41, 37,31,33
30000 DATA "SUN", "J2000.0", 0.017202099#
30010 DATA 3
40000 REM # SUNDAT #
```

8.5. COMPUTER PROGRAMS

```
40010 REM
40020 'DATA 2446369.5#,-0.7853213#,-0.5571657#,-0.2415830#
40030 'DATA 2446370.5#,-0.7744047#,-0.5695064#,-0.2469331#
40040 'DATA 2446371.5#,-0.7632541#,-0.5816759#,-0.2522092#
40050 REM
40060 'DATA 2446372.5#,-0.7518724#,-0.5936705#,-0.2574095#
40070 DATA 2446373.5#,-0.7402626#,-0.6054865#,-0.2625325#
40080 DATA 2446374.5#,-0.7284277#,-0.6171201#,-0.2675766#
40090 DATA 2446375.5#,-0.7163707#,-0.6285675#,-0.2725401#
40100 'DATA 2446376.5#,-0.7040949#,-0.6398250#,-0.2774214#
40110 REM
40120 DATA 2446377.5#,-0.6916037#,-0.6508886#,-0.2822188#
40130 DATA 2446378.5#,-0.6789003#,-0.6617546#,-0.2869306#
40140 DATA 2446379.5#,-0.6659885#,-0.6724192#,-0.2915553#
40150 REM
40160 DATA 2446389.5#,-0.5263183#,-0.7672058#,-0.3326577#
40170 DATA 2446390.5#,-0.5113954#,-0.7754313#,-0.3362239#
40180 DATA 2446391.5#,-0.4963170#,-0.7834185#,-0.3396865#
```

8.6 Numerical Examples

8.6.1 Regression of Angular Data for Satellite GEOS

Problem

Topocentric position data for the artificial satellite GEOS are shown in the table below. The angular positions are referred to the equinox J2000.0, and all times are expressed in minutes of coordinated universal time (UTC) on 1985 November 1. The observation site is 100 meters above mean sea level at west longitude 77°35'41" and north latitude 37°31'33". Use a fourth order polynomial regression to compute the angular position, velocity, and acceleration of satellite GEOS at the epoch 99.08167 minutes UTC.

t_i	α_i	δ_i
95.08167	2.09624	2.84935
96.08167	2.73443	2.39724
97.08167	3.25454	1.78343
98.08167	3.68088	1.08683
99.08167	4.03544	0.35237
100.08167	4.33550	−0.39552
101.08167	4.59395	−1.14379
102.08167	4.82023	−1.88538
103.08167	5.02129	−2.61638

Solution

Use the given data to write lines 30000 to 40090 as shown at the end of program ADGRESS. Run the program and answer the prompts as follows:

```
>>>ORDER OF POLYNOMIAL? 4

>>>EPOCH TIME FOR TT=0? 99.08167#
```

8.6. NUMERICAL EXAMPLES

Results

```
POLYNOMIAL REGRESSION OF RA AND DEC
GEOS
J2000.0

INITIAL DATA *************************
```

T(I)	A(I)	D(I)
95.0816700	2.0962400	2.8493500
96.0816700	2.7344300	2.3972400
97.0816700	3.2545400	1.7834300
98.0816700	3.6808800	1.0868300
99.0816700	4.0354400	0.3523700
100.0816700	4.3355000	-0.3955200
101.0816700	4.5939500	-1.1437900
102.0816700	4.8202300	-1.8853800
103.0816700	5.0212900	-2.6163800

```
POLYNOMIAL COEFFICIENTS ***************

EPOCH    99.0816700
```

C(0)	4.0355860	0.3518241
C(1)	4.3647689	-10.0716627
C(2)	-4.9249166	-1.0069243
C(3)	6.2345235	9.9327202
C(4)	-5.2442789	-18.6632081

```
POSITION AND MOTION ******************

EPOCH    99.0816700
```

F	4.0355860	0.3518241
F'	4.3647689	-10.0716627
F''	-9.8498332	-2.0138485

ADGRESS

8.6.2 Regression of Angular Data for Comet Rebek-Jewel

Problem

Topocentric position data for comet Rebek-Jewel are shown in the table below. The angular positions are referred to the equinox J2000.0, and all times are Julian dates expressed in terrestrial dynamical time (TDT). The observation site is at west longitude 77°35'41" and north latitude 37°31'33". Use a fourth order polynomial regression to compute the angular position, velocity, and acceleration of comet Rebek-Jewel for the epoch 2446374.57284.

t_i	α_i	δ_i
2446370.57744	5.41652	21.85272
2446371.57632	5.35088	21.92988
2446372.57518	5.28090	22.00440
2446373.57402	5.20632	22.07521
2446374.57284	5.12686	22.14104
2446375.57163	5.04222	22.20041
2446376.57040	4.95212	22.25161
2446377.56916	4.85627	22.29266
2446378.56789	4.75436	22.32127
2446379.56659	4.64614	22.33487

Solution

Use the given data to write lines 50000 to 60100 as shown at the end of program ADGRESS, where the first three digits of the Julian dates have been omitted for convenience. Run the program and answer the prompts as follows:

>>>ORDER OF POLYNOMIAL? 4

>>>EPOCH TIME FOR TT=0? 6374.57284#

8.6. NUMERICAL EXAMPLES

Results

```
POLYNOMIAL REGRESSION OF RA AND DEC
REBEK-JEWEL
J2000.0

INITIAL DATA ************************

        T(I)           A(I)          D(I)
   6370.5774400     5.4165200    21.8527200
   6371.5763200     5.3508800    21.9298800
   6372.5751800     5.2809000    22.0044000
   6373.5740200     5.2063200    22.0752100
   6374.5728400     5.1268600    22.1410400
   6375.5716300     5.0422200    22.2004100
   6376.5704000     4.9521200    22.2516100
   6377.5691600     4.8562700    22.2926600
   6378.5678900     4.7543600    22.3212700
   6379.5665900     4.6461400    22.3348700

POLYNOMIAL COEFFICIENTS ***************

EPOCH 6374.5728400

C( 0 )             5.1268591     22.1410258
C( 1 )            -4.7728391      3.6603069
C( 2 )            -8.7666583    -10.8714791
C( 3 )            -9.4305944    -53.2837279
C( 4 )            -0.9340459   -120.7175944

POSITION AND MOTION ******************

EPOCH 6374.5728400

F                  5.1268591     22.1410258
F'                -4.7728391      3.6603069
F''              -17.5333166    -21.7429581

ADGRESS
```

8.6.3 Topocentric Vector to the Geocenter

Problem

An observer is located 100 meters above mean sea level at west longitude 77°35'41" and north latitude 37°31'33". Compute the topocentric position and motion of the center of the Earth on 1985 November 1 at $1^h39^m4^s.9$ UTC.

Solution

Use the given information to write data lines 20000 to 30020 as shown at the end of program GEO. Run the program.

Results

```
TOPOCENTRIC RR-VECTOR TO THE GEOCENTER
GEOS
J2000.0
DATE(UT)     1985 / 11 / 1
WL              77.59472
BB              37.52583
SS             347.47184
JO         2446370.50000
UT(MIN)         99.08167

RR(K)        -0.7748777    0.1697957   -0.6068707
RR(K)'       -0.0099896   -0.0456380   -0.0000139
RR(K)''       0.0026850   -0.0005877    0.0000037

GEO
```

8.6. NUMERICAL EXAMPLES

8.6.4 Topocentric Vector to the Heliocenter

Problem

The geocentric rectangular coordinates of the Sun listed at the end of program HELO were taken from *The Astronautical Almanac 1985*. Given an observation site at west longitude 77°35'41" and north latitude 37°31'33", use a second order interpolation to compute the topocentric position and motion of the Sun for the Julian date 2446374.57284.

Solution

Use the data in lines 20000 to 30010 and lines 40070 to 40090 as listed at the end of program HELO. Run the program and answer the prompt with the complete Julian date as follows:

```
>>>INTERPOLATION JD? 2446374.57284#
```

Results

```
TOPOCENTRIC RR-VECTOR TO THE HELIOCENTER
SUN
J2000.0
JD(TDT)  2446374.57284
WL          77.59472
BB          37.52583
SS         352.87039
2446373.5   -0.7402626  -0.6054865  -0.2625325
2446374.5   -0.7284277  -0.6171201  -0.2675766
2446375.5   -0.7163707  -0.6285675  -0.2725401

GG(K)       -0.7275570  -0.6179602  -0.2679409
G(K)        -0.0000336   0.0000042  -0.0000260

RR(K)       -0.7275905  -0.6179560  -0.2679668
RR(K)'       0.6953877  -0.6700890  -0.2905418
RR(K)''      0.7505604   0.6292407   0.2723781

ORDER 2

HELO
```

References

[1] Moulton, *An Introduction to Celestial Mechanics*, Dover Publications, Inc., 1970.

[2] Herget, *The Computation of Orbits*, Published by Author, 1948.

[3] Dubyago, *The Determination of Orbits*, The Rand Corporation, 1961.

[4] Baker, *Astrodynamics: Applications and Advanced Topics*, Academic Press, 1967.

[5] Escobal, *Methods of Orbit Determination*, Krieger Publishing Co., 1976.

[6] Roy, *Orbital Motion*, Adam Hilger Ltd., 1978.

[7] Taff, *Celestial Mechanics*, John Wiley and Sons, 1985.

[8] Danby, *Fundamentals of Celestial Mechanics*, Willmann-Bell, Inc., 1988.

[9] King-Hele, *Observing Earth Satellites*, Van Nostrand Reinhold Co., 1983.

[10] Sidgwick, *Amateur Astronomer's Handbook*, 4th edition by Muirden, Enslow Publishers, 1980.

[11] Roth, *Astronomy: a Handbook*, Springer-Verlag, 1975.

[12] Boulet, "Making a Simple Chronograph," *Sky & Telescope*, April 1983, pp. 369-372.

[13] Boulet, "A Simple Photochronograph," *Sky & Telescope*, July 1984, p. 76.

[14] Everhart, "Constructing a Measuring Engine," *Sky & Telescope*, September 1982, pp. 279-282.

[15] Sinnott, "Editor's Note," *Sky & Telescope*, September 1982, p. 283.

[16] Marsden, "How to reduce Plate Measurements," *Sky & Telescope*, September 1982, p. 284.

[17] Montenbruck, *Practical Ephemeris Calculations*, Springer-Verlag, 1989.

Chapter 9

The Method of Laplace

9.1 Introduction

Of all the initial orbit determination methods, that of Laplace follows most directly from fundamental physical principles and the elementary processes of applied calculus. It is general, and any number of observations may be included in the solution. The method does have a potential disadvantage, however. It depends heavily on information obtained by numerical differentiation of the initial position data, and, in practice, it is sometimes difficult to attain the required precision by such numerical techniques. Consequently, more than three angular positions should be used if they are available, and it is best if they are fairly close together and evenly spaced in time. Furthermore, the magnitude of the coefficient D_0 in Equations 9.10 and 9.13 is related to the curvature of the celestial body's apparent path in the sky. If the available position data trace a path that is nearly straight, the coefficient D_0 will approach zero, and the method of Laplace may fail [1,2,3,4].

9.2 Solution by Successive Differentiation

Assume that the initial angular and vector position data have been reduced to the following values for the epoch t by the analytical and numerical procedures described in Chapter 8:

$$\{t, \mathbf{L}, \dot{\mathbf{L}}, \ddot{\mathbf{L}}, \mathbf{R}, \dot{\mathbf{R}}, \ddot{\mathbf{R}}\}.$$

It is also known that the following general conditions must be satisfied at the epoch time:

$$\ddot{\mathbf{r}} = -u\mathbf{r} \tag{9.1}$$

$$\mathbf{r} = p\mathbf{L} - \mathbf{R}, \tag{9.2}$$

where

$$u = \frac{\mu}{r^3}. \tag{9.3}$$

Furthermore, by differentiating Equation 9.2 with respect to modified time we have

$$\dot{\mathbf{r}} = \dot{p}\mathbf{L} + p\dot{\mathbf{L}} - \dot{\mathbf{R}}, \tag{9.4}$$

which can be used along with Equation 9.2 to compute the orbital elements $\mathbf{r}$ and $\dot{\mathbf{r}}$ at the epoch time t if values for the unknown quantities p and $\dot{p}$ can be found. Differentiating once more with respect to modified time and collecting terms we obtain

$$\ddot{\mathbf{r}} = \ddot{p}\mathbf{L} + 2\dot{p}\dot{\mathbf{L}} + p\ddot{\mathbf{L}} - \ddot{\mathbf{R}}. \tag{9.5}$$

Substituting Equation 9.1 into Equation 9.5 and replacing $\mathbf{r}$ by Equation 9.2 yields

$$-u(p\mathbf{L} - \mathbf{R}) = \ddot{p}\mathbf{L} + 2\dot{p}\dot{\mathbf{L}} + p\ddot{\mathbf{L}} - \ddot{\mathbf{R}}. \tag{9.6}$$

Rearranging Equation 9.6 to form scalar coefficients for the vectors, the result is

$$(up + \ddot{p})\mathbf{L} + 2\dot{p}\dot{\mathbf{L}} + p\ddot{\mathbf{L}} = \ddot{\mathbf{R}} + u\mathbf{R}. \tag{9.7}$$

9.3 The Scalar Equations for the Range and Rate

We solve Equation 9.7 for the range p by first taking the cross product $\dot{\mathbf{L}}\times$, followed by the dot product with $\mathbf{L}$. Accordingly, we have

$$p\mathbf{L} \cdot (\dot{\mathbf{L}} \times \ddot{\mathbf{L}}) = \mathbf{L} \cdot (\dot{\mathbf{L}} \times \ddot{\mathbf{R}}) + u\mathbf{L} \cdot (\dot{\mathbf{L}} \times \mathbf{R}), \tag{9.8}$$

where the $\dot{p}$ term was eliminated because $\dot{\mathbf{L}} \times \dot{\mathbf{L}} = \mathbf{0}$, and the $up + \ddot{p}$ term was eliminated because the vectors $\mathbf{L}$ and $\dot{\mathbf{L}} \times \mathbf{L}$ are perpendicular, causing their dot product to be zero. Since all the vectors are part of the given data, we have an expression for p in terms of the unknown quantity u. For convenience, let Equation 9.8 be rewritten as follows:

$$p = A + \frac{\mu B}{r^3}, \tag{9.9}$$

where u has been replaced by Equation 9.3 and

$$A = \frac{\mathbf{L} \cdot (\dot{\mathbf{L}} \times \ddot{\mathbf{R}})}{D_0}$$

$$B = \frac{\mathbf{L} \cdot (\dot{\mathbf{L}} \times \mathbf{R})}{D_0}$$

$$D_0 = \mathbf{L} \cdot (\dot{\mathbf{L}} \times \ddot{\mathbf{L}}). \tag{9.10}$$

9.4. THE SCALAR EQUATION FOR THE RADIAL DISTANCE

In a somewhat similar fashion, Equation 9.7 is solved for the rate $\dot{p}$ by first taking the cross product $\times \dot{\mathbf{L}}$ followed by the dot product with $\mathbf{L}$. Consequently,

$$2\dot{p}\mathbf{L} \cdot (\dot{\mathbf{L}} \times \dot{\mathbf{L}}) = \mathbf{L} \cdot (\ddot{\mathbf{R}} \times \dot{\mathbf{L}}) + u\mathbf{L} \cdot (\mathbf{R} \times \dot{\mathbf{L}}), \qquad (9.11)$$

where the p term was eliminated because $\dot{\mathbf{L}} \times \dot{\mathbf{L}} = \mathbf{0}$, and the $up + \ddot{p}$ term was eliminated because the vectors $\mathbf{L}$ and $\mathbf{L} \times \dot{\mathbf{L}}$ are perpendicular, causing their dot product to be zero. All vectors in Equations 9.11 are known. Therefore, we have an expression for $\dot{p}$ in terms of the same unknown, u. For convenience, let Equation 9.11 be rewritten as follows:

$$\dot{p} = C + \frac{\mu D}{r^3}, \qquad (9.12)$$

where u has been replaced by Equation 9.3 and

$$C = \frac{\mathbf{L} \cdot (\ddot{\mathbf{R}} \times \dot{\mathbf{L}})}{2D_0}$$

$$D = \frac{\mathbf{L} \cdot (\mathbf{R} \times \dot{\mathbf{L}})}{2D_0}$$

$$D_0 = \mathbf{L} \cdot (\dot{\mathbf{L}} \times \dot{\mathbf{L}}). \qquad (9.13)$$

9.4 The Scalar Equation for the Radial Distance

We have derived two scalar equations containing a total of three unknowns: p, $\dot{p}$, and r. One more scalar equation, relating p and r, can be obtained by taking the dot product of Equation 9.2 with itself. Thus,

$$\mathbf{r} \cdot \mathbf{r} = (p\mathbf{L} - \mathbf{R}) \cdot (p\mathbf{L} - \mathbf{R}), \qquad (9.14)$$

which simplifies to

$$r^2 = p^2 - 2p(\mathbf{L} \cdot \mathbf{R}) + R^2 \qquad (9.15)$$

since $L^2 = 1$ because $\mathbf{L}$ is a unit vector. For convenience, let Equation 9.15 be written

$$r^2 = p^2 + pE + F, \qquad (9.16)$$

where

$$E = -2(\mathbf{L} \cdot \mathbf{R})$$

$$F = R^2. \qquad (9.17)$$

We now have three scalar equations containing the three unknowns p, $\dot{p}$, and r.

9.5 The Scalar Equation of Lagrange

An expression containing only r can now be formed by substituting Equation 9.9 for the variable p in Equation 9.16. When this is done, the result is

$$r^2 = A^2 + \frac{2\mu AB}{r^3} + \frac{\mu^2 B^2}{r^6} + AE + \frac{\mu BE}{r^3} + F. \tag{9.18}$$

Multiplying through by r^6 to clear the fractions and collecting terms, we obtain the eight-degree equation

$$r^8 + ar^6 + br^3 + c = 0 \tag{9.19}$$

known as the *equation of Lagrange*. The coefficients are defined as follows:

$$\begin{aligned} a &= -(A^2 + AE + F) \\ b &= -\mu(2AB + BE) \\ c &= -\mu^2 B^2. \end{aligned} \tag{9.20}$$

If we let the symbol x represent the unknown distance r in Equation 9.19, we can write

$$f(x) = x^8 + ax^6 + bx^3 + c.$$

Differentiating $f(x)$ with respect to x yields

$$f'(x) = 8x^7 + 6ax^5 + 3bx^3.$$

It is easy to solve these last two equations for x by the Newton-Raphson method described in Section 7.2.2. The result will be the magnitude of the radius vector needed to complete the solution of the orbit problem.

The equation of Lagrange can have as many as three positive roots. When this occurs, it is usually possible to eliminate one or two of these roots on the grounds that they are unreasonably large or small, or that they correspond to values of p which are negative. In spite of this, there will be situations where two roots appear realistic. In this case, one must compute two sets of orbital elements and two ephemerides. The best solution must then be determined on the basis of how well the computed motion compares with the observed motion of the celestial body [5].

9.6 The Vector Orbital Elements

Armed with a value for r, the magnitude of the radius vector at the epoch time, we can now compute p and $\dot{p}$ from Equations 9.9 and 9.12, namely

$$p = A + \frac{\mu B}{r^3}$$

9.6. THE VECTOR ORBITAL ELEMENTS

$$\dot{p} = C + \frac{\mu D}{r^3}.$$

The values of p and $\dot{p}$ may then be used in Equations 9.2 and 9.4 to compute the vector orbital elements at the epoch time t as follows:

$$\mathbf{r} = p\mathbf{L} - \mathbf{R}$$
$$\dot{\mathbf{r}} = \dot{p}\mathbf{L} + p\dot{\mathbf{L}} - \dot{\mathbf{R}}.$$

Therefore, the element set is

$$\{t, \mathbf{r}, \dot{\mathbf{r}}\},$$

and the preliminary orbit is determined. In the case of heliocentric orbits, the time t should be corrected for light-time so that the elements are defined for an epoch t_c when light left the celestial body on its way toward the observer. Thus, according to the discussion in Section 2.8.1, the value t in a heliocentric element set should be replaced by a revised epoch

$$t_c = t - \frac{p}{c}, \tag{9.21}$$

where

$$c = 173.1446 \text{ au/day}. \tag{9.22}$$

An orbit computed by the method of Laplace will produce agreement with the angular position of the celestial body which was used, along with its angular motion, to determine the element set. If computed positions are compared with the observations made at times other than the epoch of the elements, the agreement may be poor. In the final chapter, we will describe a process which can be used to improve the elements so that they better represent the observed motion.

394   CHAPTER 9. THE METHOD OF LAPLACE

9.7 Program LAPLACE

Program LAPLACE computes the position and velocity elements of a preliminary orbit by the method of Laplace. LAPLACE uses angular position data reduced by ADGRESS and vector position data reduced by GEO or HELO.

Program Algorithm

Define	Line Number
FNVS(X,Y,Z): vector squaring function	1070
FNMG(X,Y,Z): vector magnitude function	1080
FNDP(X1,Y1,Z1,X2,Y2,Z2): dot product function	1090
FNDX(X2,Y2,Z2,X3,Y3,Z3): triple scalar product function	1100
FNF(X): Lagrange equation function	1110
FNDF(X): derivative of Lagrange equation function	1120
FNP(X): range equation function	1130
FNU(X): rate equation function	1140

Given	Line Number
μ: combined mass	1160
angle to radian conversion factor	1170
$1/c$: light-time constant	1180
name of the object for information only	1220
equinox for information only	1220
k: gravitational constant	1220
t: epoch time uncorrected for light-time	1230

9.7. PROGRAM LAPLACE

α, δ: angular position at epoch	1240
$\dot{\alpha}, \dot{\delta}$: angular velocity at epoch	1250
$\ddot{\alpha}, \ddot{\delta}$: angular acceleration at epoch	1260
R_x, R_y, R_z: $\mathbf{R}$ at epoch	1270
$\dot{R}_x, \dot{R}_y, \dot{R}_z$: $\dot{\mathbf{R}}$ at epoch	1280
$\ddot{R}_x, \ddot{R}_y, \ddot{R}_z$: $\ddot{\mathbf{R}}$ at epoch	1290
no light-time correction for geocentric orbits	1310

Compute	Line Number
α in radian measure	1490
$\dot{\alpha}$ in radian measure	1500
$\ddot{\alpha}$ in radian measure	1510
δ in radian measure	1530
$\dot{\delta}$ in radian measure	1540
$\ddot{\delta}$ in radian measure	1550
L_x: Equations 9.4	1570
L_y: Equations 9.4	1580
L_z: Equations 9.4	1590
$\mathcal{X}$: Equations 9.7	1610
$\mathcal{Y}$: Equations 9.7	1620
$\mathcal{Z}$: Equations 9.7	1630
$\dot{L}_x$: Equations 9.6	1650

$\dot{L}_y$: Equations 9.6	1660
$\dot{L}_z$: Equations 9.6	1670
$\dot{\mathcal{X}}$: Equations 9.9	1690
$\dot{\mathcal{Y}}$: Equations 9.9	1700
$\dot{\mathcal{Z}}$: Equations 9.9	1710
$\ddot{L}_x$: Equations 9.8	1730
$\ddot{L}_y$: Equations 9.8	1740
$\ddot{L}_z$: Equations 9.8	1750
D_0: Equations 9.10	2010
A: Equations 9.10	2020
B: Equations 9.10	2030
C: Equations 9.13	2040
D: Equations 9.13	2050
E: Equations 9.17	2060
F: Equations 9.17	2070
a: Equations 9.20	2090
b: Equations 9.20	2100
c: Equations 9.20	2110
Choose	Line Number
$r_{\min}$ and $r_{\max}$ which bound the unknown value of r	2260

9.7. PROGRAM LAPLACE

This prompt requests the user to choose minimum and maximum values for r, the magnitude of the radius vector, between which the unknown solution lies. Several trials may be necessary to find appropriate values.

Compute	Line Number
a table containing a series of values for r, p, and $f(r)$	2300-2380

Choose	Line Number
x: an approximate value for the unknown r	2400

This prompt requests the user to choose an approximate value for r as a starting point for the Newton-Raphson solution of the equation of Lagrange. From the table, choose a value of r which corresponds to a positive value of p and the smallest value of $|f(r)|$ at a point where the function changes sign. If the table does not contain any such point, the program must be run again using different values for r_{min} and r_{max}.

Compute	Line Number
solution by the Newton-Raphson method	2420-2440
r: magnitude of the radius vector at epoch	2460
p: range at epoch by Equation 9.9	2470
$\dot{p}$: rate at epoch by Equation 9.12	2480
r: position elements at epoch by Equation 9.2	2530
$\dot{\mathbf{r}}$: velocity elements at epoch by Equation 9.4	2540
t_c: epoch corrected for light-time by Equation 9.21	2570

End.

Program Listing

```
1000 CLS
1010 PRINT"# LAPLACE # PRELIMINARY ORBITAL ELEMENTS"
1020 PRINT"# BY THE METHOD OF LAPLACE"
1030 REM
1040 DEFDBL A-Z
1050 DEFINT I,K,N
1060 REM
1070 DEF FNVS(X,Y,Z)=X*X+Y*Y+Z*Z
1080 DEF FNMG(X,Y,Z)=SQR(X*X+Y*Y+Z*Z)
1090 DEF FNDP(X1,Y1,Z1,X2,Y2,Z2)=X1*X2+Y1*Y2+Z1*Z2
1100 DEF FNDX(X2,Y2,Z2,X3,Y3,Z3)=LL(1)*(Y2*Z3-Z2*Y3)-
     LL(2)*(X2*Z3-Z2*X3)+LL(3)*(X2*Y3-Y2*X3)
1110 DEF FNF(X)=C+X*X*X*(B+X*X*X*(A+X*X))
1120 DEF FNDF(X)=X*X*(3*B+X*X*X*(6*A+8*X*X))
1130 DEF FNP(X)=AA+M*BB/(X*X*X)
1140 DEF FNU(X)=CC+M*DD/(X*X*X)
1150 REM
1160 M=1
1170 Q1=.0174532925#
1180 AB=1/173.1446#
1190 G$="####.#######"
1200 S$="########.#####"
1210 REM
1220 READ N$, E$, K#
1230 READ TA
1240 READ A, D
1250 READ A1,D1
1260 READ A2,D2
1270 READ RR(1),RR(2),RR(3)
1280 READ R1(1),R1(2),R1(3)
1290 READ R2(1),R2(2),R2(3)
1300 REM
1310 IF K#>.07 THEN AB=0
1320 REM
1330 LINE INPUT"";L$
1340 CLS
1350 PRINT"PRELIMINARY ORBIT METHOD OF LAPLACE"
1360 PRINT N$
1370 PRINT E$
1380 PRINT
```

9.7. PROGRAM LAPLACE

```
1390 PRINT"TA";TAB(9);:PRINT USING G$;TA
1400 PRINT
1410 PRINT"A";TAB(9);:PRINT USING G$;A
1420 PRINT"A'";TAB(9);:PRINT USING G$;A1
1430 PRINT"A''";TAB(9);:PRINT USING G$;A2
1440 PRINT
1450 PRINT"D";TAB(9);:PRINT USING G$;D
1460 PRINT"D'";TAB(9);:PRINT USING G$;D1
1470 PRINT"D''";TAB(9);:PRINT USING G$;D2
1480 REM
1490 A=A*15*Q1
1500 A1=A1*15*Q1
1510 A2=A2*15*Q1
1520 REM
1530 D=D*Q1
1540 D1=D1*Q1
1550 D2-D2*Q1
1560 REM
1570 LL(1)=COS(D)*COS(A)
1580 LL(2)=COS(D)*SIN(A)
1590 LL(3)=SIN(D)
1600 REM
1610 XX=SIN(D)*COS(A)
1620 YY=SIN(D)*SIN(A)
1630 ZZ=COS(D)
1640 REM
1650 L1(1)=-XX*D1-LL(2)*A1
1660 L1(2)=-YY*D1+LL(1)*A1
1670 L1(3)=+ZZ*D1
1680 REM
1690 X1=+LL(1)*D1-YY*A1
1700 Y1=+LL(2)*D1+XX*A1
1710 Z1=-LL(3)*D1
1720 REM
1730 L2(1)=-X1*D1-XX*D2-L1(2)*A1-LL(2)*A2
1740 L2(2)=-Y1*D1-YY*D2+L1(1)*A1+LL(1)*A2
1750 L2(3)=+Z1*D1+ZZ*D2
1760 REM
1770 LINE INPUT"";L$
1780 CLS
1790 PRINT"LL(K)";TAB(9);
1800 PRINT USING G$;LL(1),LL(2),LL(3)
```

```
1810 REM
1820 PRINT"LL(K)'";TAB(9);
1830 PRINT USING G$;L1(1),L1(2),L1(3)
1840 REM
1850 PRINT"LL(K)''";TAB(9);
1860 PRINT USING G$;L2(1),L2(2),L2(3)
1870 PRINT
1880 PRINT"RR(K)";TAB(9);
1890 PRINT USING G$;RR(1),RR(2),RR(3)
1900 REM
1910 PRINT"RR(K)'";TAB(9);
1920 PRINT USING G$;R1(1),R1(2),R1(3)
1930 REM
1940 PRINT"RR(K)''";TAB(9);
1950 PRINT USING G$;R2(1),R2(2),R2(3)
1960 REM
1970 LINE INPUT"";L$
1980 CLS
1990 PRINT"*** NUMERICAL COEFFICIENTS ***"
2000 PRINT
2010 D0=FNDX(L1(1),L1(2),L1(3),L2(1),L2(2),L2(3))
2020 AA=FNDX(L1(1),L1(2),L1(3),R2(1),R2(2),R2(3))/D0
2030 BB=FNDX(L1(1),L1(2),L1(3),RR(1),RR(2),RR(3))/D0
2040 CC=FNDX(R2(1),R2(2),R2(3),L2(1),L2(2),L2(3))/(2*D0)
2050 DD=FNDX(RR(1),RR(2),RR(3),L2(1),L2(2),L2(3))/(2*D0)
2060 EE=FNDP(LL(1),LL(2),LL(3),RR(1),RR(2),RR(3))*(-2)
2070 FF=FNVS(RR(1),RR(2),RR(3))
2080 REM
2090 A=-(AA*AA+AA*EE+FF)
2100 B=-(M)*(2*AA*BB+BB*EE)
2110 C=-(M*M)*(BB*BB)
2120 REM
2130 PRINT"D0";TAB(9);:PRINT USING G$;D0
2140 PRINT
2150 PRINT"AA";TAB(9);:PRINT USING G$;AA
2160 PRINT"BB";TAB(9);:PRINT USING G$;BB
2170 PRINT"CC";TAB(9);:PRINT USING G$;CC
2180 PRINT"DD";TAB(9);:PRINT USING G$;DD
2190 PRINT"EE";TAB(9);:PRINT USING G$;EE
2200 PRINT"FF";TAB(9);:PRINT USING G$;FF
2210 PRINT
2220 PRINT"A";TAB(9);:PRINT USING G$;A
```

9.7. PROGRAM LAPLACE

```
2230 PRINT"B";TAB(9);:PRINT USING G$;B
2240 PRINT"C";TAB(9);:PRINT USING G$;C
2250 PRINT
2260 INPUT">>>LIMITS: RMIN, RMAX";XL,XH
2270 CLS
2280 PRINT"*** SOLUTION OF THE ORBIT EQUATION ***"
2290 PRINT
2300 PRINT TAB(10)"R";TAB(24)"P";TAB(38)"F(R)"
2310 PRINT
2320 DX=(XH-XL)/10
2330 FOR N=0 TO 10
2340   X=XL+N*DX
2350   IF X=0 THEN X=.1
2360   REM
2370   PRINT USING S$;X,FNP(X),FNF(X)
2380 NEXT N
2390 PRINT
2400 INPUT">>>APPROXIMATE R";X
2410 REM
2420 IF ABS(FNF(X))<.0000001# THEN 2460
2430   X=X-FNF(X)/FNDF(X)
2440 GOTO 2420
2450 REM
2460 R=X
2470 P=FNP(X)
2480 U=FNU(X)
2490 CLS
2500 PRINT"*** PRELIMINARY ORBITAL ELEMENTS ***
2510 PRINT
2520 FOR K=1 TO 3
2530   R(K)=P*LL(K)-RR(K)
2540   V(K)=U*LL(K)+P*L1(K)-R1(K)
2550 NEXT K
2560 REM
2570 T=TA-AB*P
2580 REM
2590 PRINT"T";TAB(9);:PRINT USING G$;T
2600 PRINT
2610 PRINT"P";TAB(9);:PRINT USING G$;P
2620 PRINT"R";TAB(9);:PRINT USING G$;R
2630 PRINT
2640 PRINT"R(K)";TAB(9);
```

```
2650 PRINT USING G$;R(1),R(2),R(3)
2660 PRINT"V(K)";TAB(9);
2670 PRINT USING G$;V(1),V(2),V(3)
2680 PRINT
2690 PRINT N$
2700 PRINT"METHOD OF LAPLACE"
2710 END
30000 DATA "GEOS", "J2000.0", 0.07436680#
30010 DATA 99.08167#
30020 DATA +4.0355860#,  +0.3518241#
30030 DATA +4.3647689#,-10.0716627#
30040 DATA -9.8498332#, -2.0138485#
30050 REM
30060 DATA -0.7748777#,+0.1697957#,-0.6068707#
30070 DATA -0.0099896#,-0.0456380#,-0.0000139#
30080 DATA +0.0026850#,-0.0005877#,+0.0000037#
30090 REM
40000 DATA "REBEK-JEWEL", "J2000.0", 0.017202099#
40010 DATA 6374.57284#
40020 DATA  +5.1268591#,+22.1410258#
40030 DATA  -4.7728391#, +3.6603069#
40040 DATA -17.5333166#,-21.7429581#
40050 REM
40060 DATA -0.7275905#,-0.6179560#,-0.2679668#
40070 DATA +0.6953877#,-0.6700890#,-0.2905418#
40080 DATA +0.7505604#,+0.6292407#,+0.2723781#
```

9.8 Numerical Examples

9.8.1 The Orbit of Satellite GEOS

Problem

The topocentric data listed below for satellite GEOS were taken from the numerical examples of Sections 8.6.1 and 8.6.3. Compute a set of preliminary position and velocity elements by the method of Laplace.

- equinox: J2000.0
- $k = 0.07436680$
- $t = 99.08167$
- $\alpha = 4.0355860$
- $\dot{\alpha} = 4.3647689$
- $\ddot{\alpha} = -9.8498332$
- $\delta = 0.3518241$
- $\dot{\delta} = -10.0716627$
- $\ddot{\delta} = -2.0138485$
- $\mathbf{R} = \{-0.7748777, +0.1697957, -0.6068707\}$
- $\dot{\mathbf{R}} = \{-0.0099896, -0.0456380, -0.0000139\}$
- $\ddot{\mathbf{R}} = \{+0.0026850, -0.0005877, +0.0000037\}$

Solution

Use the given data to write lines 30000 to 30080 as shown at the end of program LAPLACE. Run the program and answer the prompts as follows:

```
>>>LIMITS: RMIN, RMAX? 1,2

>>>APPROXIMATE R? 1.3
```

Results

```
PRELIMINARY ORBIT METHOD OF LAPLACE
GEOS
J2000.0

TA          99.0816700

A            4.0355860
A'           4.3647689
A''         -9.8498332

D            0.3518241
D'         -10.0716627
D''         -2.0138485

LL(K)        0.4919009    0.8706295    0.0061405
LL(K)'      -0.9943320    0.5630319   -0.1757804
LL(K)''      1.5855348   -2.4307800   -0.0353374

RR(K)       -0.7748777    0.1697957   -0.6068707
RR(K)'      -0.0099896   -0.0456380   -0.0000139
RR(K)''      0.0026850   -0.0005877    0.0000037

*** NUMERICAL COEFFICIENTS ***

D0          -0.4838491

AA           0.0009573
BB           1.1543485
CC          -0.0000703
DD           1.6329968
EE           0.4741207
FF           0.9975581

A           -0.9980129
B           -0.5495107
C           -1.3325204

>>>LIMITS: RMIN, RMAX? 1,2
```

9.8. NUMERICAL EXAMPLES

*** SOLUTION OF THE ORBIT EQUATION ***

R	P	F(R)
1.00000	1.15531	-1.88004
1.10000	0.86824	-1.68837
1.20000	0.66898	-0.96231
1.30000	0.52638	0.80029
1.40000	0.42164	4.40294
1.50000	0.34299	11.07380
1.60000	0.28278	22.62248
1.70000	0.23592	41.63570
1.80000	0.19889	71.71770
1.90000	0.16925	117.78162
2.00000	0.14525	186.39857

\>\>\>APPROXIMATE R? 1.3

*** PRELIMINARY ORBITAL ELEMENTS ***

T	99.0816700		
P	0.5728052		
R	1.2638207		
R(K)	1.0566411	0.3289055	0.6103880
V(K)	-0.1616737	1.0723920	-0.0957070

GEOS
METHOD OF LAPLACE

Discussion of Results

When the preliminary position and velocity elements computed by LAPLACE are used in program CLASSEL, the following set of classical elements is obtained for satellite GEOS:

- epoch: 99.08167
- $a = 2.5177492$
- $e = 0.5040773$
- $M = 4.42163°$
- $i = 30.04136°$
- $\Omega = 269.79587°$
- $\omega = 89.85937°$

9.8. NUMERICAL EXAMPLES

9.8.2 The Orbit of Comet Rebek-Jewel

Problem

The topocentric data listed below for comet Rebek-Jewel were taken from the numerical examples of Sections 8.6.2 and 8.6.4. Compute a set of preliminary position and velocity elements by the method of Laplace.

- equinox: J2000.0
- $k = 0.017202099$
- $t = 2446374.57284$
- $\alpha = 5.1268591$
- $\dot{\alpha} = -4.7728391$
- $\ddot{\alpha} = -17.5333166$
- $\delta = 22.1410258$
- $\dot{\delta} = 3.6603069$
- $\ddot{\delta} = -21.7429581$
- $\mathbf{R} = \{-0.7275905, -0.6179560, -0.2679668\}$
- $\dot{\mathbf{R}} = \{+0.6953877, -0.6700890, \ 0.2905418\}$
- $\ddot{\mathbf{R}} = \{+0.7505604, +0.6292407, +0.2723781\}$

Solution

Use the given data to write lines 40000 to 40080 as shown at the end of program LAPLACE. Run the program and answer the prompts as follows:

```
>>>LIMITS: RMIN, RMAX? 1,3

>>>APPROXIMATE R? 1.8
```

Results

```
PRELIMINARY ORBIT METHOD OF LAPLACE
REBEK-JEWEL
J2000.0

TA         6374.5728400

A             5.1268591
A'           -4.7728391
A''         -17.5333166

D            22.1410258
D'            3.6603069
D''         -21.7429581

LL(K)         0.2098924    0.9021646    0.3768876
LL(K)'        1.1218225   -0.2857170    0.0591735
LL(K)''       3.7863657   -2.2227584   -0.3530407

RR(K)        -0.7275905   -0.6179560   -0.2679668
RR(K)'        0.6953877   -0.6700890   -0.2905418
RR(K)''       0.7505604    0.6292407    0.2723781

*** NUMERICAL COEFFICIENTS ***

D0            0.0761557

AA            1.1439649
BB           -1.0966718
CC           -1.8172614
DD            1.7401839
EE            1.6224141
FF            0.9830638

A            -4.1477045
B             4.2883639
C            -1.2026889

>>>LIMITS: RMIN, RMAX? 1,3
```

9.8. NUMERICAL EXAMPLES

*** SOLUTION OF THE ORBIT EQUATION ***

R	P	F(R)
1.00000	0.04729	-0.06203
1.20000	0.50932	-1.87756
1.40000	0.74430	-5.90782
1.60000	0.87622	-10.27481
1.80000	0.95592	-7.06600
2.00000	1.00688	23.65114
2.20000	1.04097	122.95221
2.40000	1.06463	366.19413
2.60000	1.08157	881.14890
2.80000	1.09401	1872.21689
3.00000	1.10335	3651.90658

>>>APPROXIMATE R? 1.8

*** PRELIMINARY ORBITAL ELEMENTS ***

T   6374.5671889

P   0.9784659
R   1.8782801

R(K)   0.9329630   1.5006932   0.6367384
V(K)   0.0759682   -1.0120252   -0.2374871

REBEK-JEWEL
METHOD OF LAPLACE

Discussion of Results

When the preliminary position and velocity elements computed by LAPLACE are used in program CLASSEL, the following classical elements are obtained for comet Rebek-Jewel:

- epoch: 2446374.56719
- $a = -46.3771628$
- $e = 1.0136589$
- $TT = 2446469.80568$
- $i = 162.68722°$
- $\Omega = 58.96917°$
- $\omega = 107.14842°$

The computed orbit is a hyperbola.

References

[1] Herget, *The Computation of Orbits*, Published by Author, 1948.

[2] Escobal, *Methods of Orbit Determination*, Krieger Publishing Co., 1976.

[3] Baker, *Astrodynamics: Applications and Advanced Topics*, Academic Press, 1967.

[4] Taff, *Celestial Mechanics*, John Wiley and Sons, 1985.

[5] Dubyago, *The Determination of Orbits*, The Rand Corporation, 1961.

Chapter 10

The Method of Gauss

10.1 Introduction

The method of Gauss is more complex than that of Laplace; however, it has features which permit the solution to be brought to a higher state of refinement. The excellent utility of this method has encouraged successive researchers to develop increasingly more elegant and powerful variations of the original procedure [1,2]. The approach described in this chapter is one of the simpler versions of the Gaussian method.

Orbit determination by the method of Gauss requires exactly three sets of position data. These need not be separated by equal intervals of time, although it is desirable, and the inclusion of parallax can be beneficial rather than problematic as is sometimes the case for Laplace's procedure. The Gaussian method is applicable to orbits of any form, but difficulties may arise when it is used to determine very eccentric orbits because the influence of the radial velocity is neglected in the first approximation. Also, if the arc of the celestial body's apparent path is very short or has too little curvature, the coefficient D_0 in Equations 10.10, 10.11, and 10.12 will approach zero, and the method of Gauss may fail [1 through 8].

10.2 Solution by f and g Expressions

Assume that from the available data three sets have been chosen and reduced to the following:
$$\{t_i, \mathbf{L}_i, \mathbf{R}_i\},$$
where $i = 1, 2, 3$. If we let t_2 be the epoch time, then the observed times can be converted to the modified time intervals
$$\tau_1 = k(t_1 - t_2)$$

$$\tau_3 = k(t_3 - t_2).$$

Now, the closed or universal f and g expressions which were developed for the solution of the two-body equation of motion,

$$\ddot{\mathbf{r}} = -\frac{\mu \mathbf{r}}{r^3},$$

may be employed to describe the dynamic constraints at τ_1 and τ_3 as follows:

$$\mathbf{r}_1 = f_1 \mathbf{r}_2 + g_1 \dot{\mathbf{r}}_2 \tag{10.1}$$
$$\mathbf{r}_3 = f_3 \mathbf{r}_2 + g_3 \dot{\mathbf{r}}_2. \tag{10.2}$$

Furthermore, the geometric constraint of the fundamental vector triangle must hold at all three times. Thus,

$$\mathbf{r}_i = p_i \mathbf{L}_i - \mathbf{R}_i, \tag{10.3}$$

for $i = 1, 2, 3$. Multiplying Equation 10.1 by g_3 and Equation 10.2 by g_1 and subtracting, we eliminate $\dot{\mathbf{r}}_2$ to obtain

$$g_3 \mathbf{r}_1 - g_1 \mathbf{r}_3 = (f_1 g_3 - f_3 g_1) \mathbf{r}_2.$$

Dividing each term by the coefficient of $\mathbf{r}_2$ and rearranging we can write

$$c_1 \mathbf{r}_1 + c_2 \mathbf{r}_2 + c_3 \mathbf{r}_3 = \mathbf{0}, \tag{10.4}$$

where

$$c_1 = +\frac{g_3}{f_1 g_3 - f_3 g_1}$$
$$c_2 = -1$$
$$c_3 = -\frac{g_1}{f_1 g_3 - f_3 g_1}. \tag{10.5}$$

Equation 10.4 reflects the fact that two-body motion requires the three radius vectors to lie in the same plane. Substituting Equation 10.3 for each of the $\mathbf{r}_i$ in Equation 10.4 and collecting terms, the result is

$$c_1 p_1 \mathbf{L}_1 + c_2 p_2 \mathbf{L}_2 + c_3 p_3 \mathbf{L}_3 = c_1 \mathbf{R}_1 + c_2 \mathbf{R}_2 + c_3 \mathbf{R}_3. \tag{10.6}$$

The solution of this equation for the unknown p_i is the key to the method of Gauss [1,2].

10.3 The Scalar Equations for the Ranges

We solve Equation 10.6 for the three p_i by taking appropriate cross and dot products with the vectors $\mathbf{L}_i$ in a fashion similar to the process used for the method of Laplace in Section 9.3. When this is accomplished, the result is

$$p_1 = \frac{c_1(\mathbf{R}_1 \times \mathbf{L}_2)\cdot\mathbf{L}_3 + c_2(\mathbf{R}_2 \times \mathbf{L}_2)\cdot\mathbf{L}_3 + c_3(\mathbf{R}_3 \times \mathbf{L}_2)\cdot\mathbf{L}_3}{c_1(\mathbf{L}_1 \times \mathbf{L}_2)\cdot\mathbf{L}_3} \quad (10.7)$$

$$p_2 = \frac{c_1(\mathbf{L}_1 \times \mathbf{R}_1)\cdot\mathbf{L}_3 + c_2(\mathbf{L}_1 \times \mathbf{R}_2)\cdot\mathbf{L}_3 + c_3(\mathbf{L}_1 \times \mathbf{R}_3)\cdot\mathbf{L}_3}{c_2(\mathbf{L}_1 \times \mathbf{L}_2)\cdot\mathbf{L}_3} \quad (10.8)$$

$$p_3 = \frac{c_1\mathbf{L}_1\cdot(\mathbf{L}_2 \times \mathbf{R}_1) + c_2\mathbf{L}_1\cdot(\mathbf{L}_2 \times \mathbf{R}_2) + c_3\mathbf{L}_1\cdot(\mathbf{L}_2 \times \mathbf{R}_3)}{c_3\mathbf{L}_1\cdot(\mathbf{L}_2 \times \mathbf{L}_3)}. \quad (10.9)$$

For convenience, we shall write these equations in the following simpler form:

$$p_1 = \frac{c_1 D_{11} + c_2 D_{12} + c_3 D_{13}}{c_1 D_0} \quad (10.10)$$

$$p_2 = \frac{c_1 D_{21} + c_2 D_{22} + c_3 D_{23}}{c_2 D_0} \quad (10.11)$$

$$p_3 = \frac{c_1 D_{31} + c_2 D_{32} + c_3 D_{33}}{c_3 D_0}, \quad (10.12)$$

where, for $j = 1, 2, 3$,

$$\begin{aligned} D_{1j} &= (\mathbf{R}_j \times \mathbf{L}_2)\cdot\mathbf{L}_3 \\ D_{2j} &= (\mathbf{L}_1 \times \mathbf{R}_j)\cdot\mathbf{L}_3 \\ D_{3j} &= \mathbf{L}_1\cdot(\mathbf{L}_2 \times \mathbf{R}_j), \end{aligned} \quad (10.13)$$

and

$$D_0 = (\mathbf{L}_1 \times \mathbf{L}_2)\cdot\mathbf{L}_3 = \mathbf{L}_1\cdot(\mathbf{L}_2 \times \mathbf{L}_3). \quad (10.14)$$

10.4 The First Approximation

Equations 10.10, 10.11, and 10.12 were developed using no assumptions beyond that of two-body motion. Unfortunately, these equations cannot be used to compute the p_i until the quantities c_1 and c_3 are known. These coefficients were defined in Equations 10.5 as follows:

$$c_1 = +\frac{g_3}{f_1 g_3 - f_3 g_1}$$

$$c_2 = -1$$

$$c_3 = -\frac{g_1}{f_1 g_3 - f_3 g_1}.$$

Thus, there is a problem because c_1 and c_3 are expressed in terms of the f_i and g_i, which, in turn, require a knowledge of the vector orbital elements. The solution to this dilemma is to assume approximate values for the f and g expressions which can be improved once initial values for the vector elements have been computed.

Consider the first few terms of the f and g series developed in Section 3.5.1. For the two instants τ_1 and τ_3, we can write

$$f_i = 1 - \frac{1}{2} u_2 \tau_i^2 + \frac{1}{2} u_2 z_2 \tau_i^3 + \cdots$$

$$g_i = \tau_i - \frac{1}{6} u_2 \tau_i^3 + \frac{1}{4} u_2 z_2 \tau_i^4 + \cdots, \qquad (10.15)$$

where $i = 1, 3$, and

$$u_2 = \frac{\mu}{r_2^3}$$

$$z_2 = \frac{\mathbf{r}_2 \cdot \dot{\mathbf{r}}_2}{r_2^2}.$$

If the f_i and g_i series are truncated to eliminate all terms beyond the second, the resulting approximations do not require the vector elements. Thus, we make the following simplifying assumptions:

$$f_i \approx 1 - \frac{1}{2} u_2 \tau_i^2$$

$$g_i \approx \tau_i - \frac{1}{6} u_2 \tau_i^3, \qquad (10.16)$$

where the contributions of the vectors have been ignored so the only unknown is the scalar u_2. Using these approximations, it is possible to form

$$f_1 g_3 - f_3 g_1 \approx \tau - \frac{u_2}{6} \tau^3, \qquad (10.17)$$

where all orders above τ_i^3 have been neglected and $\tau_3 - \tau_1$ has been replaced by τ. Therefore, substituting Equations 10.16 and 10.17 into the expressions for c_1 and c_3, we obtain

$$c_1 \approx \frac{\tau_3 - u_2 \tau_3^3/6}{\tau - u_2 \tau^3/6}$$

$$c_3 \approx \frac{-\tau_1 + u_2 \tau_1^3/6}{\tau - u_2 \tau^3/6}.$$

10.5. THE SCALAR EQUATIONS RELATING P AND R AT EPOCH

Carrying out the division and neglecting all terms with order greater than three, the result is

$$c_1 \approx +\frac{\tau_3}{\tau} + \frac{u_2\tau_3}{6\tau}(\tau^2 - \tau_3^2)$$

$$c_3 \approx -\frac{\tau_1}{\tau} - \frac{u_2\tau_1}{6\tau}(\tau^2 - \tau_1^2). \qquad (10.18)$$

For convenience, we rewrite Equations 10.18 as follows:

$$c_1 \approx A_1 + \frac{\mu B_1}{r_2^3}$$

$$c_3 \approx A_3 + \frac{\mu B_3}{r_2^3}, \qquad (10.19)$$

where u_2 has been replaced by μ/r_2^3 and

$$A_1 = +\frac{\tau_3}{\tau}$$

$$B_1 = +\frac{1}{6}A_1(\tau^2 - \tau_3^2)$$

$$A_3 = -\frac{\tau_1}{\tau}$$

$$B_3 = +\frac{1}{6}A_3(\tau^2 - \tau_1^2). \qquad (10.20)$$

10.5 The Scalar Equations Relating p and r at Epoch

Equation 10.11 can now be solved for the range p_2 by letting $c_2 = -1$ and replacing c_1 and c_3 by Equations 10.19. Making these substitutions and collecting terms, we can obtain the following expression:

$$p_2 = A + \frac{\mu B}{r_2^3}, \qquad (10.21)$$

where

$$A = -\frac{A_1 D_{21} - D_{22} + A_3 D_{23}}{D_0}$$

$$B = -\frac{B_1 D_{21} + B_3 D_{23}}{D_0}. \qquad (10.22)$$

A second equation containing the unknowns p_2 and r_2 can be derived by taking the dot product of Equation 10.3 with itself when $i = 2$. Thus,

$$\mathbf{r}_2 \cdot \mathbf{r}_2 = (p_2 \mathbf{L}_2 - \mathbf{R}_2) \cdot (p_2 \mathbf{L}_2 - \mathbf{R}_2).$$

Making use of the fact that $L_2^2 = 1$, since $\mathbf{L}_2$ is a unit vector, the above equation can be reduced to

$$r_2^2 = p_2^2 + p_2 E + F, \qquad (10.23)$$

where

$$\begin{aligned} E &= -2(\mathbf{L}_2 \cdot \mathbf{R}_2) \\ F &= R_2^2. \end{aligned} \qquad (10.24)$$

We now have two independent scalar equations relating p_2 and r_2.

10.6 The Scalar Equation of Lagrange

Substituting Equation 10.21 for the range p_2 in Equation 10.23, we obtain the *equation of Lagrange*, that is

$$r_2^8 + a r_2^6 + b r_2^3 + c = 0, \qquad (10.25)$$

where

$$\begin{aligned} a &= -(A^2 + AE + F) \\ b &= -\mu(2AB + BE) \\ c &= -\mu^2 B^2. \end{aligned} \qquad (10.26)$$

Note that although Equations 10.26 are similar in form to Equations 9.20, the quantities A and B are defined differently.

If we let the symbol x represent the unknown value of r_2 which will satisfy Equation 10.25, we can write

$$f(x) = x^8 + ax^6 + bx^3 + c.$$

Differentiating $f(x)$ with respect to x produces

$$f'(x) = 8x^7 + 6ax^5 + 3bx^2.$$

When the above equations are solved for x by the Newton-Raphson method described in Section 7.2.2, the result will be the value of r_2 needed to complete Gauss' method.

As was the case for the method of Laplace, the equation of Lagrange can have as many as three positive roots. When this occurs, it is usually possible to rule out one or two roots on the grounds that they are unreasonably large or small, or that one or both correspond to values of p_2 which are negative. In spite of this, there will be situations where two roots appear realistic. In those cases, one must compute two sets of orbital elements and ephemerides. The best solution must then be determined on the basis of how well the computed motion compares with the observed motion of the celestial body [4].

10.7 The Vector Orbital Elements

10.7.1 Initial Position Vector

The f_i and g_i can now be computed by Equations 10.16 using the value of r_2 found from Equation 10.25. Thus,

$$f_i \approx 1 - \frac{1}{2} u_2 \tau_i^2$$
$$g_i \approx \tau_i - \frac{1}{6} u_2 \tau_i^3, \qquad (10.27)$$

where $i = 1, 3$, and

$$u_2 = \frac{\mu}{r_2^3}. \qquad (10.28)$$

Employing these values of f_i and g_i, the quantities c_1, c_2, and c_3 are obtained from Equations 10.5:

$$c_1 = +\frac{g_3}{f_1 g_3 - f_3 g_1}$$
$$c_2 = -1$$
$$c_3 = -\frac{g_1}{f_1 g_3 - f_3 g_1}. \qquad (10.29)$$

Finally, substituting c_1, c_2, and c_3 into

$$p_1 = \frac{c_1 D_{11} + c_2 D_{12} + c_3 D_{13}}{c_1 D_0}$$
$$p_2 = \frac{c_1 D_{21} + c_2 D_{22} + c_3 D_{23}}{c_2 D_0}$$
$$p_3 = \frac{c_1 D_{31} + c_2 D_{32} + c_3 D_{33}}{c_3 D_0} \qquad (10.30)$$

produces the three p_i which can be used in the general geometric constraint,

$$\mathbf{r}_i = p_i \mathbf{L}_i - \mathbf{R}_i, \qquad (10.31)$$

to determine the vector element $\mathbf{r}_2$ along with the other two radius vectors.

10.7.2 Initial Velocity Vector

We now have more than enough information to determine the vector element $\dot{\mathbf{r}}_2$. Consider again the dynamic constraints at τ_1 and τ_3:

$$\mathbf{r}_1 = f_1 \mathbf{r}_2 + g_1 \dot{\mathbf{r}}_2$$
$$\mathbf{r}_3 = f_3 \mathbf{r}_2 + g_3 \dot{\mathbf{r}}_2.$$

Multiplying the first equation by f_3 and the second equation by f_1, we subtract and eliminate $\mathbf{r}_2$ to obtain

$$f_3 \mathbf{r}_1 - f_1 \mathbf{r}_3 = -(f_1 g_3 - f_3 g_1) \dot{\mathbf{r}}_2 .$$

Dividing each term by the coefficient of $\dot{\mathbf{r}}_2$ and rearranging, we can write

$$\dot{\mathbf{r}}_2 = d_1 \mathbf{r}_1 + d_3 \mathbf{r}_3 , \tag{10.32}$$

where

$$d_1 = -\frac{f_3}{f_1 g_3 - f_3 g_1}$$

$$d_3 = +\frac{f_1}{f_1 g_3 - f_3 g_1} . \tag{10.33}$$

Therefore, since approximate values for the f_i and g_i are known, d_1 and d_3 can be calculated, and Equation 10.32 yields the velocity $\dot{\mathbf{r}}_2$.

In the case of heliocentric orbits, the correction for light-time should be introduced before going on to refine the initial approximations of $\mathbf{r}_2$ and $\dot{\mathbf{r}}_2$. Thus, for all three observation times t_i, the corrected values are

$$t_{ci} = t_i - \frac{\rho_i}{c} , \tag{10.34}$$

where

$$c = 173.1446 \text{ au/day} .$$

Consequently, the modified time intervals should also be computed anew from

$$\tau_1 = k(t_{c1} - t_{c2})$$
$$\tau_3 = k(t_{c3} - t_{c2})$$
$$\tau = \tau_3 - \tau_1 . \tag{10.35}$$

10.7.3 Refinement of the Elements

The vector orbital elements may be refined by using the initial values of $\mathbf{r}_2$ and $\dot{\mathbf{r}}_2$ to recompute the f_i and g_i from their universal formulations. When this has been accomplished, improved values of c_i and d_i can be determined. The results permit better values of the ρ_i to be found from

$$\rho_1 = \frac{c_1 D_{11} + c_2 D_{12} + c_3 D_{13}}{c_1 D_0}$$

$$\rho_2 = \frac{c_1 D_{21} + c_2 D_{22} + c_3 D_{23}}{c_2 D_0}$$

$$\rho_3 = \frac{c_1 D_{31} + c_2 D_{32} + c_3 D_{33}}{c_3 D_0} ,$$

10.7. THE VECTOR ORBITAL ELEMENTS

which may then be used with the geometric constraint,

$$\mathbf{r}_i = p_i \mathbf{L}_i - \mathbf{R}_i,$$

to improve the element $\mathbf{r}_2$ along with the other two position vectors. Finally, the new d_i and $\mathbf{r}_i$ are used in

$$\dot{\mathbf{r}}_2 = d_1 \mathbf{r}_1 + d_3 \mathbf{r}_3$$

to find an improved $\dot{\mathbf{r}}_2$. The whole process is repeated until the magnitudes of the p_i stop changing and converge to stable values. When this occurs, we have the final vector element set, and the preliminary orbit is determined.

The orbital elements derived from the method of Gauss will represent all three observations satisfactorily because the corresponding $\mathbf{L}_i$ were used to determine the $\mathbf{r}_i$ when the geometric constraint was applied. Therefore, the accuracy of the elements must be tested by comparing computed positions with observed positions not used in the solution.

10.8 Program GAUSS

Program GAUSS computes the position and velocity elements of a preliminary orbit by the method of Gauss. The f and g expressions are based on the universal formulation described in Section 5.4. GAUSS uses angular position data from observations and vector position data obtained from program GEO or HELO.

Program Algorithm

Define	Line Number
FNVS(X,Y,Z): vector squaring function	1070
FNMG(X,Y,Z): vector magnitude function	1080
FNDP(X1,Y1,Z1,X2,Y2,Z2): dot product function	1090
FNX(A1,A2,B1,B2): cross product function	1100
FNF(X): Lagrange equation function	1110
FNDF(X): derivative of Lagrange equation function	1120
FNP(X): range equation function	1130

Given	Line Number
μ: combined mass	1170
angle to radian conversion factor	1180
$1/c$: light-time constant	1190
name of the object for information only	1270
equinox for information only	1270
k: gravitational constant	1270
t_i: three observation times uncorrected for light-time	1290
α_i, δ_i: three angular positions	1300

10.8. PROGRAM GAUSS

R_{ix}, R_{iy}, R_{iz}: three $\mathbf{R}_i$	1310
no light-time correction for geocentric orbits	1340
Compute	Line Number
α_i in radians	1430
δ_i in radians	1440
L_{ix}: Equations 8.4	1450
L_{iy}: Equations 8.4	1460
L_{iz}: Equations 8.4	1470
$\tau_1 = k(t_1 - t_2)$	1650
$\tau_3 = k(t_3 - t_2)$	1660
$\tau = \tau_3 - \tau_1$	1670
D_0: Equation 10.14	1690-1720
D_{1j}: Equations 10.13	1750-1780
D_{2j}: Equations 10.13	1800-1830
D_{3j}: Equations 10.13	1850-1880
E: Equations 10.24	1910
F: Equations 10.24	1920
A_1: Equations 10.20	1940
B_1: Equations 10.20	1950
A_3: Equations 10.20	1960

B_3: Equations 10.20	1970
A: Equations 10.22	1990
B: Equations 10.22	2000
a: Equations 10.26	2020
b: Equations 10.26	2030
c: Equations 10.26	2040

Choose	Line Number
$r_{\min}$ and $r_{\max}$ which bound the unknown value of r	2180

This prompt requests the user to choose minimum and maximum values for r, the magnitude of the radius vector at epoch, between which the unknown solution lies. Several trials may be necessary to find appropriate values.

Compute	Line Number
a table containing a series of values for r, p, and $f(r)$	2220-2300

Choose	Line Number
x: an approximate value for the unknown r	2320

This prompt requests the user to choose an approximate value for r as a starting point for the Newton-Raphson solution of the equation of Lagrange. From the table, choose a value of r which corresponds to a positive value of p and the smallest value of $|f(r)|$ at a point where the function changes sign. If the table does not contain any such point, the program must be run again using different values for $r_{\min}$ and $r_{\max}$.

Compute	Line Number
solution by the Newton-Raphson method	2340-2360
u_2: Equation 10.28	2400

10.8. PROGRAM GAUSS

f_i: Equations 10.27	2420-2430
g_i: Equations 10.27	2440-2450
$f_1 g_3 - f_3 g_1$	2470
c_1: Equations 10.29	2490
c_2: Equations 10.29	2500
c_3: Equations 10.29	2510
d_1: Equations 10.33	2530
d_3: Equations 10.33	2540
B_n: coefficients for the universal f and g expressions	2560-2590
p_i: Equations 10.30	2660
$\mathbf{r}_i$: Equation 10.31	2680
r_i: magnitudes of the three radius vectors	2700
$\dot{\mathbf{r}}_2$: Equation 10.32	2740
Δp: difference between present and previous values of the p_i	2770-2790
t_i: observation times corrected for light-time by Equation 10.34	2840
τ_1 corrected for light-time by Equations 10.35	2870
τ_3 corrected for light-time by Equations 10.35	2880
τ corrected for light-time by Equations 10.35	2890

Choose	Line Number
refine solution: yes or no	3090

CHAPTER 10. THE METHOD OF GAUSS

The first approximation can be successively refined by repeating the computational steps between lines 2620 and 3500, using the results of a previous calculation to begin the next. This process is continued until the values of the p_i no longer change significantly. This point is reached when the three values of Δp equal zero or cannot be further reduced.

number of refinement loops completed	3150
store present values of the p_i in three dummy variables	3170-3190
store present values of the f_i in two dummy variables	3210-3220
store present values of the g_i in two dummy variables	3230-3240
f and g for each τ_i by universal variables	3260-3340
average of the present and previous values of f_i	3360-3370
average of the present and previous values of g_i	3380-3390

Averages are used for the new values of f_i and g_i because, in practice, those taken directly form lines 3320 and 3330 may cause the solution to oscillate rather than converge smoothly.

$f_1 g_3 - f_3 g_1$	3410
c_1: Equations 10.29	3430
c_3: Equations 10.29	3440
d_1: Equations 10.33	3460
d_3: Equations 10.33	3470
refined values of the elements	3500

End.

10.8. PROGRAM GAUSS

Program Listing

```
1000 CLS
1010 PRINT"# GAUSS # PRELIMINARY ORBITAL ELEMENTS"
1020 PRINT"# BY THE METHOD OF GAUSS"
1030 REM
1040 DEFDBL A-Z
1050 DEFINT I,J,K,N
1060 REM
1070 DEF FNVS(X,Y,Z)=X*X+Y*Y+Z*Z
1080 DEF FNMG(X,Y,Z)=SQR(X*X+Y*Y+Z*Z)
1090 DEF FNDP(X1,Y1,Z1,X2,Y2,Z2)=X1*X2+Y1*Y2+Z1*Z2
1100 DEF FNX(A1,A2,B1,B2)=A1*B2-A2*B1
1110 DEF FNF(X)=C+X*X*X*(B+X*X*X*(A+X*X))
1120 DEF FNDF(X)=X*X*(3*B+X*X*X*(6*A+8*X*X))
1130 DEF FNP(X)=AA+M*BB/(X*X*X)
1140 REM
1150 DIM B(19)
1160 REM
1170 M=1
1180 Q1=.0174532925#
1190 AB=1/173.1446#
1200 NR=0
1210 P1=0
1220 P2=0
1230 P3=0
1240 G$="####.#######"
1250 S$="########.#####"
1260 REM
1270 READ N$, E$, K#
1280 FOR I=1 TO 3
1290    READ TA(I)
1300    READ A(I), D(I)
1310    READ RR(I,1),RR(I,2),RR(I,3)
1320 NEXT I
1330 REM
1340 IF K#>.07 THEN AB=0
1350 REM
1360 LINE INPUT"";L$
1370 CLS
1380 PRINT"PRELIMINARY ORBIT METHOD OF GAUSS"
1390 PRINT N$
```

```
1400 PRINT E$
1410 PRINT
1420 FOR I=1 TO 3
1430   A=A(I)*15*Q1
1440   D=D(I)*Q1
1450   LL(I,1)=COS(D)*COS(A)
1460   LL(I,2)=COS(D)*SIN(A)
1470   LL(I,3)=SIN(D)
1480 NEXT I
1490 REM
1500 FOR I=1 TO 3
1510   PRINT"TA(I)";TAB(9);:PRINT USING G$;TA(I)
1520   PRINT"A(I)";TAB(9);:PRINT USING G$;A(I)
1530   PRINT"D(I)";TAB(9);:PRINT USING G$;D(I)
1540   PRINT"LL(I,K)";TAB(9);
1550   PRINT USING G$;LL(I,1),LL(I,2),LL(I,3)
1560   PRINT"RR(I,K)";TAB(9);
1570   PRINT USING G$;RR(I,1),RR(I,2),RR(I,3)
1580   REM
1590   LINE INPUT"";L$
1600   CLS
1610 NEXT I
1620 REM
1630 PRINT"*** NUMERICAL COEFFICIENTS ***"
1640 PRINT
1650 TT(1)=K#*(TA(1)-TA(2))
1660 TT(3)=K#*(TA(3)-TA(2))
1670 TT=TT(3)-TT(1)
1680 REM
1690 E1=LL(1,1)*FNX(LL(2,2),LL(2,3),LL(3,2),LL(3,3))
1700 E2=LL(1,2)*FNX(LL(2,1),LL(2,3),LL(3,1),LL(3,3))
1710 E3=LL(1,3)*FNX(LL(2,1),LL(2,2),LL(3,1),LL(3,2))
1720 D0=E1-E2+E3
1730 REM
1740 FOR J=1 TO 3
1750   E1=RR(J,1)*FNX(LL(2,2),LL(2,3),LL(3,2),LL(3,3))
1760   E2=RR(J,2)*FNX(LL(2,1),LL(2,3),LL(3,1),LL(3,3))
1770   E3=RR(J,3)*FNX(LL(2,1),LL(2,2),LL(3,1),LL(3,2))
1780   DD(1,J)=E1-E2+E3
1790   REM
1800   E1=LL(1,1)*FNX(RR(J,2),RR(J,3),LL(3,2),LL(3,3))
1810   E2=LL(1,2)*FNX(RR(J,1),RR(J,3),LL(3,1),LL(3,3))
```

10.8. PROGRAM GAUSS

```
1820    E3=LL(1,3)*FNX(RR(J,1),RR(J,2),LL(3,1),LL(3,2))
1830    DD(2,J)=E1-E2+E3
1840    REM
1850    E1=LL(1,1)*FNX(LL(2,2),LL(2,3),RR(J,2),RR(J,3))
1860    E2=LL(1,2)*FNX(LL(2,1),LL(2,3),RR(J,1),RR(J,3))
1870    E3=LL(1,3)*FNX(LL(2,1),LL(2,2),RR(J,1),RR(J,2))
1880    DD(3,J)=E1-E2+E3
1890 NEXT J
1900 REM
1910 EE=FNDP(LL(2,1),LL(2,2),LL(2,3),
     RR(2,1),RR(2,2),RR(2,3))*(-2)
1920 FF=FNVS(RR(2,1),RR(2,2),RR(2,3))
1930 REM
1940 A1=+TT(3)/TT
1950 B1=+A1*(TT*TT-TT(3)*TT(3))/6
1960 A3=-TT(1)/TT
1970 B3=+A3*(TT*TT-TT(1)*TT(1))/6
1980 REM
1990 AA=(A1*DD(2,1)-DD(2,2)+A3*DD(2,3))/(-D0)
2000 BB=(B1*DD(2,1)+B3*DD(2,3))/(-D0)
2010 REM
2020 A=-(AA*AA+AA*EE+FF)
2030 B=-(M)*(2*AA*BB+BB*EE)
2040 C=-(M*M)*(BB*BB)
2050 REM
2060 PRINT"D0";TAB(9);:PRINT USING G$;D0
2070 PRINT
2080 PRINT"AA";TAB(9);:PRINT USING G$;AA
2090 PRINT"BB";TAB(9);:PRINT USING G$;BB
2100 PRINT
2110 PRINT"EE";TAB(9);:PRINT USING G$;EE
2120 PRINT"FF";TAB(9);:PRINT USING G$;FF
2130 PRINT
2140 PRINT"A";TAB(9);:PRINT USING G$;A
2150 PRINT"B";TAB(9);:PRINT USING G$;B
2160 PRINT"C";TAB(9);:PRINT USING G$;C
2170 PRINT
2180 INPUT">>>LIMITS: RMIN, RMAX";XL,XH
2190 CLS
2200 PRINT"*** SOLUTION OF THE ORBIT EQUATION ***"
2210 PRINT
2220 PRINT TAB(10)"R";TAB(24)"P";TAB(38)"F(R)"
```

```
2230 PRINT
2240 DX=(XH-XL)/10
2250 FOR N=0 TO 10
2260   X=XL+N*DX
2270   IF X=0 THEN X=.1
2280   REM
2290   PRINT USING S$;X,FNP(X),FNF(X)
2300 NEXT N
2310 PRINT
2320 INPUT">>>APPROXIMATE R";X
2330 REM
2340 IF ABS(FNF(X))<.0000001# THEN 2400
2350   X=X-FNF(X)/FNDF(X)
2360 GOTO 2340
2370 REM
2380 REM
2390 REM
2400 U2=M/(X*X*X)
2410 REM
2420 F(1)=1-U2*TT(1)*TT(1)/2
2430 F(3)=1-U2*TT(3)*TT(3)/2
2440 G(1)=TT(1)*(1-U2*TT(1)*TT(1)/6)
2450 G(3)=TT(3)*(1-U2*TT(3)*TT(3)/6)
2460 REM
2470 FG=F(1)*G(3)-F(3)*G(1)
2480 REM
2490 C(1)=+G(3)/FG
2500 C(2)=-1
2510 C(3)=-G(1)/FG
2520 REM
2530 D(1)=-F(3)/FG
2540 D(3)=+F(1)/FG
2550 REM
2560 B(1)=1
2570 FOR J=2 TO 19
2580   B(J)=B(J-1)/J
2590 NEXT J
2600 REM
2610 REM
2620 CLS
2630 PRINT"*** PRELIMINARY ORBITAL ELEMENTS ***
2640 PRINT
```

10.8. PROGRAM GAUSS

```
2650 FOR I=1 TO 3
2660   P(I,0)=(C(1)*DD(I,1)+C(2)*DD(I,2)+
       C(3)*DD(I,3))/(C(I)*D0)
2670   FOR K=1 TO 3
2680     R(I,K)=P(I,0)*LL(I,K)-RR(I,K)
2690   NEXT K
2700   R(I,0)=FNMG(R(I,1),R(I,2),R(I,3))
2710 NEXT I
2720 REM
2730 FOR K=1 TO 3
2740   V(2,K)=D(1)*R(1,K)+D(3)*R(3,K)
2750 NEXT K
2760 REM
2770 DP(1)=P(1,0)-P1
2780 DP(2)=P(2,0)-P2
2790 DP(3)=P(3,0)-P3
2800 REM
2810 REM
2820 REM
2830 FOR I=1 TO 3
2840   T(I)=TA(I)-AB*P(I,0)
2850 NEXT I
2860 REM
2870 TT(1)=K#*(T(1)-T(2))
2880 TT(3)=K#*(T(3)-T(2))
2890 TT=TT(3)-TT(1)
2900 REM
2910 REM
2920 REM
2930 PRINT"P(1)";TAB(9);:PRINT USING G$;P(1,0)
2940 PRINT"P(2)";TAB(9);:PRINT USING G$;P(2,0)
2950 PRINT"P(3)";TAB(9);:PRINT USING G$;P(3,0)
2960 PRINT
2970 PRINT"T(2)";TAB(9);:PRINT USING G$;T(2)
2980 REM
2990 PRINT"R(2)";TAB(9);:PRINT USING G$;R(2,0)
3000 PRINT
3010 PRINT"R(2,K)";TAB(9);
3020 PRINT USING G$;R(2,1),R(2,2),R(2,3)
3030 PRINT"V(2,K)";TAB(9);
3040 PRINT USING G$;V(2,1),V(2,2),V(2,3)
3050 PRINT
```

```
3060 PRINT"DP(";NR;")";TAB(9);
3070 PRINT USING G$;DP(1),DP(2),DP(3)
3080 REM
3090 INPUT">>>REFINE: Y/N";A$
3100 PRINT
3110 IF A$="N" THEN 3520
3120 REM
3130 REM
3140 REM
3150 NR=NR+1
3160 REM
3170 P1=P(1,0)
3180 P2=P(2,0)
3190 P3=P(3,0)
3200 REM
3210 F1=F(1)
3220 F3=F(3)
3230 G1=G(1)
3240 G3=G(3)
3250 REM
3260 FOR I=1 TO 3 STEP 2
3270    REM
3280    H=TT(I)
3290    REM
3300    GOSUB 16010 REM UFG/SUB
3310    REM
3320    F(I)=F
3330    G(I)=G
3340 NEXT I
3350 REM
3360 F(1)=(F(1)+F1)/2
3370 F(3)=(F(3)+F3)/2
3380 G(1)=(G(1)+G1)/2
3390 G(3)=(G(3)+G3)/2
3400 REM
3410 FG=F(1)*G(3)-F(3)*G(1)
3420 REM
3430 C(1)=+G(3)/FG
3440 C(3)=-G(1)/FG
3450 REM
3460 D(1)=-F(3)/FG
3470 D(3)=+F(1)/FG
```

10.8. PROGRAM GAUSS

```
3480 REM
3490 REM
3500 GOTO 2620
3510 REM
3520 PRINT N$
3530 PRINT"METHOD OF GAUSS"
3540 END
16000 STOP
16010 REM # UFG/SUB # UNIVERSAL F&G EXPRESSIONS
16020 REM
16030 RO=FNMG(R(2,1),R(2,2),R(2,3))
16040 DO=FNDP(R(2,1),R(2,2),R(2,3),V(2,1),V(2,2),V(2,3))/SQR(M)
16050 AI=2/RO-FNVS(V(2,1),V(2,2),V(2,3))/M
16060 CO=1-RO*AI
16070 WW=H*SQR(M)
16080 XX=WW/RO
16090    X2=XX*XX
16100    XA=X2*AI
16110    X3=X2*XX
16120    CC=X2*(B(2)-XA*(B(4)-XA*(B(6)-XA*(B(8)-XA*(B(10)-
         XA*(B(12)-XA*(B(14)-XA*(B(16)-XA*(B(18)))))))))
16130    UU=X3*(B(3)-XA*(B(5)-XA*(B(7)-XA*(B(9)-XA*(B(11)-
         XA*(B(13)-XA*(B(15)-XA*(B(17)-XA*(B(19)))))))))
16140    SS=XX-UU*AI
16150    FX=RO*XX+CO*UU+DO*CC-WW
16160    IF ABS(FX)<.0000001# THEN 16200
16170    DF=RO+CO*CC+DO*SS
16180    XX=XX-FX/DF
16190    GOTO 16090
16200 F=1-CC/RO
16210 G=(RO*SS+DO*CC)/SQR(M)
16220 RETURN
30000 DATA "PALLAS", "J2000.0", 0.017202099#
30010 REM
30020 DATA   6370.57744#
30030 DATA   6.38029#, -24.25104#
30040 DATA -0.7735829#,-0.5704494#,-0.2473703#
30050 REM
30060 DATA   6378.56789#
30070 DATA   6.40793#, -26.48060#
30080 DATA -0.6780640#,-0.6624821#,-0.2872733#
30090 REM
```

```
30100 DATA   6390.65113#
30110 DATA   6.38762#, -29.48400#
30120 DATA  -0.5091536#,-0.7766740#,-0.3367798#
30130 REM
40000 DATA  "REBEK-JEWEL", "J2000.0", 0.017202099#
40010 REM
40020 DATA   6370.57744#
40030 DATA   5.41652#, 21.85272#
40040 DATA  -0.7735829#,-0.5704494#,-0.2473703#
40050 REM
40060 DATA   6374.57284#
40070 DATA   5.12686#, 22.14104#
40080 DATA  -0.7275905#,-0.6179560#,-0.2679668#
40090 REM
40100 DATA   6378.56789#
40110 DATA   4.75436#, 22.32127#
40120 DATA  -0.6780640#,-0.6624821#,-0.2872733#
```

10.9 Numerical Examples

10.9.1 The Orbit of Pallas

Problem

Topocentric angular position data for minor planet Pallas are shown in the table below. The positions are referred to the equinox J2000.0, and all Julian dates are expressed in terrestrial dynamical time (TDT). The observation site is at west longitude $77°35'41''$ and north latitude $37°31'33''$.

JD	α	δ
2446370.57744	6.38029	−24.25104
2446372.57518	6.39017	−24.82014
2446374.57284	6.39810	−25.38232
2446376.57040	6.40403	−25.93626
2446378.56789	6.40793	−26.48060
2446390.65113	6.38762	−29.48400
2446398.63867	6.33411	−31.08535

Compute a set of preliminary position and velocity elements by the method of Gauss, using the three sets of position data listed below. The $\mathbf{R}_i$ vectors were obtained by employing HELO to perform second-order interpolations of the data listed at the end of that program.

- equinox: J2000.0

- $k = 0.017202099$

- $t_1 = 6370.57744$

  $\alpha_1 = 6.38029$

  $\delta_1 = -24.25104$

  $\mathbf{R}_1 = \{-0.7735829, -0.5704494, -0.2473703\}$

- $t_2 = 6378.56789$

  $\alpha_2 = 6.40793$

  $\delta_2 = -26.48060$

  $\mathbf{R}_2 = \{-0.6780640, -0.6624821, -0.2872733\}$

- $t_3 = 6390.65113$

  $\alpha_3 = 6.38762$

  $\delta_3 = -29.48400$

  $\mathbf{R}_3 = \{-0.5091536, -0.7766740, -0.3367798\}$

Solution

Use the given data to write lines 30000 to 30120 as shown at the end of program GAUSS. Run the program and answer the first two prompts as follows:

```
>>>LIMITS: RMIN, RMAX? 2,3

>>>APPROXIMATE R? 2.3
```

Once the first approximation of the orbital elements has been computed, continue to refine the solution by responding

```
>>>REFINE: Y/N? Y
```

until the Δp are all equal to zero. When this occurs, answer the prompt as follows:

```
>>>REFINE: Y/N? N
```

In this example, 17 refinement loops were required to complete the solution. For brevity, only the results of the first two and last two refinements are shown.

10.9. NUMERICAL EXAMPLES

Results

```
PRELIMINARY ORBIT METHOD OF GAUSS
PALLAS
J2000.0

TA(I)    6370.5774400
A(I)        6.3802900
D(I)      -24.2510400
LL(I,K)   -0.0906241   0.9072396  -0.4107354
RR(I,K)   -0.7735829  -0.5704494  -0.2473703

TA(I)    6378.5678900
A(I)        6.4079300
D(I)      -26.4806000
LL(I,K)   -0.0954098   0.8899859  -0.4458948
RR(I,K)   -0.6780640  -0.6624821  -0.2872733

TA(I)    6390.6511300
A(I)        6.3876200
D(I)      -29.4840000
LL(I,K)   -0.0881850   0.8660149  -0.4921805
RR(I,K)   -0.5091536  -0.7766740  -0.3367798

*** NUMERICAL COEFFICIENTS ***

D0        -0.0005257

AA         1.8284753
BB        -1.7448400

EE         0.7936242
FF         0.9811793

A         -5.7756235
B          7.7655411
C         -3.0444667

>>>LIMITS: RMIN, RMAX?   2,3
```

*** SOLUTION OF THE ORBIT EQUATION ***

R	P	F(R)
2.00000	1.61037	-54.56004
2.10000	1.64007	-48.25203
2.20000	1.66461	-26.43789
2.30000	1.68507	19.54916
2.40000	1.70226	101.32067
2.50000	1.71681	234.10668
2.60000	1.72920	437.53210
2.70000	1.73983	736.50515
2.80000	1.74899	1162.22769
2.90000	1.75693	1753.33787
3.00000	1.76385	2557.19558

\>\>\>APPROXIMATE R?   2.3

*** PRELIMINARY ORBITAL ELEMENTS ***

P(1)   1.7354488
P(2)   1.6781037
P(3)   1.6042997

T(2)   6378.5581981
R(2)   2.2647285

R(2,K)   0.5179565   2.1559707   -0.4609844
V(2,K)   -0.7187995   0.0676815   0.0351518

DP(0)   1.7354488   1.6781037   1.6042997
\>\>\>REFINE: Y/N? Y

10.9. NUMERICAL EXAMPLES

*** PRELIMINARY ORBITAL ELEMENTS ***

```
P(1)        1.7352649
P(2)        1.6779370
P(3)        1.6041482

T(2)     6378.5581990
R(2)        2.2645758

R(2,K)      0.5179724    2.1558223   -0.4609100
V(2,K)     -0.7187796    0.0678558    0.0351332

DP( 1 )    -0.0001839   -0.0001667   -0.0001515
>>>REFINE: Y/N? Y
```

*** PRELIMINARY ORBITAL ELEMENTS ***

```
P(1)        1.7351573
P(2)        1.6778387
P(3)        1.6040578

T(2)     6378.5581996
R(2)        2.2644857

R(2,K)      0.5179818    2.1557348   -0.4608662
V(2,K)     -0.7187699    0.0679482    0.0351259

DP( 2 )    -0.0001076   -0.0000983   -0.0000903
>>>REFINE: Y/N? Y
```

*** PRELIMINARY ORBITAL ELEMENTS ***

```
P(1)        1.7350043
P(2)        1.6776970
P(3)        1.6039254

T(2)     6378.5582004
R(2)        2.2643559

R(2,K)      0.5179953     2.1556087    -0.4608030
V(2,K)     -0.7187611     0.0680558     0.0351243

DP( 16 )  -0.0000001    -0.0000001    -0.0000001
>>>REFINE: Y/N? Y
```

*** PRELIMINARY ORBITAL ELEMENTS ***

```
P(1)        1.7350042
P(2)        1.6776970
P(3)        1.6039253

T(2)     6378.5582004
R(2)        2.2643558

R(2,K)      0.5179953     2.1556087    -0.4608030
V(2,K)     -0.7187611     0.0680558     0.0351243

DP( 17 )  -0.0000000    -0.0000000    -0.0000000
>>>REFINE: Y/N? N
```

PALLAS
METHOD OF GAUSS

10.9. NUMERICAL EXAMPLES

Discussion of Results

When the preliminary position and velocity elements computed by GAUSS are used in program CLASSEL, the following set of classical elements is obtained for minor planet Pallas:

- epoch: 2446378.55820
- $a = 2.7718447$
- $e = 0.2336961$
- $M = 329.89800°$
- $i = 34.79457°$
- $\Omega = 173.34362°$
- $\omega = 309.92515°$

These elements compare fairly well with those for the epoch 2446400.5 given in the numerical example of Section 4.7.5.

CHAPTER 10. THE METHOD OF GAUSS

10.9.2 The Orbit of Comet Rebek-Jewel

Problem

The angular position data listed below for comet Rebek-Jewel were taken from the numerical example of Section 8.6.2. The values of the $\mathbf{R}_i$ vectors were taken from the numerical examples of Sections 8.6.4 and 10.9.1. Use the given information to compute a set of preliminary position and velocity elements by the method of Gauss.

- equinox: J2000.0

- $k = 0.017202099$

- $t_1 = 6370.57744$
  $\alpha_1 = 5.41652$
  $\delta_1 = 21.85272$
  $\mathbf{R}_1 = \{-0.7735829, -0.5704494, -0.2473703\}$

- $t_2 = 6374.57284$
  $\alpha_2 = 5.12686$
  $\delta_2 = 22.14104$
  $\mathbf{R}_2 = \{-0.7275905, -0.6179560, -0.2679668\}$

- $t_3 = 6378.56789$
  $\alpha_3 = 4.75436$
  $\delta_3 = 22.32127$
  $\mathbf{R}_3 = \{-0.6780640, -0.6624821, -0.2872733\}$

Solution

Transfer the given data to lines 40000 to 40120 as shown at the end of program GAUSS. Run the program and answer the first two prompts as follows:

```
>>>LIMITS: RMIN, RMAX? 1,3

>>>APPROXIMATE R? 1.8
```

10.9. NUMERICAL EXAMPLES

Once the first approximation of the orbital elements has been computed, continue to refine the solution by responding

>>>REFINE: Y/N? Y

until the Δp are all equal to zero. When this occurs, answer the prompt as follows:

>>>REFINE: Y/N? N

In this example, 20 refinement loops were required to complete the solution. For brevity, only the results of the first two and last two refinements are shown.

Results

```
PRELIMINARY ORBIT METHOD OF GAUSS
REBEK-JEWEL
J2000.0

TA(I)      6370.5774400
A(I)          5.4165200
D(I)         21.8527200
LL(I,K)       0.1412276    0.9173361    0.3722220
RR(I,K)      -0.7735829   -0.5704494   -0.2473703

TA(I)      6374.5728400
A(I)          5.1268600
D(I)         22.1410400
LL(I,K)       0.2098921    0.9021645    0.3768878
RR(I,K)      -0.7275905   -0.6179560   -0.2679668

TA(I)      6378.5678900
A(I)          4.7543600
D(I)         22.3212700
LL(I,K)       0.2963535    0.8763143    0.3797996
RR(I,K)      -0.6780640   -0.6624821   -0.2872733

*** NUMERICAL COEFFICIENTS ***

D0            0.0000250

AA            1.1292240
BB           -1.0828889

EE            1.6224139
FF            0.9830638

A            -4.0902792
B             4.2025423
C            -1.1726484

>>>LIMITS: RMIN, RMAX?   1,3
```

10.9. NUMERICAL EXAMPLES

*** SOLUTION OF THE ORBIT EQUATION ***

```
          R           P           F(R)

       1.00000      0.04634      -0.06039
       1.20000      0.50255      -1.82435
       1.40000      0.73459      -5.68089
       1.60000      0.86485      -9.63286
       1.80000      0.94354      -5.58331
       2.00000      0.99386      26.66982
       2.20000      1.02753     128.57929
       2.40000      1.05089     376.01191
       2.60000      1.06761     897.41010
       2.80000      1.07989    1898.03565
       3.00000      1.08912    3691.48244
```

\>\>\>APPROXIMATE R?   1.8

*** PRELIMINARY ORBITAL ELEMENTS ***

```
P(1)        1.0701660
P(2)        0.9614570
P(3)        0.8615434

T(2)     6374.5672871
R(2)        1.8620807

R(2,K)      0.9293928    1.4853484    0.6303282
V(2,K)      0.0630514   -0.9799957   -0.2271792

DP( 0 )     1.0701660    0.9614570    0.8615434
```
\>\>\>REFINE: Y/N? Y

*** PRELIMINARY ORBITAL ELEMENTS ***

P(1)	1.0704216
P(2)	0.9616812
P(3)	0.8617373

T(2)	6374.5672858
R(2)	1.8622941

R(2,K)	0.9294398	1.4855506	0.6304127
V(2,K)	0.0632776	-0.9802711	-0.2272673

DP(1)	0.0002556	0.0002241	0.0001938

\>>>REFINE: Y/N? Y

*** PRELIMINARY ORBITAL ELEMENTS ***

P(1)	1.0705815
P(2)	0.9618220
P(3)	0.8618601

T(2)	6374.5672850
R(2)	1.8624282

R(2,K)	0.9294694	1.4856777	0.6304658
V(2,K)	0.0634135	-0.9804582	-0.2273269

DP(2)	0.0001599	0.0001409	0.0001228

\>>>REFINE: Y/N? Y

10.9. NUMERICAL EXAMPLES

```
*** PRELIMINARY ORBITAL ELEMENTS ***

P(1)        1.0708503
P(2)        0.9620605
P(3)        0.8620705

T(2)        6374.5672836
R(2)        1.8626552

R(2,K)      0.9295194    1.4858929    0.6305557
V(2,K)      0.0636263   -0.9808128   -0.2274390

DP( 19 )    0.0000001    0.0000001    0.0000000
>>>REFINE: Y/N? Y

*** PRELIMINARY ORBITAL ELEMENTS ***

P(1)        1.0708503
P(2)        0.9620605
P(3)        0.8620705

T(2)        6374.5672836
R(2)        1.8626552

R(2,K)      0.9295194    1.4858929    0.6305557
V(2,K)      0.0636264   -0.9808129   -0.2274390

DP( 20 )    0.0000000    0.0000000    0.0000000
>>>REFINE: Y/N? N

REBEK-JEWEL
METHOD OF GAUSS
```

Discussion of Results

When the preliminary position and velocity elements computed by GAUSS are used in program CLASSEL, the following classical elements are obtained for comet Rebek-Jewel:

- epoch: 2446374.56728
- $a = 17.8682311$
- $e = 0.9671563$
- $M = 358.74248°$
- $i = 162.24048°$
- $\Omega = 58.85989°$
- $\omega = 111.88551°$

This result indicates that the orbit of this comet may be a very eccentric ellipse rather than a hyperbola as computed by the method of Laplace. Aside from the shape of the orbit, notice the close agreement with the orientation angles found previously.

References

[1] Danby, *Fundamentals of Celestial Mechanics*, Willmann-Bell, Inc., 1988.

[2] Marsden, "Initial Orbit Determination: the Pragmatist's Point of View," *The Astronomical Journal*, **90**, 1541-1547, 1985.

[3] Moulton, *An Introduction to Celestial Mechanics*, Dover Publications, Inc., 1970.

[4] Dubyago, *The Determination of Orbits*, The Rand Corporation, 1961.

[5] Herget, *The Computation of Orbits*, Published by Author, 1948.

[6] Baker, *Astrodynamics: Applications and Advanced Topics*, Academic Press, 1967.

[7] Escobal, *Methods of Orbit Determination*, Krieger Publishing Co., 1976.

[8] Taff, *Celestial Mechanics*, John Wiley and Sons, 1985.

Chapter 11

The Method of Olbers

11.1 Introduction

The method of Olbers is not general. It was specifically designed to compute preliminary orbits for newly discovered comets. Because the orbits of most non-periodic comets closely resemble parabolas, the method of Olbers assumes parabolic motion in order to simplify the computation. Using this method, it is often possible to compute a satisfactory preliminary orbit from three observations separated by only one or two days. However, there are situations where Olbers' solution becomes indeterminate. If the apparent track of the comet on the celestial sphere lies close to the ecliptic, causing the vectors $\mathbf{L}_i$ and $\mathbf{R}_i$ to be nearly coplanar, the coefficients C_1, C_3, D_1, and D_3 in Equations 11.7 approach zero. This will also occur if the direction toward the second position of the Sun is nearly opposite the direction of the second observation of the comet. In this case, the magnitude of the vector $\mathbf{O} = \mathbf{L}_2 \times \mathbf{R}_2$ approaches zero, and the method of Olbers may fail [1,2,3].

11.2 Solution by Euler's Equation

Assume that from the available data three sets have been chosen and reduced to the following:
$$\{t_i, \mathbf{L}_i, \mathbf{R}_i\},$$
where $i = 1, 2, 3$. If we let t_2 be the epoch time, then the observed times can be converted to the modified time intervals

$$\begin{aligned} \tau_1 &= k(t_1 - t_2) \\ \tau_3 &= k(t_3 - t_2) \\ \tau &= \tau_3 - \tau_1. \end{aligned}$$

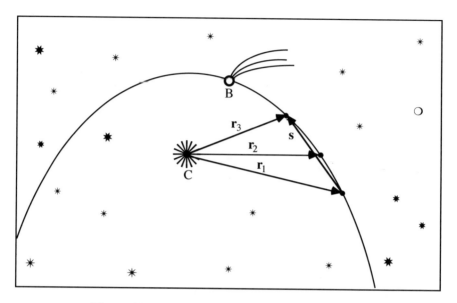

Figure 11.1: Heliocentric parabolic orbital motion.

Now, consider the orbital motion illustrated in Figure 11.1, where a comet B is following a parabolic path about the heliocentric dynamical center C. Let the radius vectors $\mathbf{r}_1$, $\mathbf{r}_2$, and $\mathbf{r}_3$ represent the positions of the comet at the respective times t_1, t_2, and t_3. Further, let $\mathbf{s}$ be the vector connecting the first and third positions of the comet as shown, so that

$$\mathbf{s} = \mathbf{r}_3 - \mathbf{r}_1. \tag{11.1}$$

During the middle of the eighteenth century, the mathematician Euler discovered an equation applicable to parabolic orbits which relates the vectors $\mathbf{s}$, $\mathbf{r}_1$, and $\mathbf{r}_3$ to the interval of time required for the comet to travel the subtended arc. *Euler's equation* may be written as follows [1,2]:

$$(r_1 + r_3 + s)^{3/2} - (r_1 + r_3 - s)^{3/2} - 6\tau = 0, \tag{11.2}$$

where

$$\begin{aligned} r_1 &= |\mathbf{r}_1| \\ r_3 &= |\mathbf{r}_3| \\ s &= |\mathbf{s}|. \end{aligned} \tag{11.3}$$

In addition to the parabolic constraint imposed by Euler's equation, we have the general geometric constraint

$$\mathbf{r}_i = p_i \mathbf{L}_i - \mathbf{R}_i \tag{11.4}$$

11.3. THE SCALAR EQUATIONS FOR THE RANGE

and the constraint of Equation 10.6, namely

$$c_1 p_1 \mathbf{L}_1 - p_2 \mathbf{L}_2 + c_3 p_3 \mathbf{L}_3 = c_1 \mathbf{R}_1 - \mathbf{R}_2 + c_3 \mathbf{R}_3, \tag{11.5}$$

where c_2 has been replaced by its numerical value of -1.

11.3 The Scalar Equations for the Range

A scalar relationship between the ranges p_1 and p_3 can be derived by taking the dot product of Equation 11.5 with the vector

$$\mathbf{O} = \mathbf{L}_2 \times \mathbf{R}_2 .$$

Since the vector $\mathbf{O}$ is perpendicular to both $\mathbf{L}_2$ and $\mathbf{R}_2$, the terms in Equation 11.5 which contain these vectors will be eliminated. The result is

$$c_1 p_1 \mathbf{L}_1 \cdot \mathbf{O} + c_3 p_3 \mathbf{L}_3 \cdot \mathbf{O} = c_1 \mathbf{R}_1 \cdot \mathbf{O} + c_3 \mathbf{R}_3 \cdot \mathbf{O}. \tag{11.6}$$

For convenience, let this equation be rewritten as follows:

$$c_1 C_1 p_1 + c_3 C_3 p_3 = c_1 D_1 + c_3 D_3, \tag{11.7}$$

where

$$\begin{aligned} C_1 &= \mathbf{L}_1 \cdot \mathbf{O} \\ C_3 &= \mathbf{L}_3 \cdot \mathbf{O} \\ D_1 &= \mathbf{R}_1 \cdot \mathbf{O} \\ D_3 &= \mathbf{R}_3 \cdot \mathbf{O} . \end{aligned} \tag{11.8}$$

Solving Equation 11.7 for p_3 in terms of p_1, we obtain

$$p_3 = \left(-\frac{c_1 C_1}{c_3 C_3}\right) p_1 + \left(\frac{c_1 D_1}{c_3 C_3} + \frac{D_3}{C_3}\right), \tag{11.9}$$

which can be written simply as

$$p_3 = Q p_1 + U, \tag{11.10}$$

where

$$\begin{aligned} Q &= -\frac{c_1 C_1}{c_3 C_3} \\ U &= +\frac{c_1 D_1}{c_3 C_3} + \frac{D_3}{C_3} . \end{aligned} \tag{11.11}$$

A second scalar relationship, which expresses p_2 in terms of p_1 and p_3, can be derived by taking the dot product of Equation 11.5 with the vector $\mathbf{L}_2$. Because $\mathbf{L}_2$ is a unit vector,
$$\mathbf{L}_2 \cdot \mathbf{L}_2 = 1,$$
and Equation 11.5 becomes
$$c_1 p_1 \mathcal{X}_1 - p_2 + c_3 p_3 \mathcal{X}_3 = c_1 \mathcal{Z}_1 - \mathcal{Z}_2 + c_3 \mathcal{Z}_3, \tag{11.12}$$
where
$$\begin{aligned} \mathcal{X}_1 &= \mathbf{L}_2 \cdot \mathbf{L}_1 \\ \mathcal{X}_3 &= \mathbf{L}_2 \cdot \mathbf{L}_3 \\ \mathcal{Z}_1 &= \mathbf{L}_2 \cdot \mathbf{R}_1 \\ \mathcal{Z}_2 &= \mathbf{L}_2 \cdot \mathbf{R}_2 \\ \mathcal{Z}_3 &= \mathbf{L}_2 \cdot \mathbf{R}_3. \end{aligned} \tag{11.13}$$

Therefore, solving Equation 11.12 for p_2, we obtain
$$p_2 = (c_1 p_1 \mathcal{X}_1 + c_3 p_3 \mathcal{X}_3) - (c_1 \mathcal{Z}_1 - \mathcal{Z}_2 + c_3 \mathcal{Z}_3). \tag{11.14}$$

As was the case for the method of Gauss, we must introduce approximate expressions for c_1 and c_3 before Equations 11.10 and 11.14 can be used to complete the method of Olbers. Consider again Equations 10.18, namely
$$\begin{aligned} c_1 &\approx +\frac{\tau_3}{\tau} + \frac{u_2 \tau_3}{6\tau}(\tau^2 - \tau_3^2) \\ c_3 &\approx -\frac{\tau_1}{\tau} - \frac{u_2 \tau_1}{6\tau}(\tau^2 - \tau_1^2), \end{aligned} \tag{11.15}$$
where $u_2 = \mu/r_2^3$. In the case of parabolic motion, the last terms in Equations 11.15 are usually not significant if the comet's distance from the Sun is not too small and the time interval between the first and third observations is not too large. Therefore, neglecting the terms which contain the unknown u_2, we take as our initial values for c_1 and c_3 the simple expressions
$$\begin{aligned} c_1 &\approx +\frac{\tau_3}{\tau} \\ c_3 &\approx -\frac{\tau_1}{\tau}. \end{aligned} \tag{11.16}$$

11.4 The Vector Orbital Elements

11.4.1 Three Radius Vectors

Enough information now exists to determine approximate values for the components of the three radius vectors $\mathbf{r}_i$. This is accomplished by solving the

11.4. THE VECTOR ORBITAL ELEMENTS

following equations simultaneously:

$$p_3 = Qp_1 + U \tag{11.17}$$
$$\mathbf{r}_1 = p_1\mathbf{L}_1 - \mathbf{R}_1 \tag{11.18}$$
$$\mathbf{r}_3 = p_3\mathbf{L}_3 - \mathbf{R}_3 \tag{11.19}$$
$$\mathbf{s} = \mathbf{r}_3 - \mathbf{r}_1 \tag{11.20}$$
$$r_1 = |\mathbf{r}_1| \tag{11.21}$$
$$r_3 = |\mathbf{r}_3| \tag{11.22}$$
$$s = |\mathbf{s}| \tag{11.23}$$
$$f = (r_1 + r_3 + s)^{3/2} - (r_1 + r_3 - s)^{3/2} - 6\tau. \tag{11.24}$$

A series of successive approximations can be used to discover the value of p_1 which produces values for p_3, r_1, r_3, and s that will yield $f = 0$ in Equation 11.24. Were it not for the microcomputer, this *brute force* approach would be extremely laborious. Fortunately, however, it is an easy matter to write a BASIC routine which will compute the function f for a given set of discrete values of p_1. From such a menu, we can pick two values of p_1 between which f passes through zero. Then, by using the bisection method described in Section 7.2.1, a particular p_1 can be found which causes the function f to be zero within some specified tolerance. When this is accomplished, p_3, $\mathbf{r}_1$, and $\mathbf{r}_3$ are also known. Equation 11.14 then yields

$$p_2 = (c_1 p_1 \mathcal{X}_1 + c_3 p_3 \mathcal{X}_3) - (c_1 \mathcal{Z}_1 - \mathcal{Z}_2 + c_3 \mathcal{Z}_3), \tag{11.25}$$

and the vector element $\mathbf{r}_2$ is determined by the geometric constraint

$$\mathbf{r}_2 = p_2 \mathbf{L}_2 - \mathbf{R}_2. \tag{11.26}$$

Finally, it follows that

$$r_2 = |\mathbf{r}_2|,$$

so that

$$u_2 = \frac{\mu}{r_2^3} \tag{11.27}$$

is also known.

The corrections for light-time should be introduced before going on to find the velocity $\dot{\mathbf{r}}_2$. Thus, for all three observation times t_i, the corrected values are

$$t_{ci} = t_i - \frac{p_i}{c}, \tag{11.28}$$

where

$$c = 173.1446 \text{ au/day}.$$

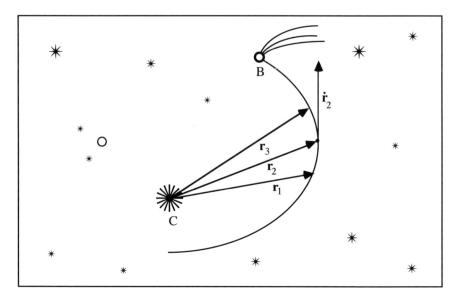

Figure 11.2: The Herrick-Gibbs transformation.

Accordingly, the modified time intervals are also computed anew from

$$\begin{aligned} \tau_1 &= k(t_{c1} - t_{c2}) \\ \tau_3 &= k(t_{c3} - t_{c2}) \\ \tau &= \tau_3 - \tau_1. \end{aligned} \qquad (11.29)$$

When the quantities found from Equations 11.27 and 11.29 are substituted into Equations 11.15, improved values for the coefficients c_1 and c_3 are obtained. These can then be used to recompute Q and U from Equations 11.11, and Equations 11.17 to 11.26 can be solved again to find improved values for the three radius vectors $\mathbf{r}_i$.

11.4.2 The Velocity Vector

The vector element $\mathbf{r}_2$ is now known along with the other two radius vectors. We shall determine the vector element $\dot{\mathbf{r}}_2$ by using a technique known as the *Herrick-Gibbs transformation*. This procedure converts three radius vectors to a velocity vector at the central date. Although the transformation is simple to use, its derivation is a little tedious [4,5].

Consider the situation depicted in Figure 11.2, where a celestial body B is orbiting the dynamical center C. Recalling the Taylor series expansion described

11.4. THE VECTOR ORBITAL ELEMENTS

in Section 3.5.2, we can write

$$\mathbf{r}_1 - \mathbf{r}_2 = \dot{\mathbf{r}}_2 \tau_1 + \ddot{\mathbf{r}}_2 \frac{\tau_1^2}{2} + \dddot{\mathbf{r}}_2 \frac{\tau_1^3}{6} + \ddddot{\mathbf{r}}_2 \frac{\tau_1^4}{24} \tag{11.30}$$

$$\mathbf{r}_3 - \mathbf{r}_2 = \dot{\mathbf{r}}_2 \tau_3 + \ddot{\mathbf{r}}_2 \frac{\tau_3^2}{2} + \dddot{\mathbf{r}}_2 \frac{\tau_3^3}{6} + \ddddot{\mathbf{r}}_2 \frac{\tau_3^4}{24}, \tag{11.31}$$

where the series has been truncated after the fourth derivative. Multiplying Equation 11.30 by τ_3 and Equation 11.31 by τ_1, we subtract to eliminate $\dot{\mathbf{r}}_2$. Thus, with

$$\tau = \tau_3 - \tau_1 \tag{11.32}$$

we have

$$\tau_3 \mathbf{r}_1 - \tau \mathbf{r}_2 - \tau_1 \mathbf{r}_3 = -\tau_1 \tau_3 \tau \left[\frac{\ddot{\mathbf{r}}_2}{2} + (\tau_1 + \tau_3) \frac{\dddot{\mathbf{r}}_2}{6} + (\tau_1^2 + \tau_1 \tau_3 + \tau_3^2) \frac{\ddddot{\mathbf{r}}_2}{24} \right]. \tag{11.33}$$

Multiplying Equation 11.30 by τ_3^2 and Equation 11.31 by τ_1^2 and then subtracting eliminates $\ddot{\mathbf{r}}_2$ so that

$$\tau_3^2 \mathbf{r}_1 - (\tau_3^2 - \tau_1^2)\mathbf{r}_2 - \tau_1^2 \mathbf{r}_3 = \tau_1 \tau_3 \tau \left[\dot{\mathbf{r}}_2 - \tau_1 \tau_3 \frac{\dddot{\mathbf{r}}_2}{6} - \tau_1 \tau_3 (\tau_1 + \tau_3) \frac{\ddddot{\mathbf{r}}_2}{24} \right]. \tag{11.34}$$

Differentiating Equation 11.33 twice and ignoring all derivatives beyond the fourth yields

$$\tau_3 \ddot{\mathbf{r}}_1 - \tau \ddot{\mathbf{r}}_2 - \tau_1 \ddot{\mathbf{r}}_3 = -\tau_1 \tau_3 \tau \frac{\ddddot{\mathbf{r}}_2}{2}. \tag{11.35}$$

Differentiating Equation 11.34 twice and ignoring all derivatives beyond the third produces

$$\tau_3^2 \ddot{\mathbf{r}}_1 - (\tau_3^2 - \tau_1^2)\ddot{\mathbf{r}}_2 - \tau_1^2 \ddot{\mathbf{r}}_3 = \tau_1 \tau_3 \tau \dddot{\mathbf{r}}_2. \tag{11.36}$$

Solving Equations 11.36 and 11.35 for $\dddot{\mathbf{r}}_2$ and $\ddddot{\mathbf{r}}_2$, respectively, we obtain

$$\dddot{\mathbf{r}}_2 = \frac{1}{\tau_1 \tau_3 \tau}[\tau_3^2 \ddot{\mathbf{r}}_1 - (\tau_3^2 - \tau_1^2)\ddot{\mathbf{r}}_2 - \tau_1^2 \ddot{\mathbf{r}}_3] \tag{11.37}$$

$$\ddddot{\mathbf{r}}_2 = -\frac{2}{\tau_1 \tau_3 \tau}[\tau_3 \ddot{\mathbf{r}}_1 - \tau \ddot{\mathbf{r}}_2 - \tau_1 \ddot{\mathbf{r}}_3]. \tag{11.38}$$

We now substitute Equations 11.37 and 11.38 into Equation 11.34 to eliminate $\dddot{\mathbf{r}}_2$ and $\ddddot{\mathbf{r}}_2$. By rearranging and collecting terms, one can arrive at the following equation after some algebra:

$$\dot{\mathbf{r}}_2 = \frac{1}{\tau_1 \tau_3 \tau}[\tau_3^2 \mathbf{r}_1 - (\tau_3^2 - \tau_1^2)\mathbf{r}_2 - \tau_1^2 \mathbf{r}_3] + \frac{1}{12}[\tau_3 \ddot{\mathbf{r}}_1 - (\tau_1 + \tau_3)\ddot{\mathbf{r}}_2 + \tau_1 \ddot{\mathbf{r}}_3]. \tag{11.39}$$

In order to simplify the above expression, let

$$u_i = \frac{\mu}{r_i^3}, \qquad (11.40)$$

so that

$$\ddot{\mathbf{r}}_i = -u_i \mathbf{r}_i, \qquad (11.41)$$

where $i = 1, 2, 3$. Thus, substituting Equation 11.41 for the appropriate $\ddot{\mathbf{r}}_i$ in Equation 11.39 and rearranging and combining terms, we obtain the following expression for the velocity vector:

$$\dot{\mathbf{r}}_2 = d_1 \mathbf{r}_1 + d_2 \mathbf{r}_2 + d_3 \mathbf{r}_3, \qquad (11.42)$$

where

$$\begin{aligned}
d_1 &= -\tau_3 \left(\frac{u_1}{12} - \frac{1}{\tau_1 \tau} \right) \\
d_2 &= +(\tau_1 + \tau_3) \left(\frac{u_2}{12} - \frac{1}{\tau_1 \tau_3} \right) \\
d_3 &= -\tau_1 \left(\frac{u_3}{12} + \frac{1}{\tau_3 \tau} \right).
\end{aligned} \qquad (11.43)$$

Therefore, we now have values for the orbital elements $\mathbf{r}_2$ and $\dot{\mathbf{r}}_2$, and the preliminary parabolic orbit is determined.

11.5 Program OLBERS

Program OLBERS computes the position and velocity elements of a preliminary heliocentric parabolic orbit by the method of Olbers. OLBERS uses angular position data from observations and vector position data obtained from program HELO.

Program Algorithm

Define	Line Number
FNVS(X,Y,Z): vector squaring function	1070
FNMG(X,Y,Z): vector magnitude function	1080
FNDP(X1,Y1,Z1,X2,Y2,Z2): dot product function	1090
FNX(A1,A2,B1,B2): cross product function	1100

Given	Line Number
μ: combined mass	1120
angle to radian conversion factor	1130
$1/c$: light-time constant	1140
name of the object for information only	1190
equinox for information only	1190
k: gravitational constant	1190
t_i: three observation times uncorrected for light-time	1210
α_i, δ_i: three angular positions	1220
R_{ix}, R_{iy}, R_{iz}: three $\mathbf{R}_i$	1230

Compute	Line Number
α_i in radians	1330

	Line Number
δ_i in radians	1340
L_{ix}: Equations 6.4	1350
L_{iy}: Equations 6.4	1360
L_{iz}: Equations 6.4	1370
$\tau_1 = k(t_1 - t_2)$	1530
$\tau_3 = k(t_3 - t_2)$	1540
$\tau = \tau_3 - \tau_1$	1550
c_1: Equations 11.15	1570
c_3: Equations 11.15	1580
C_i: Equations 11.8	1610-1640
D_i: Equations 11.8	1660-1690
$\mathcal{X}_i$: Equations 11.13	1730
$\mathcal{Z}_i$: Equations 11.13	1740
Q: Equations 11.11	1790
U: Equations 11.11	1800

Choose   Line Number

$p_{\min}$ and $p_{\max}$ which bound the unknown value of p_1 1940

This prompt requests the user to choose minimum and maximum values for p_1, the range at t_1, between which the unknown solution lies. Several trials may be necessary to find appropriate values.

Compute   Line Number

11.5. PROGRAM OLBERS

a table containing a series of values for p_1, p_3, and f 1980-2060

The quantity f is the numerical value of Equation 11.24 which corresponds to each successive value of p_1. The computations are performed in subroutine EULER/SUB.

p_3: Equation 11.17	10020
$\mathbf{r}_1$: Equation 11.18	10040
$\mathbf{r}_3$: Equation 11.19	10050
s: Equation 11.20	10060
r_1: Equation 11.21	10080
r_3: Equation 11.22	10090
s: Equation 11.23	10100
f: Equation 11.24	10110-10130

Choose	Line Number
minimum and maximum values of p_1 which are close to the solution	2100

This prompt requests the user to choose two values of p_1 which closely bracket the unknown value as a starting point for the bisection method solution of Euler's equation. From the table, choose two values which span the point where f changes sign. If the table does not contain any such points, the program must be run again using different values of p_{min} and p_{max} in line 1940.

Compute	Line Number
solution by the bisection method	2130-2310
p_2: Equation 11.25	2350 2370
$\mathbf{r}_2$: Equation 11.26	2400

$r_2 = |\mathbf{r}_2|$ 2430

u_2: Equation 11.27   2450

t_i: observation times corrected for light-time by Equation 11.28   2490

τ_1 corrected for light-time by Equations 11.29   2520

τ_3 corrected for light-time by Equations 11.29   2530

τ corrected for light-time by Equations 11.29   2540

c_1: Equations 11.15   2350

c_3: Equations 11.15   2360

The first time line 2600 is reached, the solution of Euler's equation has not been corrected for light-time. Therefore, at line 2640, the program goes back to line 1760 for another solution using the new values of c_1 and c_3 calculated above. The second time line 2600 is reached, the program jumps to line 2670, and the computation of the elements continues.

u_1: Equations 11.40   2700

u_3: Equations 11.40   2710

d_1: Equations 11.43   2730

d_2: Equations 11.43   2740

d_3: Equations 11.43   2750

$\dot{\mathbf{r}}_2$: Equation 11.42   2780

End.

11.5. PROGRAM OLBERS

Program Listing

```
1000 CLS
1010 PRINT"# OLBERS # PRELIMINARY ORBITAL ELEMENTS"
1020 PRINT"# BY THE PARABOLIC METHOD OF OLBERS"
1030 REM
1040 DEFDBL A-Z
1050 DEFINT I,K,N
1060 REM
1070 DEF FNVS(X,Y,Z)=X*X+Y*Y+Z*Z
1080 DEF FNMG(X,Y,Z)=SQR(X*X+Y*Y+Z*Z)
1090 DEF FNDP(X1,Y1,Z1,X2,Y2,Z2)=X1*X2+Y1*Y2+Z1*Z2
1100 DEF FNX(A1,A2,B1,B2)=A1*B2-A2*B1
1110 REM
1120 M=1
1130 Q1=.0174532925#
1140 AB=1/173.1446#
1150 G$="####.#######"
1160 S$="########.#####"
1170 AP$="FIRST APPROXIMATION"
1180 REM
1190 READ N$,E$,K#
1200 FOR I=1 TO 3
1210    READ TA(I)
1220    READ A(I), D(I)
1230    READ RR(I,1),RR(I,2),RR(I,3)
1240 NEXT I
1250 REM
1260 LINE INPUT"";L$
1270 CLS
1280 PRINT"PARABOLIC ORBIT METHOD OF OLBERS"
1290 PRINT N$
1300 PRINT E$
1310 PRINT
1320 FOR I=1 TO 3
1330    A=A(I)*15*Q1
1340    D=D(I)*Q1
1350    LL(I,1)=COS(D)*COS(A)
1360    LL(I,2)=COS(D)*SIN(A)
1370    LL(I,3)=SIN(D)
1380 NEXT I
1390 REM
```

```
1400 FOR I=1 TO 3
1410   PRINT"TA(I)";TAB(9);:PRINT USING G$;TA(I)
1420   PRINT"A(I)";TAB(9);:PRINT USING G$;A(I)
1430   PRINT"D(I)";TAB(9);:PRINT USING G$;D(I)
1440   PRINT"LL(I,K)";TAB(9);
1450   PRINT USING G$;LL(I,1),LL(I,2),LL(I,3)
1460   PRINT"RR(I,K)";TAB(9);
1470   PRINT USING G$;RR(I,1),RR(I,2),RR(I,3)
1480   REM
1490   LINE INPUT"";L$
1500   CLS
1510 NEXT I
1520 REM
1530 TT(1)=K#*(TA(1)-TA(2))
1540 TT(3)=K#*(TA(3)-TA(2))
1550 TT=TT(3)-TT(1)
1560 REM
1570 C(1)=+TT(3)/TT
1580 C(3)=-TT(1)/TT
1590 REM
1600 FOR I=1 TO 3 STEP 2
1610   E1=LL(I,1)*FNX(LL(2,2),LL(2,3),RR(2,2),RR(2,3))
1620   E2=LL(I,2)*FNX(LL(2,1),LL(2,3),RR(2,1),RR(2,3))
1630   E3=LL(I,3)*FNX(LL(2,1),LL(2,2),RR(2,1),RR(2,2))
1640   CC(I)=E1-E2+E3
1650   REM
1660   E1=RR(I,1)*FNX(LL(2,2),LL(2,3),RR(2,2),RR(2,3))
1670   E2=RR(I,2)*FNX(LL(2,1),LL(2,3),RR(2,1),RR(2,3))
1680   E3=RR(I,3)*FNX(LL(2,1),LL(2,2),RR(2,1),RR(2,2))
1690   DD(I)=E1-E2+E3
1700 NEXT I
1710 REM
1720 FOR I=1 TO 3
1730   LX(I)=FNDP(LL(2,1),LL(2,2),LL(2,3),
       LL(I,1),LL(I,2),LL(I,3))
1740   LZ(I)=FNDP(LL(2,1),LL(2,2),LL(2,3),
       RR(I,1),RR(I,2),RR(I,3))
1750 NEXT I
1760 CLS
1770 PRINT"*** NUMERICAL COEFFICIENTS ***"
1780 PRINT
1790 QQ=(-C(1)*CC(1))/(C(3)*CC(3))
```

11.5. PROGRAM OLBERS

```
1800 UU=(+C(1)*DD(1))/(C(3)*CC(3))+DD(3)/CC(3)
1810 REM
1820 PRINT"CC(1)";TAB(9);:PRINT USING G$;CC(1)
1830 PRINT"CC(3)";TAB(9);:PRINT USING G$;CC(3)
1840 PRINT
1850 PRINT"DD(1)";TAB(9);:PRINT USING G$;DD(1)
1860 PRINT"DD(3)";TAB(9);:PRINT USING G$;DD(3)
1870 PRINT
1880 PRINT"C(1)";TAB(9);:PRINT USING G$;C(1)
1890 PRINT"C(3)";TAB(9);:PRINT USING G$;C(3)
1900 PRINT
1910 PRINT"QQ";TAB(9);:PRINT USING G$;QQ
1920 PRINT"UU";TAB(9);:PRINT USING G$;UU
1930 PRINT
1940 INPUT">>>TABLE LIMITS: PMIN, PMAX";XL,XH
1950 CLS
1960 PRINT"*** SOLUTION OF THE ORBIT EQUATION ***"
1970 PRINT
1980 PRINT TAB(10)"P(1)";TAB(24)"P(3)";TAB(38)"F"
1990 REM
2000 DX=(XH-XL)/10
2010 FOR N=0 TO 10
2020   P(1,0)=XL+N*DX
2030   GOSUB 10010 REM EULER/SUB
2040   REM
2050   PRINT USING S$;P(1,0),P(3,0),F
2060 NEXT N
2070 PRINT
2080 PRINT TAB(4) AP$
2090 REM
2100 INPUT">>>BRACKET: P(1)MIN, P(1)MAX";XL,XH
2110 REM
2120 REM
2130   XS=(XL+XH)/2
2140   P(1,0)=XS
2150   REM
2160   GOSUB 10010 REM EULER/SUB
2170   REM
2180   FS=F
2190   REM
2200   IF ABS(FS)<.0000001# THEN 2350
2210   REM
```

```
2220    P(1,0)=XL
2230    REM
2240    GOSUB 10010 REM EULER/SUB
2250    REM
2260    SN=F*FS
2270    REM
2280    IF SN<0 THEN XH=XS
2290    IF SN>0 THEN XL=XS
2300    REM
2310    GOTO 2130
2320 REM
2330 REM
2340 REM
2350 PX=C(1)*P(1,0)*LX(1)+C(3)*P(3,0)*LX(3)
2360 PZ=C(1)*LZ(1)-LZ(2)+C(3)*LZ(3)
2370 P(2,0)=PX-PZ
2380 REM
2390 FOR K=1 TO 3
2400    R(2,K)=P(2,0)*LL(2,K)-RR(2,K)
2410 NEXT K
2420 REM
2430 R(2,0)=FNMG(R(2,1),R(2,2),R(2,3))
2440 REM
2450 U2=M/(R(2,0)*R(2,0)*R(2,0))
2460 REM
2470 REM
2480 FOR I=1 TO 3
2490    T(I)=TA(I)-AB*P(I,0)
2500 NEXT I
2510 REM
2520 TT(1)=K#*(T(1)-T(2))
2530 TT(3)=K#*(T(3)-T(2))
2540 TT=TT(3)-TT(1)
2550 REM
2560 REM
2570 C(1)=(+TT(3)/TT)*(1+(TT*TT-TT(3)*TT(3))*U2/6)
2580 C(3)=(-TT(1)/TT)*(1+(TT*TT-TT(1)*TT(1))*U2/6)
2590 REM
2600 IF AP$="" THEN 2670
2610 REM
2620 AP$=""
2630 REM
```

11.5. PROGRAM OLBERS

```
2640 GOTO 1760
2650 REM
2660 REM
2670 CLS
2680 PRINT"*** PRELIMINARY ORBITAL ELEMENTS ***
2690 PRINT
2700 U1=M/(R(1,0)*R(1,0)*R(1,0))
2710 U3=M/(R(3,0)*R(3,0)*R(3,0))
2720 REM
2730 D(1)=-TT(3)*(U1/12-1/(TT(1)*TT))
2740 D(2)=+(TT(1)+TT(3))*(U2/12-1/(TT(1)*TT(3)))
2750 D(3)=-TT(1)*(U3/12+1/(TT(3)*TT))
2760 REM
2770 FOR K=1 TO 3
2780    V(2,K)=D(1)*R(1,K)+D(2)*R(2,K)+D(3)*R(3,K)
2790 NEXT K
2800 REM
2810 PRINT"P(1)";TAB(9);:PRINT USING G$;P(1,0)
2820 PRINT"P(2)";TAB(9);:PRINT USING G$;P(2,0)
2830 PRINT"P(3)";TAB(9);:PRINT USING G$;P(3,0)
2840 PRINT
2850 PRINT"T(2)";TAB(9);:PRINT USING G$;T(2)
2860 REM
2870 PRINT"R(2)";TAB(9);:PRINT USING G$;R(2,0)
2880 PRINT
2890 PRINT"R(2,K)";TAB(9);
2900 PRINT USING G$;R(2,1),R(2,2),R(2,3)
2910 PRINT"V(2,K)";TAB(9);
2920 PRINT USING G$;V(2,1),V(2,2),V(2,3)
2930 PRINT
2940 PRINT N$
2950 PRINT"METHOD OF OLBERS"
2960 END
10000 STOP
10010 REM # EULER/SUB # EULER'S EQUATION
10020 P(3,0)=QQ*P(1,0)+UU
10030 FOR K=1 TO 3
10040    R(1,K)=P(1,0)*LL(1,K)-RR(1,K)
10050    R(3,K)=P(3,0)*LL(3,K)-RR(3,K)
10060    S(K)=R(3,K)-R(1,K)
10070 NEXT K
10080 R(1,0)=FNMG(R(1,1),R(1,2),R(1,3))
```

```
10090 R(3,0)=FNMG(R(3,1),R(3,2),R(3,3))
10100 S=FNMG(S(1),S(2),S(3))
10110 E1=SQR(R(1,0)+R(3,0)+S)
10120 E2=SQR(R(1,0)+R(3,0)-S)
10130 F=E1*E1*E1-E2*E2*E2-6*TT
10140 RETURN
30000 DATA "COMET Z", "J2000.0", 0.017202099#
30010 REM
30020 DATA   8127.41234#
30030 DATA   20.61033#,  8.24998#
30040 DATA   -0.8812239#,+0.4548668#,+0.1971999#
30050 REM
30060 DATA   8128.42345#
30070 DATA   20.38154#,  5.66773#
30080 DATA -0.8893383#,+0.4409433#,+0.1911635#
30090 REM
30100 DATA   8129.43456#
30110 DATA   20.14891#,  2.98179#
30120 DATA   -0.8971907#,+0.4268914#,+0.1850716#
30130 REM
30140 REM
50000 DATA "REBEK-JEWEL", "J2000.0", 0.017202099#
50010 REM
50020 DATA   6373.57402#
50030 DATA   5.20632#,  22.07521#
50040 DATA -0.7394278#,-0.6063494#,-0.2629346#
50050 REM
50060 DATA 6374.57284#
50070 DATA   5.12686#,  22.14104#
50080 DATA -0.7275905#,-0.6179561#,-0.2679669#
50090 REM
50100 DATA 6375.57163#
50110 DATA   5.04222#,  22.20041#
50120 DATA -0.7155322#,-0.6293764#,-0.2729185#
```

11.6 Numerical Example

11.6.1 The Orbit of Comet Z

Problem

Topocentric angular position data for comet Z are shown in the table below. The positions are referred to the equinox J2000.0, and all Julian dates are expressed in terrestrial dynamical time (TDT). The observation site is at west longitude $77°35'41''$ and north latitude $37°31'33''$.

JD	α	δ
2448125.39012	21.04982	12.97929
2448126.40123	20.83358	10.69527
2448127.41234	20.61033	8.24998
2448128.42345	20.38154	5.66773
2448129.43456	20.14891	2.98179
2448130.44567	19.91429	0.23273

Compute a set of preliminary position and velocity elements by the method of Olbers, using the specific data listed below. The $\mathbf{R}_i$ vectors were obtained by employing HELO to perform second-order interpolations of data published in *The Astronomical Almanac 1990*.

- equinox: J2000.0
- $k = 0.017202099$
- $t_1 = 8127.41234$

  $\alpha_1 = 20.61033$

  $\delta_1 = 8.24998$

  $\mathbf{R}_1 = \{-0.8812239, +0.4548668, +0.1971999\}$

- $t_2 = 8128.42345$

  $\alpha_2 = 20.38154$

  $\delta_2 = 5.66773$

  $\mathbf{R}_2 = \{-0.8893383, +0.4409433, +0.1911635\}$

- $t_3 = 8129.43456$

  $\alpha_3 = 20.14891$

  $\delta_3 = 2.98179$

  $\mathbf{R}_3 = \{-0.8971907, +0.4268914, +0.1850716\}$

Solution

Use the given data to write lines 30000 to 30120 as shown at the end of program OLBERS. Run the program and answer the first two prompts as follows:

>>>TABLE LIMITS: PMIN, PMAX? 0,1

>>>BRACKET: P(1)MIN, P(1)MAX? 0.4,0.5

Once the first approximation of the orbital elements has been computed, answer the two prompts again as follows:

>>>TABLE LIMITS: PMIN, PMAX? 0,1

>>>BRACKET: P(1)MIN, P(1)MAX? 0.4,0.5

11.6. NUMERICAL EXAMPLE

Results

```
PARABOLIC ORBIT METHOD OF OLBERS
COMET Z
J2000.0

TA(I)      8127.4123400
A(I)         20.6103300
D(I)          8.2499800
LL(I,K)       0.6248855  -0.7674165   0.1434923
RR(I,K)      -0.8812239   0.4548668   0.1971999

TA(I)      8128.4234500
A(I)         20.3815400
D(I)          5.6677300
LL(I,K)       0.5810142  -0.8078794   0.0987593
RR(I,K)      -0.8893383   0.4409433   0.1911635

TA(I)      8129.4345600
A(I)         20.1489100
D(I)          2.9817900
LL(I,K)       0.5326510  -0.8447348   0.0520186
RR(I,K)      -0.8971907   0.4268914   0.1850716

*** NUMERICAL COEFFICIENTS ***

CC(1)        -0.0374132
CC(3)         0.0385132

DD(1)        -0.0071664
DD(3)         0.0071657

C(1)          0.5000000
C(3)          0.5000000

QQ            0.9714387
UU           -0.0000176

>>>TABLE LIMITS: PMIN, PMAX?   0,1
```

*** SOLUTION OF THE ORBIT EQUATION ***

P(1)	P(3)	F
0.00000	-0.00002	-0.06187
0.10000	0.09713	-0.10468
0.20000	0.19427	-0.11959
0.30000	0.29141	-0.08608
0.40000	0.38856	-0.02343
0.50000	0.48570	0.05221
0.60000	0.58285	0.13569
0.70000	0.67999	0.22518
0.80000	0.77713	0.31989
0.90000	0.87428	0.41934
1.00000	0.97142	0.52325

FIRST APPROXIMATION
\>\>\>BRACKET: P(1)MIN, P(1)MAX?   0.4,0.5

*** NUMERICAL COEFFICIENTS ***

CC(1)   -0.0374132
CC(3)   0.0385132

DD(1)   -0.0071664
DD(3)   0.0071657

C(1)   0.5000244
C(3)   0.5000321

QQ   0.9714237
UU   -0.0000147

\>\>\>TABLE LIMITS: PMIN, PMAX?   0,1

11.6. NUMERICAL EXAMPLE

*** SOLUTION OF THE ORBIT EQUATION ***

P(1)	P(3)	F
0.00000	-0.00001	-0.06188
0.10000	0.09713	-0.10469
0.20000	0.19427	-0.11960
0.30000	0.29141	-0.08608
0.40000	0.38855	-0.02343
0.50000	0.48570	0.05221
0.60000	0.58284	0.13569
0.70000	0.67998	0.22519
0.80000	0.77712	0.31990
0.90000	0.87427	0.41935
1.00000	0.97141	0.52326

>>>BRACKET: P(1)MIN, P(1)MAX? 0.4,0.5

*** PRELIMINARY ORBITAL ELEMENTS ***

P(1)	0.4323721		
P(2)	0.4249150		
P(3)	0.4200018		
T(2)	8128.4209959		
R(2)	1.3886189		
R(2,K)	1.1362199	-0.7842233	-0.1491992
V(2,K)	-0.8767848	0.1435572	-0.8067960

COMET Z
METHOD OF OLBERS

Discussion of Results

When the preliminary position and velocity elements computed by OLBERS are used in program CLASSEL, the following set of classical elements is obtained for comet Z:

- epoch: 2448128.42100
- $q = 0.9001239$
- $e = 1.0000000$
- $T = 2448189.48913$
- $i = 132.00491°$
- $\Omega = 138.99852°$
- $\omega = 242.98905°$

For comparison, the classical elements which correspond to the vector elements used to generate the observed positions for this example are as follows:

- epoch: 2448128.5
- $q = 0.9$
- $e = 1$
- $T = 2448189.5$
- $i = 132°$
- $\Omega = 139°$
- $\omega = 243°$

11.6. NUMERICAL EXAMPLE

11.6.2 The Orbit of Comet Rebek-Jewel

Problem

The angular position data listed below for comet Rebek-Jewel were taken from the numerical example of Section 8.6.2. The values of the $\mathbf{R}_i$ vectors were obtained by employing HELO to perform second-order interpolations of the data listed at the end of that program. Use the given information to compute a set of preliminary position and velocity elements by the method of Olbers.

- equinox: J2000.0

- $k = 0.017202099$

- $t_1 = 6373.57402$

  $\alpha_1 = 5.20632$

  $\delta_1 = 22.07521$

  $\mathbf{R}_1 = \{-0.7394278, -0.6063494, -0.2629346\}$

- $t_2 = 6374.57284$

  $\alpha_2 = 5.12686$

  $\delta_2 = 22.14104$

  $\mathbf{R}_2 = \{-0.7275905, -0.6179561, -0.2679669\}$

- $t_3 = 6375.57163$

  $\alpha_3 = 5.04222$

  $\delta_3 = 22.20041$

  $\mathbf{R}_3 = \{-0.7155322, -0.6293764, -0.2729185\}$

Solution

Use the given data to write lines 50000 to 50120 as shown at the end of program OLBERS. Run the program and answer the first two prompts as follows:

```
>>>TABLE LIMITS: PMIN, PMAX? 0,2

>>>BRACKET: P(1)MIN, P(1)MAX? 1.0,1.2
```

Once the first approximation of the numerical coefficients has been computed, answer the two prompts again as follows:

```
>>>TABLE LIMITS: PMIN, PMAX? 0,2

>>>BRACKET: P(1)MIN, P(1)MAX? 1.0,1.2
```

11.6. NUMERICAL EXAMPLE

Results

```
PARABOLIC ORBIT METHOD OF OLBERS
REBEK-JEWEL
J2000.0

TA(I)     6373.5740200
A(I)         5.2063200
D(I)        22.0752100
LL(I,K)      0.1911699    0.9067584    0.3758233
RR(I,K)     -0.7394278   -0.6063494   -0.2629346

TA(I)     6374.5728400
A(I)         5.1268600
D(I)        22.1410400
LL(I,K)      0.2098921    0.9021645    0.3768878
RR(I,K)     -0.7275905   -0.6179561   -0.2679669

TA(I)     6375.5716300
A(I)         5.0422200
D(I)        22.2004100
LL(I,K)      0.2297327    0.8969137    0.3778474
RR(I,K)     -0.7155322   -0.6293764   -0.2729185

*** NUMERICAL COEFFICIENTS ***

CC(1)       -0.0013963
CC(3)        0.0014744

DD(1)        0.0002253
DD(3)       -0.0002254

C(1)         0.4999925
C(3)         0.5000075

QQ           0.9470309
UU          -0.0000600

>>>TABLE LIMITS: PMIN, PMAX?  0,2
```

*** SOLUTION OF THE ORBIT EQUATION ***

P(1)	P(3)	F
0.00000	-0.00006	-0.05992
0.20000	0.18935	-0.10217
0.40000	0.37875	-0.13949
0.60000	0.56816	-0.13313
0.80000	0.75756	-0.07661
1.00000	0.94697	-0.00066
1.20000	1.13638	0.08568
1.40000	1.32578	0.18005
1.60000	1.51519	0.28144
1.80000	1.70460	0.38928
2.00000	1.89400	0.50316

FIRST APPROXIMATION
>>>BRACKET: P(1)MIN, P(1)MAX?   1.0,1.2

*** NUMERICAL COEFFICIENTS ***

| CC(1) | -0.0013963 |
| CC(3) | 0.0014744 |

| DD(1) | 0.0002253 |
| DD(3) | -0.0002254 |

| C(1) | 0.5000028 |
| C(3) | 0.5000196 |

| QQ | 0.9470274 |
| UU | -0.0000606 |

>>>TABLE LIMITS: PMIN, PMAX?   0,2

11.6. NUMERICAL EXAMPLE

*** SOLUTION OF THE ORBIT EQUATION ***

P(1)	P(3)	F
0.00000	-0.00006	-0.05995
0.20000	0.18934	-0.10220
0.40000	0.37875	-0.13952
0.60000	0.56816	-0.13315
0.80000	0.75756	-0.07662
1.00000	0.94697	-0.00067
1.20000	1.13637	0.08568
1.40000	1.32578	0.18005
1.60000	1.51518	0.28144
1.80000	1.70459	0.38928
2.00000	1.89399	0.50317

>>>BRACKET: P(1)MIN, P(1)MAX? 1.0,1.2

*** PRELIMINARY ORBITAL ELEMENTS ***

P(1)	1.0016418
P(2)	0.9748056
P(3)	0.9485217

| T(2) | 6374.5672100 |
| R(2) | 1.8747929 |

| R(2,K) | 0.9321945 | 1.4973911 | 0.6353593 |
| V(2,K) | 0.0735395 | -1.0031844 | -0.2345139 |

REBEK-JEWEL
METHOD OF OLBERS

Discussion of Results

When the preliminary position and velocity elements computed by OLBERS are used in program CLASSEL, the following set of classical elements is obtained for comet Rebek-Jewel:

- epoch: 2446374.56721
- $q = 0.6224694$
- $e = 0.9999995$
- $T = 2446470.23977$
- $i = 162.57596°$
- $\Omega = 58.94610°$
- $\omega = 108.33491°$

References

[1] Herget, *The Computation of Orbits*, Published by Author, 1948.

[2] Dubyago, *The Determination of Orbits*, The Rand Corporation, 1961.

[3] Roy, *Orbital Motion*, Adam Hilger Ltd., 1978.

[4] Baker, *Astrodynamics: Applications and Advanced Topics*, Academic Press, 1967.

[5] Escobal, *Methods of Orbit Determination*, Krieger Publishing Co., 1976.

Chapter 12

Orbit Improvement

12.1 Introduction

When preliminary orbital elements are used to predict the motion of a celestial body, it is normally found that the observed and computed motion are not in satisfactory agreement. Thus, observed-minus computed (O-C) residuals can be determined for each time of observation by subtracting the computed coordinates from the measured coordinates. When there exist at least three reliable observations which cover a significant part of the orbital motion, yet are not too distant from the epoch of the preliminary elements, it is often possible to improve the elements by a straightforward differential correction process which ignores perturbations. If the situation does not permit this simple approach, then perturbations should be taken into account each time the residuals are determined.

In contrast to the dynamical problem of determining the orbital parameters initially, differential correction is primarily a numerical procedure which uses a multiple linear least squares regression to make small changes to the elements in order to minimize the O-C residuals. There is no guarantee that the resulting element set will ultimately prove to be better than the preliminary one. Only time and additional observations can finally decide. However, given an initial element set which is not too far off the mark and observations sufficiently accurate and spaced to get a representative sample of the conic section of the orbit, the least squares differential correction process can be very effective [1,2,3].

12.2 The Differential Equations of Condition

A celestial body's right ascension α and declination δ are complex functions of the orbital elements and the components of **R**, the position of the dynamical center. However, since the vector **R** may be regarded as accurately known, in need of no improvement, we can simplify our problem by considering only the functional dependence of α and δ on the position and velocity elements

$$\mathbf{r}_0 = \{x_0, y_0, z_0\}$$
$$\mathbf{v}_0 = \{\dot{x}_0, \dot{y}_0, \dot{z}_0\}$$

at the arbitrary epoch t_0. Thus, we let

$$\alpha = \alpha(x_0, y_0, z_0, \dot{x}_0, \dot{y}_0, \dot{z}_0) \qquad (12.1)$$
$$\delta = \delta(x_0, y_0, z_0, \dot{x}_0, \dot{y}_0, \dot{z}_0), \qquad (12.2)$$

and apply the definition of the total derivative of a function to obtain [2,4]

$$d\alpha = \frac{\partial \alpha}{\partial x_0}dx_0 + \frac{\partial \alpha}{\partial y_0}dy_0 + \frac{\partial \alpha}{\partial z_0}dz_0 + \cdots + \frac{\partial \alpha}{\partial \dot{z}_0}d\dot{z}_0 \qquad (12.3)$$

$$d\delta = \frac{\partial \delta}{\partial x_0}dx_0 + \frac{\partial \delta}{\partial y_0}dy_0 + \frac{\partial \delta}{\partial z_0}dz_0 + \cdots + \frac{\partial \delta}{\partial \dot{z}_0}d\dot{z}_0. \qquad (12.4)$$

Equations 12.3 and 12.4 express the amount of change produced in α and δ in response to independent changes in one or more of the scalar components of the position and velocity vectors. The partial derivatives represent the individual *rates* at which α and δ change with respect to each of the orbital elements. In practice, the differentials can be replaced by finite differences, so that we can write

$$\Delta\alpha = \frac{\partial \alpha}{\partial x_0}\Delta x_0 + \frac{\partial \alpha}{\partial y_0}\Delta y_0 + \frac{\partial \alpha}{\partial z_0}\Delta z_0 + \cdots + \frac{\partial \alpha}{\partial \dot{z}_0}\Delta \dot{z}_0 \qquad (12.5)$$

$$\Delta\delta = \frac{\partial \delta}{\partial x_0}\Delta x_0 + \frac{\partial \delta}{\partial y_0}\Delta y_0 + \frac{\partial \delta}{\partial z_0}\Delta z_0 + \cdots + \frac{\partial \delta}{\partial \dot{z}_0}\Delta \dot{z}_0, \qquad (12.6)$$

where $\Delta\alpha$ and $\Delta\delta$ are the measured O-C residuals in right ascension and declination, respectively, and $\Delta x_0, \Delta y_0, \ldots, \Delta \dot{z}_0$ are the small changes needed to improve the orbital elements so that the residuals in α and δ are eliminated.

The residuals $\Delta\alpha$ and $\Delta\delta$ in Equations 12.5 and 12.6 are known quantities obtained from measurements of the orbiting body's position on the celestial sphere. Three such observations would enable us to write six independent linear equations of condition. However, since there may be more than three sets of residuals available, we have in general

$$\Delta\alpha_1 = \frac{\partial \alpha_1}{\partial x_0}\Delta x_0 + \frac{\partial \alpha_1}{\partial y_0}\Delta y_0 + \frac{\partial \alpha_1}{\partial z_0}\Delta z_0 + \cdots + \frac{\partial \alpha_1}{\partial \dot{z}_0}\Delta \dot{z}_0$$

$$\Delta\alpha_2 = \frac{\partial\alpha_2}{\partial x_0}\Delta x_0 + \frac{\partial\alpha_2}{\partial y_0}\Delta y_0 + \frac{\partial\alpha_2}{\partial z_0}\Delta z_0 + \cdots + \frac{\partial\alpha_2}{\partial \dot{z}_0}\Delta \dot{z}_0$$

$$\Delta\alpha_3 = \frac{\partial\alpha_3}{\partial x_0}\Delta x_0 + \frac{\partial\alpha_3}{\partial y_0}\Delta y_0 + \frac{\partial\alpha_3}{\partial z_0}\Delta z_0 + \cdots + \frac{\partial\alpha_3}{\partial \dot{z}_0}\Delta \dot{z}_0$$

$$\vdots$$

$$\Delta\alpha_n = \frac{\partial\alpha_n}{\partial x_0}\Delta x_0 + \frac{\partial\alpha_n}{\partial y_0}\Delta y_0 + \frac{\partial\alpha_n}{\partial z_0}\Delta z_0 + \cdots + \frac{\partial\alpha_n}{\partial \dot{z}_0}\Delta \dot{z}_0$$

$$\Delta\delta_1 = \frac{\partial\delta_1}{\partial x_0}\Delta x_0 + \frac{\partial\delta_1}{\partial y_0}\Delta y_0 + \frac{\partial\delta_1}{\partial z_0}\Delta z_0 + \cdots + \frac{\partial\delta_1}{\partial \dot{z}_0}\Delta \dot{z}_0$$

$$\Delta\delta_2 = \frac{\partial\delta_2}{\partial x_0}\Delta x_0 + \frac{\partial\delta_2}{\partial y_0}\Delta y_0 + \frac{\partial\delta_2}{\partial z_0}\Delta z_0 + \cdots + \frac{\partial\delta_2}{\partial \dot{z}_0}\Delta \dot{z}_0$$

$$\Delta\delta_3 = \frac{\partial\delta_3}{\partial x_0}\Delta x_0 + \frac{\partial\delta_3}{\partial y_0}\Delta y_0 + \frac{\partial\delta_3}{\partial z_0}\Delta z_0 + \cdots + \frac{\partial\delta_3}{\partial \dot{z}_0}\Delta \dot{z}_0$$

$$\vdots$$

$$\Delta\delta_n = \frac{\partial\delta_n}{\partial x_0}\Delta x_0 + \frac{\partial\delta_n}{\partial y_0}\Delta y_0 + \frac{\partial\delta_n}{\partial z_0}\Delta z_0 + \cdots + \frac{\partial\delta_n}{\partial \dot{z}_0}\Delta \dot{z}_0, \quad (12.7)$$

where $n \geq 3$. Now, if we are somehow able to obtain reasonable values for the partial derivatives, then these equations can be solved by multiple linear least squares regression to yield values for the corrections $\Delta x_0, \Delta y_0, \Delta z_0, \Delta \dot{x}_0, \Delta \dot{y}_0$, and $\Delta \dot{z}_0$ which best fit all the data.

12.3 Numerical Evaluation of the Partial Derivatives

Values for the partial derivatives in Equations 12.7 can be determined to sufficient accuracy by a simple numerical process which is accomplished by the computer. If ϵ represents any one of the six orbital elements, and $\Delta\epsilon$ is some small change introduced in that typical element, then the partial derivatives of α_i and δ_i with respect to ϵ can be approximated as follows [2]:

$$\frac{\partial\alpha_i}{\partial\epsilon} \approx \frac{\alpha_i(x_0,\ldots,\epsilon_0+\Delta\epsilon,\ldots,\dot{z}_0) - \alpha_i(x_0,\ldots,\epsilon_0,\ldots,\dot{z}_0)}{\Delta\epsilon} \quad (12.8)$$

$$\frac{\partial\delta_i}{\partial\epsilon} \approx \frac{\delta_i(x_0,\ldots,\epsilon_0+\Delta\epsilon,\ldots,\dot{z}_0) - \delta_i(x_0,\ldots,\epsilon_0,\ldots,\dot{z}_0)}{\Delta\epsilon}, \quad (12.9)$$

where $i = 1$ to n. Therefore, by incrementing each element in turn while the others maintain their original values, Equations 12.8 and 12.9 will produce the six partial derivatives of α_i and δ_i at each observation time t_i. In most cases, selecting an incrementation $\Delta\epsilon$ which is equal to a few percent of ϵ will produce a small change in α_i and δ_i which is sufficient to yield satisfactory approximations of the partial derivatives [2,3,5].

Once the partial derivatives have been determined, Equations 12.7 are solved for the corrections $\Delta x_0, \Delta y_0, \ldots, \Delta \dot{z}_0$. When these are added to the original elements, a new set

$$\begin{aligned} \mathbf{r}_0 &= \{x_0 + \Delta x_0, y_0 + \Delta y_0, z_0 + \Delta z_0\} \\ \mathbf{v}_0 &= \{\dot{x}_0 + \Delta \dot{x}_0, \dot{y}_0 + \Delta \dot{y}_0, \dot{z}_0 + \Delta \dot{z}_0\} \end{aligned} \qquad (12.10)$$

is obtained for the arbitrary epoch t_0. Finally, new angular positions are generated and compared to the observations. If significant residuals remain, these may be used to compute further improvements to the elements and the entire process repeated until the observed and computed α_i and δ_i agree within the limits of the accuracy of each angular measurement.

12.4 Comparing Observation with Theory

Two least-squares orbit improvement programs are described in the following section. Program IMPROVE is based on the simplifying assumption that all observations are not too distant in time from the epoch of the preliminary elements. This approach neglects the effects of perturbations to produce a computer routine which successively improves its own results in a single *self-contained* program. However, when perturbations are present and the observations are spread over several years, the improvement process is much more lengthy and better handled in two separate operations.

Program CORRECT is a modification of IMPROVE which can be used along with COWELL or ENCKE to refine the orbital elements when perturbations must be taken into account. Each time the least-squares program is applied to reduce the latest group of residuals, the resulting set of improved elements must be taken over to the special perturbation program to determine the next group of residuals. These, in turn, are used again in the least-squares program, and the process continued until the residuals in α and δ are within the uncertainties of the observed positions. One should keep in mind that any such mathematical procedure does not insure that the improved elements are the true orbital elements of the celestial body. The new elements are simply ones that have been forced into a form which will produce agreement with a specific set of observations using a specific set of perturbations. However, if the observations are reliable, distributed fairly evenly over a sizable portion of the orbit, and all significant perturbations are accounted for, then the improved elements will probably be a much better representation of the orbit than the original set.

12.5 Computer Programs

12.5.1 Program IMPROVE

Program IMPROVE computes improvements to geocentric or heliocentric position and velocity elements. IMPROVE uses the universal formulation to predict orbital motion and assumes that perturbations are not significant during time interval of the computation.

Program Algorithm

Define	Line Number
FNVS(X,Y,Z): vector squaring function	1080
FNMG(X,Y,Z): vector magnitude function	1090
FNDP(X1,Y1,Z1,X2,Y2,Z2): dot product function	1100
FNASN(X): inverse sine function	1110
FNACN(X): inverse cosine function	1120

Given	Line Number
angle to radian conversion factor	1160
$1/c$: light-time constant	1220
name of the object for information only	1240
equinox for information only	1240
k: gravitational constant	1240
μ: combined mass	1250
t_0: epoch time	1260
$\mathbf{r}_0$: position elements	1270
$\mathbf{v}_0$: velocity elements	1280

number of ephemeris positions	1300
t_i: time for each ephemeris position	1330
α_i, δ_i: observed angular coordinates	1330
t_i: time for each position of the dynamical center	1380
$\mathbf{R}_i$: position of the dynamical center for each t_i	1380
percent change to the elements for partial derivatives	1470

Compute	Line Number
B_n: coefficients for the universal f and g expressions	1510-1540
α_i and δ_i in hours and degrees, respectively	1920-1950
$\Delta\alpha_i$ in hours	1970
$\Delta\delta_i$ in degrees	1980
For each element:	2140
$\quad \epsilon_0 + \Delta\epsilon$	2160-2240
$\quad \alpha_i(x_0,\ldots,\epsilon_0+\Delta\epsilon,\ldots,\dot{z}_0)$ for Equation 12.8	2260-2300
$\quad \delta_i(x_0,\ldots,\epsilon_0+\Delta\epsilon,\ldots,\dot{z}_0)$ for Equation 12.9	2260-2310
For each element and observation:	2350
$\quad \partial\alpha_i/\partial\epsilon$: Equation 12.8	2370
$\quad \partial\delta_i/\partial\epsilon$: Equation 12.9	2380
coefficients of the equations of condition	2420-2490
Multiple linear least squares regression:	2510

12.5. COMPUTER PROGRAMS

coefficients of the normal equations	17090-17160
$\Delta x_0, \Delta y_0, \ldots, \Delta \dot{z}_0$ by Gauss elimination	17220-17560
least squares residuals	17600-17660
$\mathbf{r}_0 = \{x_0 + \Delta x_0, y_0 + \Delta y_0, z_0 + \Delta z_0\}$: Equations 12.10	2650
$\mathbf{v}_0 = \{\dot{x}_0 + \Delta \dot{x}_0, \dot{y}_0 + \Delta \dot{y}_0, \dot{z}_0 + \Delta \dot{z}_0\}$: Equations 12.10	2650
residuals in α_i and δ_i using improved elements	2770
continued improvement, if desired	2060

End.

Program Listing

```
1000 CLS
1010 PRINT"# IMPROVE # TWO-BODY ORBIT IMPROVEMENT"
1020 PRINT"# UNIVERSAL VARIABLES"
1030 PRINT"# METHOD OF LEAST SQUARES"
1040 PRINT
1050 DEFDBL A-Z
1060 DEFINT I,J,K,N
1070 REM
1080 DEF FNVS(X,Y,Z)=X*X+Y*Y+Z*Z
1090 DEF FNMG(X,Y,Z)=SQR(X*X+Y*Y+Z*Z)
1100 DEF FNDP(X1,Y1,Z1,X2,Y2,Z2)=X1*X2+Y1*Y2+Z1*Z2
1110 DEF FNASN(X)=ATN(X/SQR(-X*X+1))
1120 DEF FNACN(X)=-ATN(X/SQR(-X*X+1))+1.5707963263#
1130 REM
1140 DIM B(19),X(20,7),SM(20),S(20)
1150 REM
1160 Q1=.017453293#
1170 G$="#####.#######"
1180 T$="####.#####"
1190 A$="#####.#####"
1200 D$="#######.#####"
1210 REM
1220 READ AB
1230 REM
1240 READ N$, E$, K#
1250 READ M
1260 READ T(0)
1270 READ E0(1),E0(2),E0(3)
1280 READ E0(4),E0(5),E0(6)
1290 REM
1300 READ NP
1310 REM
1320 FOR I=1 TO NP
1330    READ TA(I),AO(I),DO(I)
1340    T(I)=TA(I)
1350 NEXT I
1360 REM
1370 FOR I=1 TO NP
1380    READ TX(I),RR(I,1),RR(I,2),RR(I,3)
1390    IF TX(I)<>TA(I) THEN 1440
```

12.5. COMPUTER PROGRAMS

```
1400 NEXT I
1410 REM
1420 GOTO 1470
1430 REM
1440 PRINT"??? TIMES DO NOT AGREE ???"
1450 STOP
1460 REM
1470 INPUT">>>PERCENT CHANGE TO ELEMENTS";PC
1480 PRINT
1490 REM B(J)=1/J!
1500 REM
1510 B(1)=1
1520 FOR J=2 TO 19
1530    B(J)=B(J-1)/J
1540 NEXT J
1550 CLS
1560 PRINT"TWO-BODY ORBIT IMPROVEMENT"
1570 PRINT N$
1580 PRINT E$
1590 PRINT
1600 PRINT"EPOCH";TAB(11);:PRINT USING G$;T(0)
1610 PRINT
1620 PRINT"ELEMENTS"
1630 PRINT
1640 FOR N=1 TO 6
1650    PRINT"E(";N;")";TAB(11);
1660    PRINT USING G$;E0(N)
1670 NEXT N
1680 LINE INPUT"";L$
1690 CLS
1700 PRINT"DYNAMICAL CENTER"
1710 PRINT
1720 PRINT TAB(6)"TA(I)";
1730 PRINT TAB(17)"RR(I,1)";TAB(30)"RR(I,2)";TAB(43)"RR(I,3)"
1740 FOR I=1 TO NP
1750    PRINT USING T$;TA(I);
1760    PRINT USING G$;RR(I,1);RR(I,2);RR(I,3)
1770 NEXT I
1780 LINE INPUT"";L$
1790 CLS
1800 PRINT"RESIDUALS IN RA AND DEC"
1810 PRINT
```

```
1820 PRINT TAB(6)"TA(I)";
1830 PRINT TAB(17)"DA(I)";TAB(30)"DD(I)"
1840 REM
1850 FOR K=1 TO 3
1860   R(0,K)=E0(K)
1870   V(0,K)=E0(K+3)
1880 NEXT K
1890 REM
1900 FOR I=1 TO NP
1910   REM
1920   GOSUB 16010 REM POS/SUB
1930   REM
1940   AC(I,0)=AX
1950   DC(I,0)=DX
1960   REM
1970   DA(I)=A0(I)-AC(I,0)
1980   DD(I)=D0(I)-DC(I,0)
1990   REM
2000   PRINT USING T$;TA(I);
2010   PRINT USING A$;DA(I);
2020   PRINT USING D$;DD(I)
2030   REM
2040 NEXT I
2050 REM
2060 INPUT">>>CONTINUE? Y/N";C$
2070 REM
2080 IF C$="N" THEN 2790 ELSE 2090
2090 CLS
2100 PRINT"PERCENT CHANGE"
2110 PRINT
2120 PRINT"PC =";PC;"%"
2130 PRINT
2140 FOR N=1 TO 6
2150   REM
2160   FOR J=1 TO 6
2170     CE(J)=E0(J)
2180   NEXT J
2190   PE(N)=ABS(E0(N)*PC/100)
2200   CE(N)=E0(N)+PE(N)
2210   FOR K=1 TO 3
2220     R(0,K)=CE(K)
2230     V(0,K)=CE(K+3)
```

12.5. COMPUTER PROGRAMS

```
2240    NEXT K
2250    REM
2260    FOR I=1 TO NP
2270      REM
2280      GOSUB 16010 REM POS/SUB
2290      REM
2300      AC(I,N)=AX
2310      DC(I,N)=DX
2320    NEXT I
2330 NEXT N
2340 REM
2350 FOR I=1 TO NP
2360   FOR N=1 TO 6
2370     PA(I,N)=Q1*15*(AC(I,N)-AC(I,0))/PE(N)
2380     PD(I,N)=Q1*(DC(I,N)-DC(I,0))/PE(N)
2390   NEXT N
2400 NEXT I
2410 REM
2420 FOR I=1 TO NP
2430   FOR N=1 TO 6
2440     X(I,N)=PA(I,N)
2450     X(I+NP,N)=PD(I,N)
2460   NEXT N
2470   X(I,7)=Q1*15*DA(I)
2480   X(I+NP,7)-Q1*DD(I)
2490 NEXT I
2500 REM
2510 GOSUB 17010 REM MGRESS/SUB
2520 REM
2530 PRINT"DIFFERENTIAL CORRECTIONS"
2540 PRINT
2550 FOR N=1 TO 6
2560   DE(N)=XU(N)
2570   PRINT"DE(";N;")";TAB(11);
2580   PRINT USING G$;DE(N)
2590 NEXT N
2600 LINE INPUT"";L$
2610 CLS
2620 PRINT"IMPROVED ELEMENTS"
2630 PRINT
2640 FOR N=1 TO 6
2650   E0(N)=E0(N)+DE(N)
```

```
2660    PRINT"E(";N;")";TAB(11);
2670    PRINT USING G$;EO(N)
2680 NEXT N
2690 REM
2700 REM
2710 REM   CONCLUDE
2720 REM   BY PRINTING
2730 REM   NEW RESIDUALS
2740 REM   USING
2750 REM   IMPROVED ELEMENTS
2760 REM
2770 GOTO 1780
2780 REM
2790 PRINT
2800 PRINT N$
2810 END
16000 STOP
16010 REM # POS/SUB # COMPUTES POSITIONS
16020 REM # BY UNIVERSAL VARIABLES
16030 REM
16040 WW=K#*SQR(M)*(T(I)-T(I-1))
16050 RO=FNMG(R(I-1,1),R(I-1,2),R(I-1,3))
16060 DO=FNDP(R(I-1,1),R(I-1,2),R(I-1,3),
      V(I-1,1),V(I-1,2),V(I-1,3))/SQR(M)
16070 AI=2/RO-FNVS(V(I-1,1),V(I-1,2),V(I-1,3))/M
16080 CO=1-RO*AI
16090 XX=WW/RO
16100    X2=XX*XX
16110    XA=X2*AI
16120    X3=X2*XX
16130    CC=X2*(B(2)-XA*(B(4)-XA*(B(6)-XA*(B(8)-XA*(B(10)-
         XA*(B(12)-XA*(B(14)-XA*(B(16)-XA*(B(18))))))))))
16140    UU=X3*(B(3)-XA*(B(5)-XA*(B(7)-XA*(B(9)-XA*(B(11)-
         XA*(B(13)-XA*(B(15)-XA*(B(17)-XA*(B(19))))))))))
16150    SS=XX-UU*AI
16160    FX=RO*XX+CO*UU+DO*CC-WW
16170    IF ABS(FX)<.0000001# THEN 16210
16180    DF=RO+CO*CC+DO*SS
16190    XX=XX-FX/DF
16200    GOTO 16100
16210 F=1-CC/RO
16220 G=(RO*SS+DO*CC)/SQR(M)
```

12.5. COMPUTER PROGRAMS

```
16230 R=R0+C0*CC+D0*SS
16240 FP=-SQR(M)*SS/(R*R0)
16250 GP=1-CC/R
16260 FOR K=1 TO 3
16270    R(I,K)=F*R(I-1,K)+G*V(I-1,K)
16280    V(I,K)=FP*R(I-1,K)+GP*V(I-1,K)
16290    P(I,K)=R(I,K)+RR(I,K)
16300 NEXT K
16310 P=FNMG(P(I,1),P(I,2),P(I,3))
16320 REM
16330 AP=AB*P
16340 IF T(I)-(TA(I)-AP)<.00001 THEN 16380
16350 T(I)=TA(I)-AP
16360 GOTO 16040
16370 REM
16380 FOR K=1 TO 3
16390    LL(I,K)=P(I,K)/P
16400 NEXT K
16410 CD=SQR(1-LL(I,3)*LL(I,3))
16420 CX=LL(I,1)/CD
16430 SX=LL(I,2)/CD
16440 REM
16450 IF ABS(SX)<=.707107 THEN X=FNASN(ABS(SX))
16460 IF ABS(CX)<=.707107 THEN X=FNACN(ABS(CX))
16470 IF CX>=0 AND SX>=0 THEN X=X
16480 IF CX<0 AND SX>=0 THEN X=180*Q1-X
16490 IF CX<0 AND SX<0 THEN X=180*Q1+X
16500 IF CX>=0 AND SX<0 THEN X=360*Q1-X
16510 REM
16520 AX=X/(Q1*15)
16530 DX=FNASN(LL(I,3))/Q1
16540 RETURN
17000 STOP
17010 REM # MGRESS/SUB # LEAST SQUARES
17020 REM # MULTIPLE LINEAR REGRESSION
17030 REM
17040 NM=6
17050 NE=2*NP
17060 REM
17070 REM NORMAL EQUATIONS
17080 REM
17090 FOR J=1 TO NM
```

```
17100    FOR K=1 TO (NM+1)
17110      A(J,K)=0
17120      FOR I=1 TO NE
17130        A(J,K)=A(J,K)+X(I,J)*X(I,K)
17140      NEXT I
17150    NEXT K
17160 NEXT J
17170 REM
17180 NN=NM
17190 REM
17200 REM GAUSS ELIMINATION
17210 REM
17220 FOR I=1 TO (NN-1)
17230    REM
17240    JP=I
17250    PE!=ABS(A(I,I))
17260    FOR J=(I+1) TO NN
17270      CE!=ABS(A(J,I))
17280      IF CE!-PE! < 0 GOTO 17330
17290      PE!=CE!
17300      JP=J
17310    NEXT J
17320    REM
17330    IF JP=I GOTO 17410
17340    REM
17350    FOR K=I TO (NN+1)
17360      HE=A(I,K)
17370      A(I,K)=A(JP,K)
17380      A(JP,K)=HE
17390    NEXT K
17400    REM
17410    FOR J=(I+1) TO NN
17420      FOR K=(I+1) TO (NN+1)
17430        A(J,K)=A(J,K)-A(J,I)*A(I,K)/A(I,I)
17440      NEXT K
17450      A(J,I)=0
17460    NEXT J
17470 NEXT I
17480 REM
17490 XU(NN)=A(NN,NN+1)/A(NN,NN)
17500 FOR I=(NN-1) TO 1 STEP -1
17510    SS=0
```

12.5. COMPUTER PROGRAMS

```
17520    FOR K=(I+1) TO NN
17530       SS=SS+A(I,K)*XU(K)
17540    NEXT K
17550    XU(I)=(A(I,NN+1)-SS)/A(I,I)
17560 NEXT I
17570 REM
17580 REM LEAST SQUARES RESIDUALS
17590 REM
17600 FOR I=1 TO NE
17610    SM(I)=0
17620    FOR K=1 TO NM
17630       SM(I)=SM(I)+X(I,K)*XU(K)
17640    NEXT K
17650    S(I)=X(I,7)-SM(I)
17660 NEXT I
17670 RETURN
20000 DATA 0
20010 'DATA 0.005775519#
20020 REM
30000 DATA   "GEOS(LAPLACE)","J2000.0",0.07436680#
30010 DATA   1
30020 DATA   99.08167#
30030 DATA +1.0566411#,+0.3289055#,+0.6103880#
30040 DATA -0.1616737#,+1.0723920#,-0.0957070#
30050 REM
30060 DATA 9
30070 REM
31000 REM # OBSERVATIONS #
31010 DATA   95.08167#, 2.09624#, 2.84935#
31020 DATA   96.08167#, 2.73443#, 2.39724#
31030 DATA   97.08167#, 3.25454#, 1.78343#
31040 DATA   98.08167#, 3.68088#, 1.08683#
31050 DATA   99.08167#, 4.03544#, 0.35237#
31060 DATA  100.08167#, 4.33550#,-0.39552#
31070 DATA  101.08167#, 4.59395#,-1.14379#
31080 DATA  102.08167#, 4.82023#,-1.88538#
31090 DATA  103.08167#, 5.02129#,-2.61638#
31100 REM
32000 REM # RR-VECTOR #
32010 DATA   95.08167#,-0.7717874#, 0.1833449#,-0.6068664#
32020 DATA   96.08167#,-0.7725822#, 0.1799627#,-0.6068675#
32030 DATA   97.08167#,-0.7733622#, 0.1765771#,-0.6068686#
```

```
32040 DATA   98.08167#,-0.7741273#, 0.1731881#,-0.6068696#
32050 DATA   99.08167#,-0.7748777#, 0.1697957#,-0.6068707#
32060 DATA  100.08167#,-0.7756131#, 0.1664002#,-0.6068717#
32070 DATA  101.08167#,-0.7763337#, 0.1630014#,-0.6068727#
32080 DATA  102.08167#,-0.7770395#, 0.1595996#,-0.6068737#
32090 DATA  103.08167#,-0.7777303#, 0.1561946#,-0.6068746#
32100 REM
32110 REM
40000 DATA "REBEK-JEWEL(LAPLACE)", "J2000.0", 0.017202099#
40010 DATA   1
40020 DATA   6374.5671889#
40030 DATA +0.9329630#,+1.5006932#,+0.6367384#
40040 DATA +0.0759682#,-1.0120252#,-0.2374871#
40050 REM
40060 DATA 5
40070 REM
41000 REM # OBSERVATIONS #
41010 DATA   6310.50000#,  6.19839#, 19.35543#
41020 DATA   6340.50000#,  6.23470#, 20.03022#
41030 DATA   6370.50000#,  5.42130#, 21.84796#
41040 DATA   6400.50000#,  1.13233#, 13.91750#
41050 DATA   6430.50000#, 22.33290#, -2.06082#
41060 REM
42000 REM # RR-VECTOR #
42010 DATA 6310.50000#,-0.9461372#, 0.3214489#, 0.1393729#
42020 DATA 6340.50000#,-0.9886643#,-0.1428416#,-0.0619377#
42030 DATA 6370.50000#,-0.7744047#,-0.5695064#,-0.2469331#
42040 DATA 6400.50000#,-0.3543651#,-0.8442170#,-0.3660427#
42050 DATA 6430.50000#, 0.1614512#,-0.8898986#,-0.3858482#
```

12.5. COMPUTER PROGRAMS

12.5.2 Program CORRECT

Program CORRECT is a modification of IMPROVE which is used along with COWELL or ENCKE to refine heliocentric position and velocity elements when perturbations are significant. The partial derivatives for the equations of conditions are computed by means of closed elliptic f and g expressions. The differential correction procedure is outlined in Section 12.4.

Program Algorithm

Define	Line Number
FNVS(X,Y,Z): vector squaring function	1080
FNMG(X,Y,Z): vector magnitude function	1090
FNDP(X1,Y1,Z1,X2,Y2,Z2): dot product function	1100
FNASN(X): inverse sine function	1110
FNACN(X): inverse cosine function	1120

Given	Line Number
angle to radian conversion factor	1160
$1/c$: light-time constant	1220
name of celestial body for information only	1240
equinox for information only	1240
k: gravitational constant	1240
m: mass of celestial body	1250
t_0: epoch time	1260
$\mathbf{r}_0$: position elements	1270
$\mathbf{v}_0$: velocity elements	1280

μ: combined mass	1300
number of pairs of residuals	1320
For each pair of residuals:	1340-1420
time of observation	1350
$\Delta\alpha$: right ascension residual in hours	1350
$\Delta\delta$: declination residual in degrees	1350
position of the Sun for each observation time	1400
percent change to the elements for partial derivatives	1480
Compute	Line Number
For each element:	1950-2190
$\epsilon_0 + \Delta\epsilon$	1970-2090
$\alpha_i(x_0,\ldots,\epsilon_0 + \Delta\epsilon,\ldots,\dot{z}_0)$ for Equation 12.8	2110-2150
$\delta_i(x_0,\ldots,\epsilon_0 + \Delta\epsilon,\ldots,\dot{z}_0)$ for Equation 12.9	2110-2160
For each element and observation:	2210-2260
$\partial\alpha_i/\partial\epsilon$: Equation 12.8	2230
$\partial\delta_i/\partial\epsilon$: Equation 12.9	2240
coefficients of the equations of condition	2300-2370
multiple linear least squares regression	2390
$\mathbf{r}_0 = \{x_0 + \Delta x_0, y_0 + \Delta y_0, z_0 + \Delta z_0\}$: Equations 12.10	2520-2560
$\mathbf{v}_0 = \{\dot{x}_0 + \Delta\dot{x}_0, \dot{y}_0 + \Delta\dot{y}_0, \dot{z}_0 + \Delta\dot{z}_0\}$: Equations 12.10	2520-2560

End.

12.5. COMPUTER PROGRAMS

Program Listing

```
1000 CLS
1010 PRINT"# CORRECT # ORBIT CORRECTION"
1020 PRINT"# CLOSED ELLIPTIC F&G EXPRESSIONS"
1030 PRINT"# METHOD OF LEAST SQUARES"
1040 PRINT
1050 DEFDBL A-Z
1060 DEFINT I,J,K,N
1070 REM
1080 DEF FNVS(X,Y,Z)=X*X+Y*Y+Z*Z
1090 DEF FNMG(X,Y,Z)=SQR(X*X+Y*Y+Z*Z)
1100 DEF FNDP(X1,Y1,Z1,X2,Y2,Z2)=X1*X2+Y1*Y2+Z1*Z2
1110 DEF FNASN(X)=ATN(X/SQR(-X*X+1))
1120 DEF FNACN(X)=-ATN(X/SQR(-X*X+1))+1.5707963263#
1130 REM
1140 DIM X(20,7),SM(20),S(20)
1150 REM
1160 Q1=.01745329a#
1170 G$="#####.#######"
1180 T$="####.#####"
1190 A$="#####.#####"
1200 D$="#######.#####"
1210 REM
1220 READ AD
1230 REM
1240 READ N$, E$, K#
1250 READ M(0)
1260 READ T(0)
1270 READ E0(1),E0(2),E0(3)
1280 READ E0(4),E0(5),E0(6)
1290 REM
1300 M=1+M(0)
1310 REM
1320 READ NP
1330 REM
1340 FOR I=1 TO NP
1350    READ TA(I),DA(I),DD(I)
1360    T(I)=TA(I)
1370 NEXT I
1380 REM
1390 FOR I=1 TO NP
```

```
1400    READ TX(I),RR(I,1),RR(I,2),RR(I,3)
1410    IF TX(I)<>TA(I) THEN 1460
1420 NEXT I
1430 REM
1440 GOTO 1480
1450 REM
1460 PRINT"??? TIMES DO NOT AGREE ???"
1470 STOP
1480 INPUT">>>PERCENT CHANGE TO ELEMENTS";PC
1490 CLS
1500 PRINT"ORBIT CORRECTION"
1510 PRINT N$
1520 PRINT E$
1530 PRINT
1540 PRINT"EPOCH";TAB(11);:PRINT USING G$;T(0)
1550 PRINT
1560 PRINT"ELEMENTS"
1570 PRINT
1584 FOR N=1 TO 6
1590    PRINT"E(";N;")";TAB(11);
1600    PRINT USING G$;EO(N)
1610 NEXT N
1620 LINE INPUT"";L$
1630 CLS
1640 PRINT"DYNAMICAL CENTER"
1650 PRINT
1660 PRINT TAB(6)"TA(I)";
1670 PRINT TAB(17)"RR(I,1)";TAB(30)"RR(I,2)";TAB(43)"RR(I,3)"
1680 FOR I=1 TO NP
1690    PRINT USING T$;TA(I);
1700    PRINT USING G$;RR(I,1);RR(I,2);RR(I,3)
1710 NEXT I
1720 LINE INPUT"";L$
1730 CLS
1740 PRINT"RESIDUALS IN RA AND DEC"
1750 PRINT
1760 PRINT TAB(6)"TA(I)";
1770 PRINT TAB(17)"DA(I)";TAB(30)"DD(I)"
1780 REM
1790 FOR I=1 TO NP
1800    PRINT USING T$;TA(I);
1810    PRINT USING A$;DA(I);
```

12.5. COMPUTER PROGRAMS

```
1820    PRINT USING D$;DD(I)
1830 NEXT I
1840 REM
1850 REM
1860 REM
1870 LINE INPUT"";L$
1880 CLS
1890 PRINT"PERCENT CHANGE"
1900 PRINT
1910 PRINT"PC =";PC;"%"
1920 PRINT
1930 REM COMPUTING PARTIAL DERIVATIVES
1940 REM
1950 FOR N=0 TO 6
1960    REM
1970    FOR J=1 TO 6
1980       CE(J)=E0(J)
1990    NEXT J
2000    REM
2010    IF N=0 THEN 2060
2020    REM
2030    PE(N)=ABS(E0(N)*PC/100)
2040    CE(N)=E0(N)+PE(N)
2050    REM
2060    FOR K=1 TO 3
2070       R(0,K)=CE(K)
2080       V(0,K)=CE(K+3)
2090    NEXT K
2100    REM
2110    FOR I=1 TO NP
2120       REM
2130       GOSUB 16010 REM POZ/SUB
2140       REM
2150       AC(I,N)=AX
2160       DC(I,N)=DX
2170    NEXT I
2180    REM
2190 NEXT N
2200 REM
2210 FOR I=1 TO NP
2220    FOR N=1 TO 6
2230       PA(I,N)=Q1*15*(AC(I,N)-AC(I,0))/PE(N)
```

```
2240     PD(I,N)=Q1*(DC(I,N)-DC(I,0))/PE(N)
2250    NEXT N
2260 NEXT I
2270 REM
2280 REM COMPUTING LEAST SQUARES REGRESSION
2290 REM
2300 FOR I=1 TO NP
2310    FOR N=1 TO 6
2320      X(I,N)=PA(I,N)
2330      X(I+NP,N)=PD(I,N)
2340    NEXT N
2350    X(I,7)=Q1*15*DA(I)
2360    X(I+NP,7)=Q1*DD(I)
2370 NEXT I
2380 REM
2390 GOSUB 17010 REM MGRESS/SUB
2400 REM
2410 PRINT"DIFFERENTIAL CORRECTIONS"
2420 PRINT
2430 FOR N=1 TO 6
2440    DE(N)=XU(N)
2450    PRINT"DE(";N;")";TAB(11);
2460    PRINT USING G$;DE(N)
2470 NEXT N
2480 LINE INPUT"";L$
2490 CLS
2500 PRINT"CORRECTED ELEMENTS"
2510 PRINT
2520 FOR N=1 TO 6
2530    EO(N)=EO(N)+DE(N)
2540    PRINT"E(";N;")";TAB(11);
2550    PRINT USING G$;EO(N)
2560 NEXT N
2570 PRINT
2580 PRINT N$
2590 END
16000 STOP
16010 REM # POZ/SUB # COMPUTES POSITIONS
16020 REM # BY CLOSED ELLIPTIC F&G EXPRESSIONS
16030 REM
16040 R0=FNMG(R(0,1),R(0,2),R(0,3))
16050 D0=FNDP(R(0,1),R(0,2),R(0,3),
```

12.5. COMPUTER PROGRAMS

```
             V(0,1),V(0,2),V(0,3))/SQR(M)
16060 AI=2/R0-FNVS(V(0,1),V(0,2),V(0,3))/M
16070 C0=1-R0*AI
16080 S0=D0*SQR(AI)
16090 N#=K#*SQR(M)*SQR(AI*AI*AI)
16100 WW=N#*(T(I)-T(0))
16110 GG=WW
16120    FG=GG-C0*SIN(GG)-S0*COS(GG)+S0-WW
16130    IF ABS(FG)<.0000001# THEN 16170
16140    DF=1-C0*COS(GG)+S0*SIN(GG)
16150    GG=GG-FG/DF
16160    GOTO 16120
16170 CC=(1-COS(GG))/AI
16180 SS=SIN(GG)/SQR(AI)
16190 F=1-CC/R0
16200 G=(R0*SS+D0*CC)/SQR(M)
16210 R=R0+C0*CC+D0*SS
16220 FP=-SQR(M)*SS/(R*R0)
16230 GP=1-CC/R
16240 FOR K=1 TO 3
16250    R(I,K)=F*R(0,K)+G*V(0,K)
16260    V(I,K)=FP*R(0,K)+GP*V(0,K)
16270    P(I,K)=R(I,K)+RR(I,K)
16280 NEXT K
16290 P=FNMG(P(I,1),P(I,2),P(I,3))
16300 REM
16310 AP=AB*P
16320 IF T(I)-(TA(I)-AP)<.00001 THEN 16360
16330 T(I)=TA(I)-AP
16340 GOTO 16040
16350 REM
16360 FOR K=1 TO 3
16370    LL(I,K)=P(I,K)/P
16380 NEXT K
16390 CD=SQR(1-LL(I,3)*LL(I,3))
16400 CX=LL(I,1)/CD
16410 SX=LL(I,2)/CD
16420 REM
16430 IF ABS(SX)<=.707107 THEN X=FNASN(ABS(SX))
16440 IF ABS(CX)<=.707107 THEN X=FNACN(ABS(CX))
16450 IF CX>=0 AND SX>=0 THEN X=X
16460 IF CX<0 AND SX>=0 THEN X=180*Q1-X
```

```
16470 IF CX<0 AND SX<0 THEN X=180*Q1+X
16480 IF CX>=0 AND SX<0 THEN X=360*Q1-X
16490 REM
16500 AX=X/(Q1*15)
16510 DX=FNASN(LL(I,3))/Q1
16520 RETURN
17000 STOP
17010 REM # MGRESS/SUB # LEAST SQUARES
17020 REM # MULTIPLE LINEAR REGRESSION
17030 REM
17040 NM=6
17050 NE=2*NP
17060 REM
17070 REM NORMAL EQUATIONS
17080 REM
17090 FOR J=1 TO NM
17100   FOR K=1 TO (NM+1)
17110     A(J,K)=0
17120     FOR I=1 TO NE
17130       A(J,K)=A(J,K)+X(I,J)*X(I,K)
17140     NEXT I
17150   NEXT K
17160 NEXT J
17170 REM
17180 NN=NM
17190 REM
17200 REM GAUSS ELIMINATION
17210 REM
17220 FOR I=1 TO (NN-1)
17230   REM
17240   JP=I
17250   PE!=ABS(A(I,I))
17260   FOR J=(I+1) TO NN
17270     CE!=ABS(A(J,I))
17280     IF CE!-PE! < 0 GOTO 17330
17290     PE!=CE!
17300     JP=J
17310   NEXT J
17320   REM
17330   IF JP=I GOTO 17410
17340   REM
17350   FOR K=I TO (NN+1)
```

12.5. COMPUTER PROGRAMS

```
17360     HE=A(I,K)
17370     A(I,K)=A(JP,K)
17380     A(JP,K)=HE
17390     NEXT K
17400     REM
17410     FOR J=(I+1) TO NN
17420       FOR K=(I+1) TO (NN+1)
17430         A(J,K)=A(J,K)-A(J,I)*A(I,K)/A(I,I)
17440       NEXT K
17450       A(J,I)=0
17460     NEXT J
17470 NEXT I
17480 REM
17490 XU(NN)=A(NN,NN+1)/A(NN,NN)
17500 FOR I=(NN-1) TO 1 STEP -1
17510     SS=0
17520     FOR K=(I+1) TO NN
17530       SS=SS+A(I,K)*XU(K)
17540     NEXT K
17550     XU(I)=(A(I,NN+1)-SS)/A(I,I)
17560 NEXT I
17570 REM
17580 REM LEAST SQUARES RESIDUALS
17590 REM
17600 FOR I=1 TO NE
17610     SM(I)=0
17620     FOR K=1 TO NM
17630       SM(I)=SM(I)+X(I,K)*XU(K)
17640     NEXT K
17650     S(I)=X(I,7)-SM(I)
17660 NEXT I
17670 RETURN
20000 DATA 0.005775519#
20010 REM
31000 DATA  "PALLAS (GAUSS FIRST CORRECTION)", "J2000.0",
      0.017202099#
31010 DATA  0
31020 DATA  6400.5#
31030 DATA +0.2440400#,+2.1677978#,-0.4446989#
31040 DATA -0.7314533#,-0.0041469#,+0.0502464#
31050 REM
32000 'DATA  "PALLAS (GAUSS SECOND CORRECTION)", "J2000.0",
```

```
              0.017202099#
32010 'DATA   0
32020 'DATA   6400.5#
32030 'DATA  +0.2440441#,+2.1678285#,-0.4447147#
32040 'DATA  -0.7314548#,-0.0041250#,+0.0502257#
32050 REM
39000 DATA 5
39010 REM
40000 REM # RESIDUALS #
41010 DATA 6400.5#,-0.00002#, 0.00001#
41020 DATA 6800.5#,-0.00054#,-0.00152#
41030 DATA 7000.5#,-0.00084#,-0.00188#
41040 DATA 7525.5#,-0.00078#, 0.00377#
41050 DATA 7765.5#,-0.00169#, 0.00604#
41060 REM
42010 'DATA 6400.5#, 0.00000#,-0.00007#
42020 'DATA 6800.5#, 0.00001#,-0.00007#
42030 'DATA 7000.5#, 0.00000#,-0.00004#
42040 'DATA 7525.5#, 0.00002#,-0.00012#
42050 'DATA 7765.5#, 0.00005#,-0.00008#
42060 REM
50000 REM # SUN #
50010 DATA 6400.5#,-0.3543651#,-0.8442170#,-0.3660427#
50020 DATA 6800.5#, 0.2426492#,-0.8742501#,-0.3790632#
50030 DATA 7000.5#,-0.5197797#, 0.8008713#, 0.3472473#
50040 DATA 7525.5#, 0.1481994#,-0.8918724#,-0.3866997#
50050 DATA 7765.5#,-0.9069833#, 0.4084881#, 0.1771108#
```

12.6 Numerical Examples

12.6.1 Improved Orbit for GEOS

Problem

Nine topocentric observations of satellite GEOS are listed below along with the corresponding coordinates of the geocenter:

GEOS
OBSERVED POSITIONS

t	α	δ
95.08167	2.09624	2.84935
96.08167	2.73443	2.39724
97.08167	3.25454	1.78343
98.08167	3.68088	1.08683
99.08167	4.03544	0.35237
100.08167	4.33550	−0.39552
101.08167	4.59395	−1.14379
102.08167	4.82023	−1.88538
103.08167	5.02129	−2.61638

GEOCENTER
TOPOCENTRIC COORDINATES

t	X	Y	Z
95.08167	−0.7717874	0.1833449	−0.6068664
96.08167	−0.7725822	0.1799627	−0.6068675
97.08167	−0.7733622	0.1765771	−0.6068686
98.08167	−0.7741273	0.1731881	−0.6068696
99.08167	−0.7748777	0.1697957	−0.6068707
100.08167	−0.7756131	0.1664002	−0.6068717
101.08167	−0.7763337	0.1630014	−0.6068727
102.08167	−0.7770395	0.1595996	−0.6068737
103.08167	−0.7777303	0.1561946	−0.6068746

Use this information to improve the following set of preliminary elements, which were computed for GEOS by the method of Laplace:

- $t_0 = 99.08167$
- $\mathbf{r}_0 = \{+1.0566411, +0.3289055, +0.6103880\}$
- $\dot{\mathbf{r}} = \{-0.1616737, +1.0723920, -0.0957070\}$

Continue to refine the result until the residuals in right ascension and declination are less than 0.00001^h and $0.0001°$, respectively.

Solution

Use the given information to write data lines 20000 and 30000 through 32090 as shown at the end of program IMPROVE. Run the program and answer the prompt as follows:

```
>>>PERCENT CHANGE TO ELEMENTS? 3
```

Results

TWO-BODY ORBIT IMPROVEMENT
GEOS(LAPLACE)
J2000.0

EPOCH   99.0816700

ELEMENTS

E(1)	1.0566411
E(2)	0.3289055
E(3)	0.6103880
E(4)	-0.1616737
E(5)	1.0723920
E(6)	-0.0957070

DYNAMICAL CENTER

TA(I)	RR(I,1)	RR(I,2)	RR(I,3)
95.08167	-0.7717874	0.1833449	-0.6068664
96.08167	-0.7725822	0.1799627	-0.6068675
97.08167	-0.7733622	0.1765771	-0.6068686
98.08167	-0.7741273	0.1731881	-0.6068696
99.08167	-0.7748777	0.1697957	-0.6068707
100.08167	-0.7756131	0.1664002	-0.6068717
101.08167	-0.7763337	0.1630014	-0.6068727
102.08167	-0.7770395	0.1595996	-0.6068737
103.08167	-0.7777303	0.1561946	-0.6068746

12.6. NUMERICAL EXAMPLES

RESIDUALS IN RA AND DEC

TA(I)	DA(I)	DD(I)
95.08167	0.00610	-0.03118
96.08167	0.00233	-0.02149
97.08167	0.00049	-0.01250
98.08167	-0.00016	-0.00514
99.08167	-0.00015	0.00054
100.08167	0.00022	0.00483
101.08167	0.00078	0.00798
102.08167	0.00140	0.01023
103.08167	0.00207	0.01181

>>>CONTINUE? Y/N? Y

PERCENT CHANGE

PC = 3 %

DIFFERENTIAL CORRECTIONS

DE(1)	0.0006068
DE(2)	0.0010164
DE(3)	0.0000129
DE(4)	-0.0033162
DE(5)	-0.0021844
DE(6)	0.0004392

IMPROVED ELEMENTS

E(1)	1.0572479
E(2)	0.3299219
E(3)	0.6104009
E(4)	-0.1649899
E(5)	1.0702076
E(6)	-0.0952678

RESIDUALS IN RA AND DEC

TA(I)	DA(I)	DD(I)
95.08167	0.00021	-0.00018
96.08167	0.00015	-0.00013
97.08167	0.00010	-0.00008
98.08167	0.00006	-0.00004
99.08167	0.00004	-0.00001
100.08167	0.00003	0.00001
101.08167	0.00003	0.00001
102.08167	0.00003	-0.00001
103.08167	0.00003	-0.00002

>>>CONTINUE? Y/N? Y

PERCENT CHANGE

PC = 3 %

DIFFERENTIAL CORRECTIONS

DE(1)	-0.0000092
DE(2)	-0.0000019
DE(3)	0.0000000
DE(4)	-0.0000098
DE(5)	-0.0000743
DE(6)	0.0000030

IMPROVED ELEMENTS

E(1)	1.0572387
E(2)	0.3299200
E(3)	0.6104009
E(4)	-0.1649997
E(5)	1.0701333
E(6)	-0.0952648

12.6. NUMERICAL EXAMPLES

RESIDUALS IN RA AND DEC

TA(I)	DA(I)	DD(I)
95.08167	-0.00000	0.00003
96.08167	0.00000	-0.00000
97.08167	0.00000	-0.00002
98.08167	-0.00000	-0.00002
99.08167	-0.00001	-0.00002
100.08167	-0.00000	-0.00000
101.08167	0.00000	0.00001
102.08167	-0.00000	0.00001
103.08167	0.00000	0.00003

>>>CONTINUE? Y/N? Y

PERCENT CHANGE

PC = 3 %

DIFFERENTIAL CORRECTIONS

DE(1)	0.0000007
DE(2)	0.0000008
DE(3)	0.0000000
DE(4)	-0.0000015
DE(5)	0.0000005
DE(6)	-0.0000002

IMPROVED ELEMENTS

E(1)	1.0572393
E(2)	0.3299208
E(3)	0.6104009
E(4)	-0.1650012
E(5)	1.0701338
E(6)	-0.0952649

RESIDUALS IN RA AND DEC

TA(I)	DA(I)	DD(I)
95.08167	-0.00000	0.00003
96.08167	0.00000	-0.00000
97.08167	0.00000	-0.00002
98.08167	-0.00000	-0.00002
99.08167	-0.00000	-0.00002
100.08167	-0.00000	-0.00000
101.08167	0.00000	0.00001
102.08167	-0.00000	0.00001
103.08167	0.00000	0.00003

>>>CONTINUE? Y/N? N

GEOS(LAPLACE)

12.6. NUMERICAL EXAMPLES

Discussion of Results

When the position and velocity element set produced by program IMPROVE is used in program CLASSEL, the following set of classical elements is obtained for satellite GEOS:

- epoch: 99.08167
- $a = 2.4996806$
- $e = 0.4999389$
- $M = 4.40076°$
- $i = 30.00015°$
- $\Omega = 269.99992°$
- $\omega = 89.99933°$

The improved values agree well with the original classical element set which was employed to generate the angular data used in the numerical examples:

- epoch: 95
- $a = 2.5$
- $e = 0.5$
- $M = 0°$
- $i = 30°$
- $\Omega = 270°$
- $\omega = 90°$

12.6.2 Improved Orbit for Rebek-Jewel

Problem

Five geocentric positions of comet Rebek-Jewel are listed below along with the corresponding coordinates of the Sun:

REBEK-JEWEL
GEOCENTRIC POSITIONS

t	α	δ
6310.5	6.19839	19.35543
6340.5	6.23470	20.03022
6370.5	5.42130	21.84796
6400.5	1.13233	13.91750
6430.5	22.33290	−2.06082

SUN
GEOCENTRIC COORDINATES

t	X	Y	Z
6310.5	−0.9461372	0.3214489	0.1393729
6340.5	−0.9886643	−0.1428416	−0.0619377
6370.5	−0.7744047	−0.5695064	−0.2469331
6400.5	−0.3543651	−0.8442170	−0.3660427
6430.5	0.1614512	−0.8898986	−0.3858482

Use this information to improve the following set of preliminary elements, which were computed for Rebek-Jewel by the method of Laplace:

- $t_0 = 6374.5671889$

- $\mathbf{r}_0 = \{+0.9329630, +1.5006932, +0.6367384\}$

- $\dot{\mathbf{r}} = \{+0.0759682, -1.0120252, -0.2374871\}$

Continue to refine the result until the differential corrections are all zero.

Solution

Use the given information to write data lines 20010 and 40000 through 42050 as shown at the end of program IMPROVE. Run the program and answer the prompt as follows:

>>>PERCENT CHANGE TO ELEMENTS? 3

12.6. NUMERICAL EXAMPLES

Results

TWO-BODY ORBIT IMPROVEMENT
REBEK-JEWEL(LAPLACE)
J2000.0

EPOCH   6374.5671889

ELEMENTS

E(1)   0.9329630
E(2)   1.5006932
E(3)   0.6367384
E(4)   0.0759682
E(5)   -1.0120252
E(6)   -0.2374871

DYNAMICAL CENTER

TA(I)	RR(I,1)	RR(I,2)	RR(I,3)
6310.50000	-0.9461372	0.3214489	0.1393729
6340.50000	-0.9886643	-0.1428416	-0.0619377
6370.50000	-0.7744047	-0.5695064	-0.2469331
6400.50000	-0.3543651	-0.8442170	-0.3660427
6430.50000	0.1614512	-0.8898986	-0.3858482

RESIDUALS IN RA AND DEC

TA(I)	DA(I)	DD(I)
6310.50000	-0.00748	0.00097
6340.50000	-0.00229	0.00032
6370.50000	-0.00004	0.00013
6400.50000	0.00959	0.03777
6430.50000	0.01689	0.09395

>>>CONTINUE? Y/N? Y

PERCENT CHANGE

PC = 3 %

DIFFERENTIAL CORRECTIONS

DE(1)	-0.0034137
DE(2)	-0.0151044
DE(3)	-0.0062567
DE(4)	-0.0123220
DE(5)	0.0315619
DE(6)	0.0101430

IMPROVED ELEMENTS

E(1)	0.9295493
E(2)	1.4855888
E(3)	0.6304817
E(4)	0.0636462
E(5)	-0.9804633
E(6)	-0.2273441

RESIDUALS IN RA AND DEC

TA(I)	DA(I)	DD(I)
6310.50000	0.00000	-0.00117
6340.50000	0.00000	-0.00151
6370.50000	0.00029	-0.00233
6400.50000	0.00120	0.00382
6430.50000	0.00019	0.00139

>>>CONTINUE? Y/N? Y

12.6. NUMERICAL EXAMPLES

PERCENT CHANGE

PC = 3 %

DIFFERENTIAL CORRECTIONS

DE(1)	-0.0000042
DE(2)	0.0004264
DE(3)	0.0001233
DE(4)	0.0000845
DE(5)	-0.0004729
DE(6)	-0.0001239

IMPROVED ELEMENTS

E(1)	0.9295451
E(2)	1.4860152
E(3)	0.6306050
E(4)	0.0637307
E(5)	-0.9809362
E(6)	-0.2274680

RESIDUALS IN RA AND DEC

TA(I)	DA(I)	DD(I)
6310.50000	-0.00000	0.00007
6340.50000	0.00001	0.00008
6370.50000	-0.00001	0.00012
6400.50000	-0.00005	-0.00023
6430.50000	0.00000	-0.00004

>>>CONTINUE? Y/N? Y

PERCENT CHANGE

PC = 3 %

DIFFERENTIAL CORRECTIONS

DE(1)   -0.0000017
DE(2)   -0.0000271
DE(3)   -0.0000087
DE(4)   -0.0000063
DE(5)   0.0000368
DE(6)   0.0000096

IMPROVED ELEMENTS

E(1)   0.9295434
E(2)   1.4859881
E(3)   0.6305963
E(4)   0.0637244
E(5)   -0.9808994
E(6)   -0.2274584

RESIDUALS IN RA AND DEC

TA(I)	DA(I)	DD(I)
6310.50000	-0.00000	-0.00000
6340.50000	0.00001	-0.00001
6370.50000	-0.00000	0.00000
6400.50000	0.00000	0.00000
6430.50000	-0.00000	0.00001

>>>CONTINUE? Y/N? Y

12.6. NUMERICAL EXAMPLES

PERCENT CHANGE

PC = 3 %

DIFFERENTIAL CORRECTIONS

DE(1)	0.0000001
DE(2)	0.0000014
DE(3)	0.0000004
DE(4)	0.0000003
DE(5)	-0.0000018
DE(6)	-0.0000005

IMPROVED ELEMENTS

E(1)	0.9295435
E(2)	1.4859895
E(3)	0.6305968
E(4)	0.0637247
E(5)	-0.9809012
E(6)	-0.2274588

RESIDUALS IN RA AND DEC

TA(I)	DA(I)	DD(I)
6310.50000	-0.00000	0.00000
6340.50000	0.00001	-0.00000
6370.50000	-0.00000	0.00001
6400.50000	0.00000	-0.00001
6430.50000	-0.00000	0.00001

>>>CONTINUE? Y/N? Y

PERCENT CHANGE

PC = 3 %

DIFFERENTIAL CORRECTIONS

DE(1)	-0.0000000
DE(2)	-0.0000001
DE(3)	-0.0000000
DE(4)	-0.0000000
DE(5)	0.0000001
DE(6)	0.0000000

IMPROVED ELEMENTS

E(1)	0.9295435
E(2)	1.4859894
E(3)	0.6305967
E(4)	0.0637247
E(5)	-0.9809011
E(6)	-0.2274588

RESIDUALS IN RA AND DEC

TA(I)	DA(I)	DD(I)
6310.50000	-0.00000	0.00000
6340.50000	0.00001	-0.00000
6370.50000	-0.00000	0.00001
6400.50000	0.00000	-0.00001
6430.50000	-0.00000	0.00001

>>>CONTINUE? Y/N? Y

12.6. NUMERICAL EXAMPLES

PERCENT CHANGE

PC = 3 %

DIFFERENTIAL CORRECTIONS

DE(1)	0.0000000
DE(2)	0.0000000
DE(3)	0.0000000
DE(4)	0.0000000
DE(5)	-0.0000000
DE(6)	-0.0000000

IMPROVED ELEMENTS

E(1)	0.9295435
E(2)	1.4859894
E(3)	0.6305968
E(4)	0.0637247
E(5)	-0.9809011
E(6)	-0.2274588

RESIDUALS IN RA AND DEC

TA(I)	DA(I)	DD(I)
6310.50000	-0.00000	0.00000
6340.50000	0.00001	-0.00000
6370.50000	-0.00000	0.00001
6400.50000	0.00000	-0.00001
6430.50000	-0.00000	0.00001

\>>>CONTINUE? Y/N? N

REBEK-JEWEL(LAPLACE)

Discussion of Results

When the position and velocity elements produced by program IMPROVE are used in program CLASSEL, the following set of classical elements is obtained for comet Rebek-Jewel:

- epoch: 2446374.5671889
- $a = 17.9496560$
- $e = 0.9672909$
- $M = 358.75099°$
- $i = 162.24240°$
- $\Omega = 58.86072°$
- $\omega = 111.86383°$

Notice that the orbit is a highly eccentric ellipse, similar to that computed by the method of Gauss, rather than a hyperbola as initially determined by the method of Laplace. The improved elements compare well with the original classical element set used to generate the angular data for the numerical examples:

- epoch: 2446470.93867
- $a = 17.939003$
- $e = 0.9672725$
- $M = 0°$
- $i = 162.24219°$
- $\Omega = 58.86066°$
- $\omega = 111.86576°$

12.6. NUMERICAL EXAMPLES

12.6.3 Improved Orbit for Pallas

Problem

This numerical example uses the motion of the minor planet Pallas to demonstrate the process of orbit improvement when perturbations are significant. Before we can do this, however, it is necessary to establish a context in which a problem can be worked.

Recall the preliminary orbital elements of Pallas which were determined by the method of Gauss in Section 10.9.1:

- $m \approx 0$
- $t_0 = 6378.5582004$
- $\mathbf{r}_0 = \{+0.5179953, +2.1556087, -0.4608030\}$
- $\dot{\mathbf{r}} = \{-0.7187611, +0.0680558, +0.0351243\}$

We shall assume that the attractions of the planets Mercury through Saturn are the only ones that need to be considered. The orbital elements of these perturbers are recorded at the end of ENCKE.

Now, assume that the following astrometric positions of Pallas published in *The Astronomical Almanac* are observations taken on five dates during the 1365 days between 1985 December 1 and 1989 August 27 [8,9,10,11]:

PALLAS
GEOCENTRIC POSITIONS

t	α			δ			α	δ
	h	m	s	°	′	″	h	°
6400.5	6	19	02.3	−31	24	04	6.31731	−31.40111
6800.5	15	09	00.5	0	23	26	15.15014	0.39056
7000.5	15	32	21.9	21	05	06	15.53942	21.08500
7525.5	21	09	14.9	− 3	05	11	21.15414	− 3.08639
7765.5	1	08	53.1	− 1	46	15	1.14808	− 1.77083

References 8 through 11 also provide the geocentric positions of the Sun on the dates corresponding to the times of the observations. These are listed in the table below.

SUN
GEOCENTRIC COORDINATES

t	X	Y	Z
6400.5	−0.3543651	−0.8442170	−0.3660427
6800.5	0.2426492	−0.8742501	−0.3790632
7000.5	−0.5197797	0.8008713	0.3472473
7525.5	0.1481994	−0.8918724	−0.3866997
7765.5	−0.9069833	0.4084881	0.1771108

526   CHAPTER 12. ORBIT IMPROVEMENT

For convenience, we shall first use program ENCKE to adjust the preliminary orbital elements forward in time to the epoch JD 2446400.5. This computation yields the following vector orbital elements for Pallas at the new epoch:

- $m \approx 0$

- $t_0 = 6400.5$

- $\mathbf{r}_0 = \{+0.2440400, +2.1677978, -0.4446989\}$

- $\dot{\mathbf{r}} = \{-0.7314533, -0.0041469, +0.0502464\}$

Finally, we use the orbital elements of Pallas for the epoch JD 2446400.5 in program ENCKE to compute the minor planet's right ascension and declination for the time of each observation and determine the corresponding O-C residuals. The results are summarized in the table below.

PALLAS
COMPUTED POSITIONS AND RESIDUALS
UNCORRECTED

t	α	δ	α_r	δ_r	α_r	δ_r
	h	o	h	o	s	"
6400.5	6.31733	−31.40112	−0.00002	0.00001	− 0.1	0
6800.5	15.15068	0.39208	−0.00054	−0.00152	− 1.9	− 5
7000.5	15.54026	21.08688	−0.00084	−0.00188	− 3.0	− 7
7525.5	21.15492	− 3.09016	−0.00078	0.00377	− 2.8	14
7765.5	1.14977	− 1.77687	−0.00169	0.00604	− 6.1	22

The columns labeled α_r and δ_r are the residuals in right ascension and declination, respectively.

Use the data provided above to improve the preliminary orbital elements of Pallas. Repeat the procedure until the residuals in right ascension and declination do not exceed $0^s.1$ and $1''$, respectively.

Solution

Use the given information to compose the data lines shown at the end of program CORRECT. The uncorrected orbital elements are written in lines 31000 to 31040, the number of residuals sets and the values of the residuals for the first correction are recorded in lines 39000 to 41050. The coordinates of the Sun are written in lines 50010 to 50050. Run the program and answer the prompt as follows:

```
>>>PERCENT CHANGE TO ELEMENTS? 1
```

12.6. NUMERICAL EXAMPLES

Results *first correction*

```
ORBIT CORRECTION
PALLAS (GAUSS FIRST CORRECTION)
J2000.0

EPOCH       6400.5000000

ELEMENTS

E( 1 )       0.2440400
E( 2 )       2.1677978
E( 3 )      -0.4446989
E( 4 )      -0.7314533
E( 5 )      -0.0041469
E( 6 )       0.0502464

DYNAMICAL CENTER
```

TA(I)	RR(I,1)	RR(I,2)	RR(I,3)
6400.50000	-0.3543651	-0.8442170	-0.3660427
6800.50000	0.2426492	-0.8742501	-0.3790632
7000.50000	-0.5197797	0.8008713	0.3472473
7525.50000	0.1481994	-0.8918724	-0.3866997
7765.50000	-0.9069833	0.4084881	0.1771108

RESIDUALS IN RA AND DEC

TA(I)	DA(I)	DD(I)
6400.50000	-0.00002	0.00001
6800.50000	-0.00054	-0.00152
7000.50000	-0.00084	-0.00188
7525.50000	-0.00078	0.00377
7765.50000	-0.00169	0.00604

PERCENT CHANGE

PC = 1 %

DIFFERENTIAL CORRECTIONS

DE(1)	0.0000041
DE(2)	0.0000307
DE(3)	-0.0000158
DE(4)	-0.0000015
DE(5)	0.0000219
DE(6)	-0.0000207

CORRECTED ELEMENTS

E(1)	0.2440441
E(2)	2.1678285
E(3)	-0.4447147
E(4)	-0.7314548
E(5)	-0.0041250
E(6)	0.0502257

PALLAS (GAUSS FIRST CORRECTION)

12.6. NUMERICAL EXAMPLES

Discussion of Initial Results

The result of the least squares differential correction is a new set of orbital elements for Pallas:

- $m \approx 0$
- $t_0 = 6400.5$
- $\mathbf{r}_0 = \{+0.2440441, +2.1678285, -0.4447147\}$
- $\dot{\mathbf{r}} = \{-0.7314548, -0.0041250, +0.0502257\}$

When these are used in program ENCKE, new computed positions and residuals are obtained as shown in the table below:

PALLAS
COMPUTED POSITIONS AND RESIDUALS
CORRECTED ONCE

t	α	δ	α_r	δ_r	α_r	δ_r
	h	o	h	o	s	"
6400.5	6.31731	−31.40104	0.00000	−0.00007	0.0	0
6800.5	15.15013	0.39063	0.00001	−0.00007	0.0	0
7000.5	15.53942	21.08504	0.00000	−0.00004	0.0	0
7525.5	21.15412	− 3.08627	0.00002	−0.00012	0.1	0
7765.5	1.14803	− 1.77075	0.00005	−0.00008	0.2	0

The residuals have been greatly reduced; however, those in right ascension for the last two dates are still significant. Therefore, we apply program CORRECT a second time using the above data. The improved orbital elements are recorded in lines 32000 to 32040 and the new residuals in lines 42010 to 42050. We run the program and answer the prompt as before:

```
>>>PERCENT CHANGE TO ELEMENTS? 1
```

Results *second correction*

ORBIT CORRECTION
PALLAS (GAUSS SECOND CORRECTION)
J2000.0

EPOCH   6400.5000000

ELEMENTS

E(1)	0.2440441
E(2)	2.1678285
E(3)	-0.4447147
E(4)	-0.7314548
E(5)	-0.0041250
E(6)	0.0502257

DYNAMICAL CENTER

TA(I)	RR(I,1)	RR(I,2)	RR(I,3)
6400.50000	-0.3543651	-0.8442170	-0.3660427
6800.50000	0.2426492	-0.8742501	-0.3790632
7000.50000	-0.5197797	0.8008713	0.3472473
7525.50000	0.1481994	-0.8918724	-0.3866997
7765.50000	-0.9069833	0.4084881	0.1771108

RESIDUALS IN RA AND DEC

TA(I)	DA(I)	DD(I)
6400.50000	0.00000	-0.00007
6800.50000	0.00001	-0.00007
7000.50000	0.00000	-0.00004
7525.50000	0.00002	-0.00012
7765.50000	0.00005	-0.00008

12.6. NUMERICAL EXAMPLES

PERCENT CHANGE

PC = 1 %

DIFFERENTIAL CORRECTIONS

DE(1)	0.0000003
DE(2)	-0.0000017
DE(3)	0.0000008
DE(4)	-0.0000002
DE(5)	0.0000004
DE(6)	-0.0000005

CORRECTED ELEMENTS

E(1)	0.2440444
E(2)	2.1678268
E(3)	-0.4447139
E(4)	-0.7314550
E(5)	-0.0041246
E(6)	0.0502252

PALLAS (GAUSS SECOND CORRECTION)

Discussion of Final Results

The result of the second application of program CORRECT is a satisfactory set of orbital elements for Pallas:

- $m \approx 0$

- $t_0 = 6400.5$

- $\mathbf{r}_0 = \{+0.2440444, +2.1678268, -0.4447139\}$

- $\dot{\mathbf{r}} = \{-0.7314550, -0.0041246, +0.0502252\}$

When these elements are used in program ENCKE, the computed positions agree closely with the observations:

PALLAS
COMPUTED POSITIONS AND RESIDUALS
CORRECTED TWICE

t	α	δ	α_r	δ_r	α_r	δ_r
	h	o	h	o	s	"
6400.5	6.31731	-31.40105	0.00000	-0.00006	0.0	0
6800.5	15.15013	0.39061	0.00001	-0.00005	0.0	0
7000.5	15.53943	21.08504	-0.00001	-0.00004	0.0	0
7525.5	21.15414	-3.08632	0.00000	-0.00007	0.0	0
7765.5	1.14808	-1.77081	0.00000	-0.00002	0.0	0

Thus, the residuals are within the required tolerances, and the correction process is completed.

When the position and velocity elements produced by program CORRECT are used in program CLASSEL, the following set of classical elements is obtained for Pallas:

- epoch: 2446400.5

- $a = 2.7720010$

- $e = 0.2337063$

- $M = 334.59382°$

- $i = 34.79480°$

- $\Omega = 173.34566°$

- $\omega = 309.90988°$

12.6. NUMERICAL EXAMPLES

The published set of classical elements, which correspond to the vector elements used to generate the angular data for the method of Gauss, are listed below for comparison [7]:

- epoch: 2446400.5
- $a = 2.7720$
- $e = 0.2337$
- $M = 334.594°$
- $i = 34.795°$
- $\Omega = 173.34566°$
- $\omega = 309.909°$

References

[1] Brouwer and Clemence, *Methods of Celestial Mechanics*, Academic Press, 1961.

[2] Escobal, *Methods of Orbit Determination*, Krieger Publishing Co., 1976.

[3] Baker, *Astrodynamics: Applications and Advanced Topics*, Academic Press, 1967.

[4] Thomas, *Calculus and Analytic Geometry*, Addison-Wesley Publishing Company, Inc., 1960.

[5] Bate, Mueller, and White, *Fundamentals of Astrodynamics*, Dover Publications Inc., 1971.

[6] Roy, *Orbital Motion*, Adam Hilger Ltd., 1978.

[7] *The Astronomical Almanac 1985*, U.S. Government Printing Office, 1984.

Appendix A

Vectors

A.1 Basic Vector Operations

Scalars and Vectors

A *scalar* is a quantity having magnitude without direction. Some common examples are mass, length, speed, time, and temperature. The variables in ordinary algebraic equations are scalars, and operations with scalars follow the familiar rules of elementary algebra.

A *vector* is a quantity which has both magnitude and direction. Common examples are position, velocity, acceleration, and force. As illustrated in Figure A.1, a vector can be graphically represented by an arrow defining its direction, with its magnitude represented by the arrow's length. The tail of the vector is called the *origin* or *initial point*, and the head of the vector is called the *terminal point*. In written text or equations, vectors are often printed in bold faced type, such as **A**. The magnitude of a vector is denoted by putting it between absolute value signs, that is $|\mathbf{A}|$. Alternately, since the magnitude of a vector is a scalar, it can be represented simply by a letter in normal type, such as A.

Fundamental Definitions of Vector Algebra

Many of the familiar operations of ordinary algebra can be extended to produce a vector algebra. The operations of vector addition, vector subtraction, and the multiplication of vectors and scalars, are based on several fundamental definitions:

1. Two vectors are *equal* if they have the same magnitude and direction, regardless of their positions in space. Figure A.1(a) illustrates a situation where $\mathbf{A} = \mathbf{B}$.

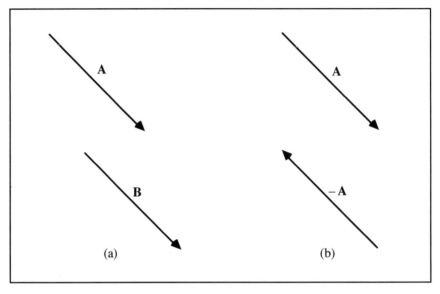

Figure A.1: Vectors.

2. Given a vector **A**, another vector having the magnitude of **A** but a direction opposite to that of **A** can be denoted by $-\mathbf{A}$. This is shown in Figure A.1(b).

3. Consider the vectors **A** and **B** shown in Figure A.2(a). The *sum* of these two vectors is a vector $\mathbf{C} = \mathbf{A} + \mathbf{B}$, found by placing the initial point of **B** on the terminal point of **A** and then constructing **C** from the initial point of **A** to the terminal point of **B** as shown in Figure A.2(b). Note that Definition 1 permits vectors to be moved in space as long as their magnitude and direction are not changed.

4. Given two vectors **A** and **B**, their *difference* is a vector $\mathbf{C} = \mathbf{A} - \mathbf{B}$, found by reversing the direction of **B** and adding the result to **A**. In other words, $\mathbf{C} = \mathbf{A} + (-\mathbf{B})$.

5. If two vectors are equal, their difference is defined to be the *null* or *zero vector* denoted by the symbol **0**. Thus, $\mathbf{A} - \mathbf{A} = \mathbf{0}$. The null vector has zero magnitude and no specific direction. Vectors which are not null vectors are called *proper vectors*. In this text, all vectors may be considered proper vectors unless otherwise noted.

6. Given a vector **A** and scalar m, their product is a vector $\mathbf{C} = m\mathbf{A}$. The direction of **C** is the same as **A** if m is positive and opposite to that of **A**

A.1. BASIC VECTOR OPERATIONS

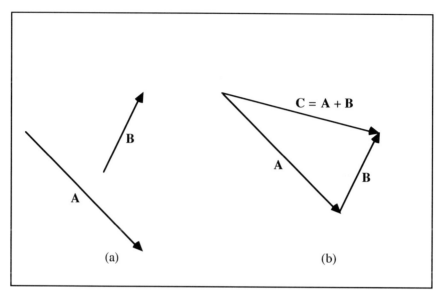

Figure A.2: Vector Addition.

if m is negative. In either case, the magnitude is $|\mathbf{C}| = |m||\mathbf{A}|$.

7. A *unit vector* is a vector having unit magnitude. Thus, if $\mathbf{u}$ is a unit vector, then $|\mathbf{u}| = 1$. If $\mathbf{A}$ is a proper vector with magnitude A, then a unit vector in the same direction as $\mathbf{A}$ can be formed by multiplying $\mathbf{A}$ by the reciprocal of its magnitude. Thus, $\mathbf{A}/A$ represents a unit vector in the direction of $\mathbf{A}$.

Rules of Vector Algebra

Given the vectors $\mathbf{A}$, $\mathbf{B}$, and $\mathbf{C}$, and the scalars m and n, then the rules of vector algebra may be stated as follows:

1. $\mathbf{A} + \mathbf{B} = \mathbf{B} + \mathbf{A}$
2. $\mathbf{A} + (\mathbf{B} + \mathbf{C}) = (\mathbf{A} + \mathbf{B}) + \mathbf{C}$
3. $m\mathbf{A} = \mathbf{A}m$
4. $m(n\mathbf{A}) = (mn)\mathbf{A}$
5. $(m + n)\mathbf{A} = m\mathbf{A} + n\mathbf{A}$
6. $m(\mathbf{A} + \mathbf{B}) = m\mathbf{A} + m\mathbf{B}$

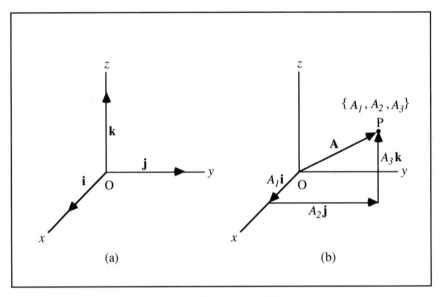

Figure A.3: Component Vectors.

These rules do not cover situations where vectors are multiplied by vectors. The rules governing the products of vectors are discussed in Section A.2.

The rules of vector algebra permit vector equations to be manipulated in much the same way as ordinary algebraic equations. For example, the vector equation

$$\mathbf{C} = \mathbf{A} + \mathbf{B}$$

can be solved for $\mathbf{A}$ to yield

$$\mathbf{A} = \mathbf{C} - \mathbf{B}.$$

Vector Components

Consider the right-handed rectangular coordinate system shown in Figure A.3(a). Let $\mathbf{i}$, $\mathbf{j}$, and $\mathbf{k}$ be three mutually perpendicular unit vectors pointing from the origin O in the directions of the positive x, y, and z axes. Since the system is right-handed, $\mathbf{k}$ points in the direction in which a right-handed screw would advance if rotated counterclockwise through the 90° angle from $\mathbf{i}$ to $\mathbf{j}$.

Any vector in space can be represented as the sum of three *component vectors* which are parallel to each of the coordinate axes. These components are formed by multiplying the unit vectors $\mathbf{i}$, $\mathbf{j}$, and $\mathbf{k}$ by appropriate scalars. Therefore, in the situation illustrated in Figure A.3(b), the vector $\mathbf{A}$ can be defined in terms

A.1. BASIC VECTOR OPERATIONS

of the rectangular unit vectors as

$$\mathbf{A} = A_1\mathbf{i} + A_2\mathbf{j} + A_3\mathbf{k}.$$

Notice that the sum of these component vectors follows the procedure of Definition 3 above. For convenience, we may also define a vector **A** as an ordered set of three scalars, that is

$$\mathbf{A} = \{A_1, A_2, A_3\},$$

where it is understood that A_1, A_2, and A_3 correspond to the **i**, **j**, and **k** directions, respectively.

When algebraic expressions contain vectors whose components are known, the components can be used to evaluate the expressions. For example, in terms of its components, the magnitude of $\mathbf{A} = \{A_1, A_2, A_3\}$ is given by

$$|\mathbf{A}| = A = \sqrt{A_1^2 + A_2^2 + A_3^2}.$$

Furthermore, if **A** is a proper vector, then a unit vector in the direction of **A** is given by

$$\frac{\mathbf{A}}{A} = \left\{\frac{A_1}{A}, \frac{A_2}{A}, \frac{A_3}{A}\right\},$$

and, if m is any scalar,

$$m\mathbf{A} = \{mA_1, mA_2, mA_3\}.$$

Finally, given another vector $\mathbf{B} = \{B_1, B_2, B_3\}$, then

$$\mathbf{A} + \mathbf{B} = \{(A_1 + B_1), (A_2 + B_2), (A_3 + B_3)\}.$$

Thus, a vector equation such as

$$\mathbf{C} = \mathbf{A} + \mathbf{B}$$

can be solved to find the components of **C** by replacing the vector expression with an equivalent set of three simultaneous scalar equations:

$$\begin{aligned} C_1 &= A_1 + B_1 \\ C_2 &= A_2 + B_2 \\ C_3 &= A_3 + B_3. \end{aligned}$$

Numerical Examples

1. Given $\mathbf{A} = \{1, 2, -2\}$, find the magnitude $A = |\mathbf{A}|$.

$$\begin{aligned} A &= \sqrt{(1)^2 + (2)^2 + (-2)^2} \\ A &= \sqrt{1 + 4 + 4} \\ A &= \sqrt{9} \\ A &= 3 \end{aligned}$$

2. Given $\mathbf{A} = \{0, 1, 2\}$ and $\mathbf{B} = \{2, 1, 0\}$, find the vector sum $\mathbf{C} = \mathbf{A} + \mathbf{B}$.

$$\begin{aligned} \mathbf{C} &= \{0+2, 1+1, 2+0\} \\ \mathbf{C} &= \{2, 2, 2\} \end{aligned}$$

3. Given $\mathbf{A} = \{1, -1, 2\}$ and $m = -3$, find the product $\mathbf{C} = m\mathbf{A}$.

$$\begin{aligned} \mathbf{C} &= -3\{1, -1, 2\} \\ \mathbf{C} &= \{-3, 3, -6\} \end{aligned}$$

4. Given $\mathbf{A} = \{1, 2, -2\}$, find a unit vector $\mathbf{a}$ in the direction of $\mathbf{A}$.

- According to the first numerical example, we have

$$|\mathbf{A}| = A = 3.$$

- Therefore, letting $\mathbf{a}$ represent the unit vector,

$$\mathbf{a} = \frac{\mathbf{A}}{A}$$

$$\mathbf{a} = \left\{\frac{1}{3}, \frac{2}{3}, -\frac{2}{3}\right\}.$$

5. Confirm that the vector $\mathbf{a} = \{1/3, 2/3, -2/3\}$ computed in the previous numerical example is a unit vector.

$$\begin{aligned} a &= \sqrt{\frac{1}{9} + \frac{4}{9} + \frac{4}{9}} \\ a &= \sqrt{\frac{9}{9}} \\ a &= \sqrt{1} \\ a &= 1 \end{aligned}$$

A.2. THE DOT AND CROSS PRODUCTS

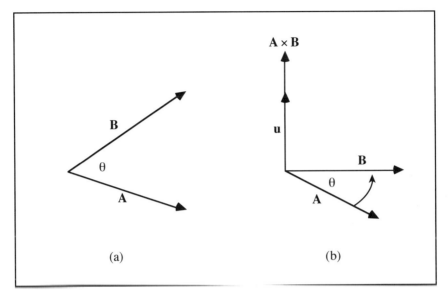

Figure A.4: Dot and Cross Products.

A.2 The Dot and Cross Products

Dot Product

Consider the situation depicted in Figure A.4(a) where the two vectors **A** and **B** form an angle $0 \leq \theta \leq 180°$. The *dot product*, represented by $\mathbf{A} \cdot \mathbf{B}$ (read as **A** dot **B**), is defined to be

$$\mathbf{A} \cdot \mathbf{B} = |\mathbf{A}||\mathbf{B}| \cos \theta.$$

Thus, the dot product produces a scalar and so is sometimes called the *scalar product*. Notice that this product equals zero when two vectors are perpendicular because $\cos 90° = 0$.

If **A**, **B**, and **C** are vectors and m is a scalar, then the following rules apply to the dot product:

1. $\mathbf{A} \cdot \mathbf{B} = \mathbf{B} \cdot \mathbf{A}$

2. $\mathbf{A} \cdot (\mathbf{B} + \mathbf{C}) = \mathbf{A} \cdot \mathbf{B} + \mathbf{A} \cdot \mathbf{C}$

3. $m(\mathbf{A} \cdot \mathbf{B}) = (m\mathbf{A}) \cdot \mathbf{B} = \mathbf{A} \cdot (m\mathbf{B})$

4. $\mathbf{i} \cdot \mathbf{i} = \mathbf{j} \cdot \mathbf{j} = \mathbf{k} \cdot \mathbf{k} = 1$

5. $\mathbf{i} \cdot \mathbf{j} = \mathbf{j} \cdot \mathbf{k} = \mathbf{k} \cdot \mathbf{i} = 0$

6. $\mathbf{A} \cdot \mathbf{B} = A_1 B_1 + A_2 B_2 + A_3 B_3$

7. $\mathbf{A} \cdot \mathbf{A} = A^2 = A_1^2 + A_2^2 + A_3^2$

Cross Product

Consider the situation shown in Figure A.4(b) where two vectors $\mathbf{A}$ and $\mathbf{B}$ form an angle $0 \leq \theta \leq 180°$, and $\mathbf{u}$ is a unit vector perpendicular to both $\mathbf{A}$ and $\mathbf{B}$. The direction of $\mathbf{u}$ is defined to be the direction in which a right-handed screw would advance if rotated from $\mathbf{A}$ to $\mathbf{B}$ through the angle θ. The *cross product* is a vector, represented by $\mathbf{A} \times \mathbf{B}$ (read as $\mathbf{A}$ cross $\mathbf{B}$), defined as follows:

$$\mathbf{A} \times \mathbf{B} = \mathbf{u}|\mathbf{A}||\mathbf{B}|\sin\theta.$$

As a consequence of this definition, $\mathbf{A} \times \mathbf{A} = \mathbf{0}$ and $\mathbf{A} \times \mathbf{B} = \mathbf{0}$ if $\mathbf{A}$ is parallel to $\mathbf{B}$ because $\sin 0° = 0$.

If $\mathbf{A}$, $\mathbf{B}$, and $\mathbf{C}$ are vectors and m is a scalar, then the following rules apply to the cross product:

1. $\mathbf{A} \times \mathbf{B} = -\mathbf{B} \times \mathbf{A}$

2. $\mathbf{A} \times (\mathbf{B} + \mathbf{C}) = \mathbf{A} \times \mathbf{B} + \mathbf{A} \times \mathbf{C}$

3. $m(\mathbf{A} \times \mathbf{B}) = (m\mathbf{A}) \times \mathbf{B} = \mathbf{A} \times (m\mathbf{B})$

4. $\mathbf{i} \times \mathbf{i} = \mathbf{j} \times \mathbf{j} = \mathbf{k} \times \mathbf{k} = 0$

5. $\mathbf{i} \times \mathbf{j} = \mathbf{k}$, $\mathbf{j} \times \mathbf{k} = \mathbf{i}$, and $\mathbf{k} \times \mathbf{i} = \mathbf{j}$

6. $\mathbf{A} \times \mathbf{B} = \begin{vmatrix} \mathbf{i} & \mathbf{j} & \mathbf{k} \\ A_1 & A_2 & A_3 \\ B_1 & B_2 & B_3 \end{vmatrix}$

   $\mathbf{A} \times \mathbf{B} = (A_2 B_3 - A_3 B_2)\mathbf{i} + (A_3 B_1 - A_1 B_3)\mathbf{j} + (A_1 B_2 - A_2 B_1)\mathbf{k}$

   $\mathbf{A} \times \mathbf{B} = \{(A_2 B_3 - A_3 B_2), (A_3 B_1 - A_1 B_3), (A_1 B_2 - A_2 B_1)\}$

7. The magnitude of $\mathbf{A} \times \mathbf{B}$ is equal to the area of a parallelogram with sides determined by $\mathbf{A}$ and $\mathbf{B}$.

A.2. THE DOT AND CROSS PRODUCTS

Triple Scalar Product

Given three vectors $\mathbf{A}$, $\mathbf{B}$, and $\mathbf{C}$, the product $\mathbf{A} \cdot (\mathbf{B} \times \mathbf{C})$ is called a *triple scalar product*. The absolute value of the triple scalar product is equal to the volume of a box whose sides are determined by the vectors $\mathbf{A}$, $\mathbf{B}$, and $\mathbf{C}$. The following rules apply:

1. $\mathbf{A} \cdot (\mathbf{B} \times \mathbf{C}) = \mathbf{B} \cdot (\mathbf{C} \times \mathbf{A}) = (\mathbf{A} \times \mathbf{B}) \cdot \mathbf{C}$

2. $\mathbf{A} \cdot (\mathbf{B} \times \mathbf{C}) = \begin{vmatrix} A_1 & A_2 & A_3 \\ B_1 & B_2 & B_3 \\ C_1 & C_2 & C_3 \end{vmatrix}$

   $\mathbf{A} \cdot (\mathbf{B} \times \mathbf{C}) = A_1(B_2C_3 - B_3C_2) + A_2(B_3C_1 - B_1C_3) + A_3(B_1C_2 - B_2C_1)$

Triple Vector Product

Given three vectors $\mathbf{A}$, $\mathbf{B}$, and $\mathbf{C}$, the product $\mathbf{A} \times (\mathbf{B} \times \mathbf{C})$ is called a *triple vector product*. The following rules apply:

1. $\mathbf{A} \times (\mathbf{B} \times \mathbf{C}) \neq (\mathbf{A} \times \mathbf{B}) \times \mathbf{C}$

2. $\mathbf{A} \times (\mathbf{B} \times \mathbf{C}) = (\mathbf{A} \cdot \mathbf{C})\mathbf{B} - (\mathbf{A} \cdot \mathbf{B})\mathbf{C}$

3. $(\mathbf{A} \times \mathbf{B}) \times \mathbf{C} = (\mathbf{A} \cdot \mathbf{C})\mathbf{B} - (\mathbf{B} \cdot \mathbf{C})\mathbf{A}$

Numerical Examples

1. Given $\mathbf{A} = \{2, -2, -7\}$ and $\mathbf{B} = \{5, -4, 4\}$, find the dot product $m = \mathbf{A} \cdot \mathbf{B}$.

$$\begin{aligned} m &= \{2, -2, -7\} \cdot \{5, -4, 4\} \\ m &= (2)(5) + (-2)(-4) + (-7)(4) \\ m &= 10 + 8 - 28 \\ m &= -10 \end{aligned}$$

2. Given $\mathbf{A} = \{1, -3, 4\}$ and $\mathbf{B} = \{4, 1, -6\}$, find the cross product $\mathbf{C} = \mathbf{A} \times \mathbf{B}$.

$$\mathbf{C} = \begin{vmatrix} \mathbf{i} & \mathbf{j} & \mathbf{k} \\ 1 & -3 & 4 \\ 4 & 1 & -6 \end{vmatrix}$$

$$\mathbf{C} = \{[(-3)(-6) - (4)(1)], [(4)(4) - (1)(-6)], [(1)(1) - (-3)(4)]\}$$
$$\mathbf{C} = \{14, 22, 14\}$$

3. Given $\mathbf{C} = \{1, 1, 0\}$, $\mathbf{A} = \{1, -3, 4\}$, and $\mathbf{B} = \{4, 1, -6\}$, find the triple scalar product $m = \mathbf{C} \cdot (\mathbf{A} \times \mathbf{B})$.

$$m = \begin{vmatrix} 1 & 1 & 0 \\ 1 & -3 & 4 \\ 4 & 1 & -6 \end{vmatrix}$$

$$m = (1)[(-3)(-6) - (4)(1)] + (1)[(4)(4) - (1)(-6)] + (0)[(1)(1) - (-3)(4)]$$

$$m = 14 + 22 + 0$$

$$m = 36$$

References

Spiegel, *Theory and Problems of Vector Analysis and an Introduction to Tensor Analysis*, Schaum Publishing Co., 1959.

Thomas, *Calculus and Analytic Geometry*, Addison-Wesley Publishing Co., 1960.

Appendix B

Elementary Calculus

B.1 Differentiation

Increments

Consider the situation illustrated in Figure B.1(a) where the points P and Q lie on a curve which is the graph of the function $y = f(x)$. The values of y at P and Q are

$$y_p = f(x_p)$$
$$y_q = f(x_q),$$

respectively. If we let Δx represent the *increment* which must be added to x_p to yield x_q, then

$$\Delta x = x_q - x_p,$$

and the increment Δy is given by

$$\Delta y = f(x_p + \Delta x) - f(x_p).$$

Slope

The *slope of a line* through points P and Q in Figure B.1(a) is defined by the ratio

$$\text{slope of a line} = \frac{\Delta y}{\Delta x} = \tan \varphi,$$

where φ is the angle which the secant line makes with the x-axis. This slope expresses the *average rate of change* of the function between points P and Q. If we now hold point P fixed and let the increment Δx decrease until it approaches zero, then point Q will move down the curve and begin to merge with point P, as shown in Figure B.1(b). Thus, the secant line becomes a tangent line, and

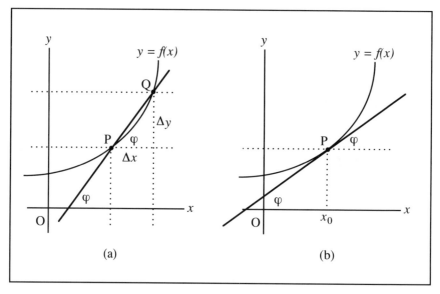

Figure B.1: Slope.

the *slope of a curve* at a point P is the slope of the tangent line at that point. This limiting process can be summarized in mathematical symbols as follows:

$$\text{slope of a curve} = \lim_{\Delta x \to 0} \frac{\Delta y}{\Delta x} = \tan \varphi,$$

where φ is the angle which the tangent line makes with the x-axis. The slope of a curve at a given point represents the *instantaneous rate of change* of the function at that point.

The slope can be used to determine if a function is increasing, decreasing, or passing through some maximum or minimum value. Consider the curve depicted in Figure B.2. Its slope is positive at all points where the value of the function $y = f(x)$ increases as the independent variable x increases, and its slope is negative in the region where the function decreases as x increases. At the maximum point M and the minimum point N, the slope is exactly zero.

The Derivative

The *derivative* of a function $y = f(x)$ with respect to x at a point $x = x_0$ is equivalent to the slope of the curve at that point. It is defined as follows:

$$f'(x_0) = \lim_{\Delta x \to 0} \frac{f(x_0 + \Delta x) - f(x_0)}{\Delta x}.$$

B.1. DIFFERENTIATION

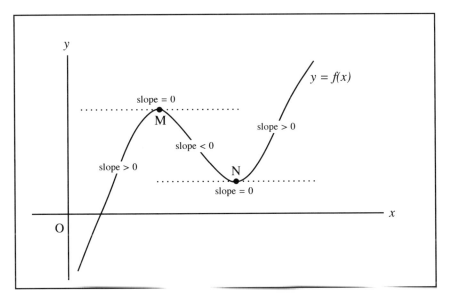

Figure B.2: The behavior of the slope.

If the limit exists for all points along the curve, then x_0 can be any value of x, and it is customary to drop the subscript 0 and simply write

$$f'(x) = \lim_{\Delta x \to 0} \frac{f(x + \Delta x) - f(x)}{\Delta x}.$$

In addition to the notation $f'(x)$, there are a number of other ways to express the derivative of the function $y = f(x)$ with respect to x. Some of the common ones are

$$f', y', \frac{dy}{dx}, \frac{d}{dx}f(x), \frac{d}{dx}(y).$$

When the notation dy/dx is used, dx is called the *differential of x* and may be any real number. The *differential of y* is a function of x and dx given by

$$dy = f'(x)dx,$$

where $f'(x) = dy/dx$.

Higher order derivatives of a function can be obtained by successive applications of the differentiation process. In other words,

$$y' = \frac{dy}{dx} = f'(x),$$

$$y'' = \frac{d^2y}{dx^2} = f''(x),$$

$$y''' = \frac{d^3y}{dx^3} = f'''(x),$$

and so forth.

The differentiation process can also be applied to vectors. As in the case of scalar functions, the derivative of a vector expresses its instantaneous rate of change with respect to the differentiation variable. The two most important vector derivatives in this text are the first and second derivatives of a position vector with respect to time. Letting **r** represent the position of a body at time t, then the velocity and acceleration of the body are

$$\mathbf{v} = \frac{d\mathbf{r}}{dt}$$

and

$$\mathbf{a} = \frac{d\mathbf{v}}{dt} = \frac{d}{dt}\left(\frac{d\mathbf{r}}{dt}\right) = \frac{d^2\mathbf{r}}{dt^2},$$

respectively.

Differentiation Formulas

It can be shown that the differentials and derivatives of many functions can be found by applying the appropriate formulas from the list below, where u and v represent functions of x, while c and n represent constants.

1. $du = \dfrac{du}{dx}dx$

2. $\dfrac{d}{dx}(c) = 0$

3. $\dfrac{d}{dx}(cu) = c\dfrac{du}{dx}$

4. $\dfrac{d}{dx}(u^n) = nu^{n-1}\dfrac{du}{dx}$

5. $\dfrac{d}{dx}(u + v + \cdots) = \dfrac{du}{dx} + \dfrac{dv}{dx} + \cdots$

6. $\dfrac{d}{dx}(uv) = u\dfrac{dv}{dx} + v\dfrac{du}{dx}$

7. $\dfrac{d}{dx}\left(\dfrac{u}{v}\right) = \dfrac{v(du/dx) - u(dv/dx)}{v^2}$

8. $\dfrac{d}{dx}(\sin u) = (\cos u)\dfrac{du}{dx}$

B.1. DIFFERENTIATION

9. $\dfrac{d}{dx}(\cos u) = (-\sin u)\dfrac{du}{dx}$

10. $\dfrac{d}{dx}(\tan u) = (\sec^2 u)\dfrac{du}{dx}$

11. $\dfrac{d}{dx}(\sinh u) = (\cosh u)\dfrac{du}{dx}$

12. $\dfrac{d}{dx}(\cosh u) = (\sinh u)\dfrac{du}{dx}$

In the special case where $u = x$ or $v = x$, the corresponding du/dx or dv/dx term in the above formulas becomes simply $dx/dx = 1$.

Partial Differentiation

The concept of differentiation can be extended to functions of more than one independent variable. In the case of a function of several variables, a *partial derivative* is found by differentiating the function with respect to one variable while treating the other variables as constants. Let $w = f(x, y, z)$ represent a function of three variables. Then the partial derivatives of w with respect to x, y, and z are defined by

$$\frac{\partial w}{\partial x} = \lim_{\Delta x \to 0} \frac{f(x + \Delta x, y, z) - f(x, y, z)}{\Delta x}$$

$$\frac{\partial w}{\partial y} = \lim_{\Delta y \to 0} \frac{f(x, y + \Delta y, z) - f(x, y, z)}{\Delta y}$$

$$\frac{\partial w}{\partial z} = \lim_{\Delta z \to 0} \frac{f(x, y, z + \Delta z) - f(x, y, z)}{\Delta z}.$$

The *total differential* of the function $w = f(x, y, z)$ is defined in terms of its partial derivatives by

$$dw = \frac{\partial w}{\partial x}dx + \frac{\partial w}{\partial y}dy + \frac{\partial w}{\partial z}dz.$$

These definitions can be generalized to functions of any number of variables.

Numerical Examples

1. Given $u = x^3 - x + 3$, find du/dx.

$$\frac{d}{dx}(x^3 - x + 3) = 3x^2 - 1$$

2. Given $u = x^2 + 3x$, find du.
$$du = \frac{du}{dx}dx$$
$$du = (2x+3)dx$$

3. Given $v = x^{-5}$, find dv/dx.
$$\frac{d}{dx}(x^{-5}) = -5x^{-6}$$

4. Given $y = t^4 + 1$, find dy/dt.
$$\frac{d}{dt}(t^4 + 1) = 4t^3$$

5. Given $u = (x^2+3)^3$, find du/dx.
$$\frac{d}{dx}(x^2+3)^3 = 3(x^2+3)^2(2x)$$

6. Given $u = x^3$ and $v = \sin x^2$, find $d(uv)/dx$.
$$\frac{d}{dx}(x^3 \sin x^2) = 3x^2(\sin x^2) + x^3(\cos x^2)(2x)$$

7. Given $y = \cosh x^7$, find dy/dx.
$$\frac{d}{dx}(\cosh x^7) = (\sinh x^7)(7x^6)$$

8. Given $y = x^2 + 2x + 2$, find the point where the function is a minimum.

- First, find dy/dx to obtain an expression for the slope at any point on the curve of the function:
$$y' = 2x + 2.$$

- Next, because the slope is zero at the minimum, set the expression for the derivative equal to zero and solve for x:
$$2x + 2 = 0$$
$$2x = -2$$
$$x = -1.$$

- Finally, substitute the value $x = -1$ into the original equation to find the corresponding value of y:
$$y = (-1)^2 + 2(-1) + 2$$
$$y = 1 - 2 + 2$$
$$y = 1.$$

- Therefore, the function reaches a minimum value at the point $\{-1, 1\}$. The fact that this is a minimum, and not a maximum, can be verified by checking the algebraic sign of the slope on either side of the point or by evaluating the function at other nearby points.

9. Given $y = x^3 + x^2 + x + 7$, find the third derivative.

$$y' = 3x^2 + 2x + 1$$
$$y'' = 6x + 2$$
$$y''' = 6$$

10. Given $w = 3x^2 + 2xy - y^4 + z^3$, find the partial derivatives with respect to the three independent variables.

$$\frac{\partial w}{\partial x} = 6x + 2y$$
$$\frac{\partial w}{\partial y} = 2x - 4y^3$$
$$\frac{\partial w}{\partial z} = 3z^2$$

11. The position of a certain body at any time t is given by the radius vector $\mathbf{r} = \{t^3, t^2, t\}$. Find expressions for the velocity $\mathbf{v}$ and acceleration $\mathbf{a}$.

- Velocity is the first derivative of position. Thus,

$$\mathbf{v} = \frac{d}{dt}(\mathbf{r})$$
$$\mathbf{v} = \frac{d}{dt}\{t^3, t^2, t\}$$
$$\mathbf{v} = \{3t^2, 2t, 1\}.$$

- Acceleration is the first derivative of velocity. Therefore,

$$\mathbf{a} = \frac{d}{dt}(\mathbf{v})$$
$$\mathbf{a} = \frac{d}{dt}\{3t^2, 2t, 1\}$$
$$\mathbf{a} = \{6t, 2, 0\}.$$

B.2 Integration

Assume a function $f(x)$ is known to be the derivative of some other function $F(x)$. Then an equation of the form

$$\frac{d}{dx}F(x) = f(x)$$

is a *differential equation*, and its solution is called an *antiderivative* or *indefinite integral* of $f(x)$ which can be denoted by

$$\int f(x)\,dx = F(x) + C.$$

The symbol $\int$ is an *integral sign*, and C is an *arbitrary constant of integration*. This constant must be included because if

$$\frac{d}{dx}F(x) = f(x),$$

then it is also true that

$$\frac{d}{dx}(F(x) + C) = f(x),$$

since the derivative of any constant is zero.

It is helpful to notice that the term inside the integral on the left side of the equation for the antiderivative is equivalent to the differential of the function $F(x)$. Recalling that

$$f(x) = \frac{d}{dx}F(x),$$

then

$$\int \frac{d}{dx}F(x)\,dx = F(x) + C.$$

Thus, employing the definition of the differential, we can write

$$\int dF(x) = F(x) + C.$$

Integration Formulas

The formulas listed below can be used to integrate many differential equations. Assume u and v are functions of x while C, c, and n represent constants.

1. $\int du = u + C$

2. $\int a\,du = a\int du$

3. $\int (du + dv + \cdots) = \int du + \int dv + \cdots$

4. $\int u^n\,du = \dfrac{u^{n+1}}{n+1} + C$, $(n \neq -1)$

5. $\int \cos u\,du = \sin u + C$

B.2. INTEGRATION

6. $\int \sin u \, du = -\cos u + C$

7. $\int \cosh u \, du = \sinh u + C$

8. $\int \sinh u \, du = \cosh u + C$

Numerical Examples

1. Given $f(x) = x^2 + x - 1$, find the antiderivative.

$$F(x) = \int (x^2 + x - 1) \, dx$$
$$F(x) = \int x^2 \, dx + \int x \, dx - \int dx$$
$$F(x) = \frac{x^3}{3} + \frac{x^2}{2} - x + C$$

2. Given $dy/dx = 6x(x^2 + 3)^2$, find the indefinite integral y.

- In order to integrate the given function, we must be able to write it in a form which fits one of the differentiation formulas. Since the fourth form seems to most closely resemble the given expression, let $u = x^2 + 3$ and find out what du will be:

$$du = \frac{du}{dx} dx$$
$$du = (2x) dx.$$

- Next, write the expression for dy/dx as follows:

$$\frac{dy}{dx} = 3(x^2 + 3)^2 (2x),$$

and multiply both sides by the differential dx to obtain

$$\frac{dy}{dx} dx = 3(x^2 + 3)^2 (2x) dx$$
$$\frac{dy}{dx} dx = 3u^2 du.$$

- The left side of this differential equation is the perfect differential dy, and the right differs from the fourth integration formula only by the constant 3.

Therefore,
$$\int dy = 3 \int u^2 du$$
$$y = 3\left(\frac{u^3}{3} + C\right)$$
$$y = u^3 + C.$$

Thus, substituting the expression for u, the result is
$$y = (x^2 + 3)^3 + C.$$

- Notice that it is not necessary to keep track of the changing value of the integration constant at each step because its value is arbitrary.

3. Integrate $y' = \sin x$.
$$\frac{dy}{dx} = \sin x$$
$$dy = \sin x \, dx$$
$$\int dy = \int \sin x \, dx$$
$$y = -\cos x + C$$

4. Assume the acceleration of a certain body is given by $\mathbf{a} = \{6t, 2, 0\}$, and that at time $t = 0$ its velocity was $\{0, 0, 1\}$, and its position was $\{0, 0, 0\}$. Find the body's velocity and position vectors as functions of time.

- Integrate the acceleration once to obtain an expression for the velocity:
$$\mathbf{v} = \int \mathbf{a} \, dt$$
$$\mathbf{v} = \left\{\int 6t \, dt, \int 2 \, dt, \int (0) \, dt\right\}$$
$$\mathbf{v} = \{3t^2 + C_x, 2t + C_y, C_z\},$$

where C_x, C_y, and C_z are the x, y, and z components of an arbitrary constant vector. Now, making use of what is given, we know that the equality
$$\mathbf{v} = \{3t^2 + C_x, 2t + C_y, C_z\} = \{0, 0, 1\}$$
must hold at $t = 0$. This will be true only if
$$C_x = 0$$
$$C_y = 0$$
$$C_z = 1.$$

B.2. INTEGRATION

Therefore, the velocity vector of the body at any time is

$$\mathbf{v} = \left\{3t^2, 2t, 1\right\}.$$

- The position vector is obtained by integrating the velocity vector:

$$\mathbf{r} = \int \mathbf{v}\, dt$$
$$\mathbf{r} = \left\{\int 3t^2\, dt, \int 2t\, dt, \int dt\right\}$$
$$\mathbf{r} = \left\{t^3 + C_x, t^2 + C_y, t + C_z\right\},$$

where C_x, C_y, and C_z are again the x, y, and z components of an arbitrary constant vector. The initial condition requires that

$$\mathbf{r} = \left\{t^3 + C_x, t^2 + C_y, t + C_z\right\} = \{0, 0, 0\}$$

at $t = 0$. Thus,

$$C_x = 0$$
$$C_y = 0$$
$$C_z = 0.$$

Therefore, the position vector of the body at any time is

$$\mathbf{r} = \left\{t^3, t^2, t\right\}.$$

- Compare these results with the last numerical example of Section B.1.

References

Spiegel, *Advanced Mathematics for Engineers and Scientists*, McGraw-Hill Book Co., 1971.

Thomas, *Calculus and Analytic Geometry*, Addison-Wesley Publishing Co., 1960.

Appendix C

Astronomical Constants

C.1 Constants Related to Units

The values listed below are based on the International Astronomical Union (IAU) 1976 system of astronomical constants. The fundamental units meter (m), kilogram (kg), and second (s) are the defining units of length, mass, and time.

- Gaussian gravitational constant, $k = 0.017\ 202\ 098\ 95$

- Astronomical unit of distance (au) $= 1.495\ 978\ 70 \times 10^{11}$ m

- Astronomical unit of mass $= 1$ solar mass $= 1.9891 \times 10^{30}$ kg

- Astronomical unit of time $= 1$ day $= 1440$ minutes $= 86\ 400$ s

- Equatorial radius of Earth $= 6.378\ 140 \times 10^{6}$ m $= 4.263\ 523 \times 10^{-5}$ au

- Solar parallax $= 8''.794\ 148$

- Speed of light, $c = 2.997\ 924\ 58 \times 10^{8}$ m/s $= 173.1446$ au/day

- Julian century $= 36\ 525$ days $= 3.155\ 760 \times 10^{9}$ s

- Obliquity of the ecliptic for epoch J2000, $\varepsilon = 23°\ 26'\ 21''.448$

- Flattening factor for Earth, $f = 1/298.257 = 0.003\ 352\ 81$

- Ratio of mass of Earth to that of Sun $= 0.000\ 003\ 003$

- Ratio of mass of Moon to that of Earth $= 0.012\ 300\ 034$

C.2 Masses of the Planets

The masses of the planets are given below in units of the mass of the Sun. The minor planets Vesta, Ceres, and Pallas are also included so that their masses can be seen in comparison to those of the major planets.

Mercury	0.000 000 166
Venus	0.000 002 448
Earth + Moon	0.000 003 040
Mars	0.000 000 323
Vesta	1.2×10^{-10}
Ceres	5.9×10^{-10}
Pallas	1.1×10^{-10}
Jupiter	0.000 954 791
Saturn	0.000 285 878
Uranus	0.000 043 554
Neptune	0.000 051 776
Pluto	0.000 000 008

References

Planetary and Lunar Coordinates for the Years 1984-2000, U.S. Government Printing Office, 1983.

The Astronomical Almanac 1989, U.S. Government Printing Office, 1988.

Index

A

Aberration, 33, 36
    light-time, 34, 35, 222, 393, 420, 455
    planetary, 33
    stellar, 34
Acceleration, 2, 4, 255
    inertial, 8
    perturbed, 82, 253
    relative, 9, 81, 253
    two-body, 81
Almanac for Computers, 80
Analytical integration, 551
Angular measurement, 1, 353, 355, 382, 483, 486
Angular momentum, 86, 137
    per unit mass, 86
    vector, 152, 155
Angular velocity, 137
Angle of inclination, 150, 153
Anomaly
    eccentric, 140
    hyperbolic eccentric, 142
    mean, 142, 150, 155
    parabolic eccentric, 147
    true, 90, 137
Apparent
    direction, 33
    motion, 1, 353
    place, 33, 35
Areal velocity, 92
Argument of perifocus, 150, 154
Aries, first point of, 16
Ascending node, 150
    longitude of, 150
    vector, 152
Astrodynamic constants, 10, 557
Astrometric position, 35, 36
Astronomical Almanac, 14, 80, 134, 200, 252, 316, 534, 558
Astronomical unit, 10, 557
Atomic time, 18, 23
Attractions, 254
    direct, 255
    indirect, 255
    net, 255
    total, 259
Autumnal equinox, 16

B

Back-substitution, 325
Baker, R.M.L., 14, 200, 252, 316, 388, 411, 449, 481, 534
Barker's equation, 149, 206
Barycenter, 24
Barycentric dynamical time, 24
Bate, R.R., 134, 316, 534
Besselian day numbers, 36
Besselian star constants, 36
Bevington, P.R., 352
Bisection method, 319
Boulet, D.L., 388
Brouwer, D., 134, 316, 534
Butcher's fifth-order method, 102

C

Calculus, 1, 545
Canale, R.P., 134, 352
Carnahan, B., 134, 352

Celestial
　ecliptic systems, 30
　equator, 15, 25
　equatorial systems, 26
Celestial mechanics, 1
Celestial sphere, 15
Center of mass, 6
Central force field, 6
Ceres, 36
Chapra, S.C., 134, 352
Classical elements, 149, 157
Classical orientation angles, 150, 153, 161
Clemence, G.M., 80, 316
Closed f and g expressions, 207, 414
Combined mass, 11
Comet orbits, 451
Components of vectors, 3, 538
Conic sections, 85, 201
　ellipse, 89, 140, 201
　hyperbola, 89, 142, 204
　parabola, 89, 146, 206
　equation of, 89
Constants
　astrodynamic, 10, 557
　flattening of Earth, 557, 29
　gravitational, 10, 557
　light-time, 34, 35, 222, 393, 420, 455
Coordinate systems, 25
　ecliptic, 30
　equatorial, 26
　orbit plane, 135, 157
　terrestrial, 28
Coordinated universal time, 24
Coordinates, 15, 25
　reduction, 33
　transformation, 30
Correction for light-time, 33
Cowell's method, 257
CRC mathematical tables, 200, 252
Cross product, 542

D

Danby, J.M.A., 134, 252, 316, 352, 388, 449

Days, units of, 10
Declination, 26
Derivatives, 3, 545
　partial, 484, 549
　total, 484, 549
Determination of orbits, 1, 353, 389, 413, 451, 483
Differential correction, 262, 483
Differentiation
　analytical, 545
　numerical, 335
Direct motion, 87, 150
Direct attractions, 255
Direction cosines, 32
Diurnal motion, 28
Diurnal rotation, 20, 360
Dot product, 541
Dubyago, A.D., 316, 388, 411, 449, 481
Duffett-Smith, P.D., 80
Dynamical time, 18, 24

E

Earth's figure, 28
Earth
　constants of, 11, 557
　rotation rate, 360
Earth masses, units of, 11
Earth-Moon system, 82
East longitude, 23
Eccentric anomaly, 140
Eccentricity, 90, 150
　vector, 151
Ecliptic, 15, 30
　coordinates, 30
　latitude, 30
　longitude, 30
　plane, 15
Ecliptic-equatorial transformations, 30
Elements of an orbit, 95, 135, 149, 157
　classical, 149
　vector, 95, 135, 157
Elements of the precession matrix, 38
Ellipse, 89, 140, 201
Elliptic orbits, 90, 140, 201

INDEX 561

Encke's method, 258
Energy of an orbit, 95
Ephemeris time, 24
Epoch, 12, 15, 18, 35, 95
Equation
    of inertial motion, 8
    of Lagrange, 392, 418
    of motion, 7
    of relative motion, 10
Equations of condition, 331, 333, 483
Equator
    celestial, 15, 25
    of date, 35
    of Earth, 15, 28
    mean, 18
    true, 18
Equinox, 16, 35
    autumnal, 16
    vernal, 16
Errors in computation, 104
Escobal, R.P., 14, 80, 134, 200, 252, 388, 411, 449, 481, 534
Euler's equation, 452
Everhart, E., 388
Explanatory Supplement, 80

F

f and g expressions
    closed, 207
    universal, 217
f and g series, 95, 99
First point of Aries, 16
Fitzpatrick, P.M., 14
fixed
    equator, 35
    equinox, 35
Flattening, 29
Force, 2, 5, 6
Forward elimination, 324
Fundamental plane, 15, 17
Fundamental vector triangle, 31, 221

G

g-radii, units of, 11

Gauss elimination, 323
Gauss' method, 413
Gaussian gravitational constant, 10, 557
General perturbations, 253
General precession, 17, 37
Geocentric
    constants, 11, 557
    coordinates, 26, 357
    parallax, 39
    system of units, 11
Geocentric gravitational constant, 11
Geodetic latitude, 28
Geometric direction, 33
Gravitation, law of, 6
Gravitational constant, 6
Greenwich mean sidereal time, 21

H

Harmonic law, 92
Height, 29
Heliocentric
    constants, 10, 557
    coordinates, 26, 361
    system of units, 10
Herget, P., 316, 388, 411, 449, 481
Herrick-Gibbs method, 456
Horizontal parallax, 40
Hour angle, 39
Hyperbola, 89, 142, 204
Hyperbolic
    eccentric anomaly, 142
    functions, 204
    orbits, 204, 212
    radian, 143

I

Inclination, angle of, 150, 153
Increment function, 102
Indirect attractions, 255
Inertia, law of, 2
Inertial coordinate system, 2, 15, 16
Inertial frame of reference, 2
Inertial origin, 2, 15
Infinite power series, 95

Integration
 analytic, 551
 numerical, 95, 100, 102, 253
international atomic time, 23
Interpolation, 326
Instantaneous
 equator of date, 35
 equinox of date, 35

J

Julian century, 21
Julian date, 19
 modified, 20
Julian day numbers, 18
Julian epoch, 18
Juno, 36
Jupiter, 13

K

Kepler's equation, 142
 for ellipse, 142
 for hyperbola, 145
 for parabola, 149
 solution of, 212, 214, 217, 220
 universal formulation, 220
Kepler's laws, 85
King-Hele, D., 388
Kinetic energy, 95

L

Lagrange interpolating polynomial, 327
Lagrange's equation, 392, 418
Lagrangian interpolation, 327
Laplace's method, 389
Latitude
 geodetic, 28
 ecliptic, 30
Law of acceleration, 5
Law of action and reaction, 5
Law of areas, 91
Law of inertia, 2
Law of gravitation, 6
Laws of motion, 1, 2

Least squares, 329, 333, 356, 485
 multiple linear regression, 333
 polynomial regression, 329
Light-time correction, 33, 34, 222, 393, 420, 455
Local mean sidereal time, 23, 28
Longitude
 east, 23
 ecliptic, 30
Longitude of ascending node, 150, 154
Longitude of perifocus, 154
Longitude along the ecliptic, 30
Luther, H.A., 134

M

Makemson, M.W., 14, 200, 252
Marsden, B.G., 388, 449
Mars, 13, 129, 184, 304, 307
Mass, combined, 11
Matrix, precession, 38
Mean
 anomaly, 142, 145, 149, 150, 155
 ecliptic, 18
 equator, 18
 equinox, 18, 358
 longitude, 156
 motion, 142, 145, 148, 150
 place, 18, 35
 sidereal time, 21, 23
 solar time, 19
Mercury, 124, 127
Meridian, Greenwich, 19, 21
Minutes, units of, 11
Modified
 Julian date, 20
 time, 12
Momentum, angular, 86, 137
Montenbruck, O., 388
Moulton, F.R., 388, 449
Mueller, D.D., 200, 316, 534
Multiple linear regression, 333

N

Naive Gauss elimination, 323

INDEX 563

Newton's law of graduation, 6
Newton's laws of motion, 1, 2
Newtonian frame of reference, 2
Newton-Raphson's method, 321
Newton-Raphson formula, 322
Node, ascending, 150
Normal equations, 332, 334
Null vector, 86, 536
Numerical differentiation, 335, 485
Numerical error, 104
Numerical integration, 95, 100, 102, 253
Nutation, 17, 33, 36

O

Oblate spheroid, 16, 28
Obliquity of ecliptic, 15
Observational data, 1, 353, 355, 486
Optical measurements, 353, 355
Orbit plane coordinate system, 135, 157
Orbital elements, 95, 135, 149, 157
Orbits
    determination, 1, 353, 389, 413, 451, 483
    differential correction, 262, 483
Orbits
    elliptic, 89, 140, 201
    hyperbolic, 89, 142, 204
    parabolic, 89, 146, 206
Orbits
    two-body, 81
    perturbed, 253
Order
    of integration, 102
    of interpolation, 327
    of polynomial, 331

P

P-matrix, 38
Pallas, 36, 192, 245, 345, 525
Parabola, 89, 146, 206
Parabolic
    orbit, 89, 146, 206
    eccentric anomaly, 147
Parallax, 39

Partial derivatives, 484, 549
Partial pivoting, 323, 325
Perifocal distance, 90, 150
Perifocus, 89, 90
Period, 93,
Perturbations, 82, 253
Pivot coefficient, 325
    equation, 325
Planetary aberration, 33, 34
Planetary attractions, 254
Planetary and Lunar Coordinates for the Years 1984-2000, 80, 316, 558
Planetary masses, 558
Pluto, 36
Potential energy, 95
Precession, 17, 37
Preliminary orbit, 81, 353, 389, 413, 451
Principle axes, 25

R

Radial distance, 26
Radius vector, 2
Range, 32
Rate of change, 3, 4, 545, 546
Rectification, 259
Recurrence equations, 335
Reduction
    of coordinates, 33
    of observations, 355
    for aberration, 33
    for precession, 37
    for nutation, 36
Reference ellipsoid, 28
Reference orbit, 258
Regression
    polynomial, 329
    multiple linear, 333
Relative acceleration, 10
Relative motion, 10, 11, 12
Residuals, 331, 483, 484
Retrograde motion, 87, 150
Right ascension, 26, 221, 355
Roots, solving an equation for, 317
Rotation axes, 161

Roth, G.D., 388
Round-off error, 104
Roy, A.E., 134, 316, 388, 481
Runge-Kutta methods, 102

S

Scalar product, 541
Scalar, 535
Semimajor axis, 90, 149
Semiminor axis, 90
Semiparameter, 90
Sidereal time, 18, 20
Sidgwick, J.B., 388
Sinnott, R.W., 352, 388
Solar coordinates, 221, 357
Solar masses, units of, 10
Solution of an equation, 317
Special perturbations, 253
Speed, 3
Spiegel, M.R., 544, 555
Spring equinox, 16
Standard Julian epoch, 35
Station coordinates, 28
Stellar aberration, 34
Step-size, 102, 104
Sun, coordinates, 221, 357

T

Taff, L.G., 80, 134, 388, 411, 449
Taylor series, 100
Terrestrial equatorial systems, 28
Terrestrial dynamical time, 24
Thomas, G.B., 200, 534, 544, 555
Time zones, 19
Time, 18
    atomic, 23
    barycentric dynamical, 24
    coordinated universal, 24
    dynamical, 24
    ephemeris, 24
    Julian date, 19
    solar, 19
    sidereal, 20
    terrestrial dynamical, 24
    universal, 19
Time of perifocal passage, 142, 145, 149, 150, 157
Topocentric coordinates, 26, 355, 357
Total derivative, 484, 549
Transformation of coordinates, 30
Triple products, 543
True
    anomaly, 90, 137
    ecliptic, 18
    equator, 18
    equinox, 18
    place, 18
Truncation error, 105
Two-body orbit, 81
Two-body equation of motion, 82

U

Uniform motion
Uniform time, 24
Unit vector, 8, 85, 537
Units, 10
Universal formulation, 217
    variables, 217
Universal gravitation, 6
Universal time, 18, 19
Upper triangular system, 325
Uranus, 311

V

Van de Kamp, P., 352
Vector algebra, 535
Vector calculus
Vector elements, 95, 135, 157
Vector product, 542
Vectors, 535
    components of, 538
    difference of, 536
    equal, 535
    null, 536
    proper, 536
    sum of, 536
    unit, 537
    zero, 536

Velocity, 3
Velocity components, 3, 538
Vernal equinox, 16
Vesta, 36
Vis-Viva equation, 93, 95

W

White, J.E., 200, 316, 534
Wilkes, J.O., 134
Woolard, E.W., 80
Working equation of motion, 12
Working units and constants, 10

Z

Zero vector, 86, 536

Some Other Books Published by Willmann-Bell, Inc.

Astronomical Algorithms by Jean Meeus, 6.00" by 9.00", 440 pages, hardbound, $24.95, ISBN 0-943396-35-2. In the last few years the International Astronomical Union has introduced subtle changes in the reference frame used for the coordinates of celestial objects, both within and far beyond our solar system. So sweeping are these revisions that a highly respected work for professional astronomers, the *Explanatory Supplement to the Astronomical Ephemeris*, published in 1961, is now seriously out of date. While the technical journals have seen a flurry of scientific papers on these issues, this book is the first to offer succinct and practical methods for coping with the changeover. With 56 chapters devoted to a wide range of subjects you will find most of the subjects of interest to the amateur and professional astronomer. By presenting these astronomical algorithms in standard mathematical notation, rather than in the form of program listings, the author has made them accessible to users of a wide variety of machines and computer languages—including those not yet invented.

Fundamentals of Celestial Mechanics, by J.M.A. Danby, 6.00" by 9.00", 466 pages, softbound, $24.95, ISBN 0-943396-20-4. This text assumes an understanding of calculus and elementary differential equations. Emphasis is on computations. Sample BASIC program listings (for a PC) are included. Covered are the problem of two bodies including the use of universal variables, several methods (including that of Laguerre) for solving Kepler's equation, and three methods for solving the two point boundary value problem. The chapter on the determination of orbits includes two versions of Gauss' method, the application of least squares and an introduction to recursive methods. The chapter on numerical methods includes three methods for the numerical integration of differential equations, one of which has full stepsize control. There are also chapters on perturbations, the three- and n-body problems plus much, much more.

Astronomical Formulae for Calculators, by Jean Meeus, 6.00" by 7.00", 201 pages, softbound, $14.95, ISBN 0-943396-22-0. Now in its 4th edition, it is the one source to which thousands turn when they need both equations and worked examples. Almost every other astronomical math book now in print credits Meeus as a reference but they do not provide you with his detailed formulae. *The Journal of The Royal Astronomical Society of Canada* said: "In just under 200 pages, he has covered most of the calculations one is likely to encounter in practical, computational astronomy. The writing is crisp and clear throughout and the examples are thoughtful and well-formatted."

Elements of Solar Eclipses 1951–2200, by Jean Meeus, 8.50" by 11.00", 112 pages, softbound, $19.95, ISBN 0-943396-21-2. This book contains Besselian elements for the 570 solar eclipses during the 250 years between 1951 and 2200. The elements were calculated using highly accurate modern theories of the Sun and Moon developed at the Bureau des Longitudes of Paris. Formulae are provided for the calculation of local circumstances, points on the central line or the northern and southern limits, etc. These algorithms can easily be programmed on a home computer and checked against numerical examples in this book.

Transits, Jean Meeus, 8.50" by 11.00", 75 pages. softbound, $14.95, ISBN 0-943396-26-3. Transits of Venus across face of the Sun rank among the rarest astronomical phenomena—only 81 occur during the 6,000 year period spanning -2000 to $+4000$. Two transits of Venus will occur early in the next century (2004 and 2012). Transits of Mercury are somewhat more frequent—117 occur during the 700 year period $+1600$ to $+2300$. Four Mercury transits will take place between 1993 and 2006: 1993, 1999, 2003 and 2006. This book presents elements, geocentric data for all transits of Venus from -2000 to $+4000$ and Mercury from $+0160$ to $+2300$. These elements allow the calculation of local circumstances and Jean Meeus has provided all necessary formulae and worked examples to do this. Also presented is a discussion (without elements) of transits seen from other planets.

Introduction to BASIC Astronomy With a PC, J.L. Lawrence, 8.5" by 11.00", 130 pages, softbound, $19.95, ISBN 0-943396-23-9. Introductory astronomy books usually follow two broad paths; descriptive or mathematical. This book is fundamentally a mathematical approach but with a difference. The author has written the text and IBM-PC computer programs to emphasize concepts rather than derivation of formulas. In order that you can begin immediately to learn the how and why of astronomy, the programs (BASICA) are provided on a diskette (IBM-PC, 5.25-inch DSDD) along with this book.

Lunar Tables and Programs from 4000 B.C to A.D. 8000 Michelle Chapront-Touzé and Jean Chapron, 8.50" by 11.00", 176 pages, softbound, $19.95, ISBN 0-943396-33-6. This book provides high precision time-dependent expansions of the longitude, latitude, and radius vector of the Moon, referred to the mean ecliptic and equinox of date. For historians and others who do not need this full precision, more compact procedures are given. Included are formulae for computing coordinates referred to other reference frames and for corrections of aberration. Microcomputer program listings in FORTRAN, BASIC and PASCAL are provided to implement the tables and formulae presented in the book. The best accuracy of the longitude for a date in ephemeris time for periods between -4000 to $+8000$ is as follows: -4000 to -2000: $0.80°$, $-2000°$ to -500: $0.37°$, -500 to $+500$: $0.136°$, $+500$ to $+1500$: $0.0474°$, $+1500$ to $+1900$: $0.0054°$, $+1900$ to $+2100$: $0.0004°$, $+2100$ to $+2500$: $0.0054°$, $+2500$ to $+3500$: $0.0474°$, $+3500$ to $+4500$: $0.136°$, $+4500$ to $+6000$: $0.37°$, $+6000$ to $+8000$: $0.80°$.

Planetary Programs and Tables from -4000 to $+2800$ by Pierre Bretagnon and Jean-Louis Simon, 8.50" by 11.00", 150 pages, softbound, $19.95, ISBN 0-943396-08-5 Included in this book are formulae, tables and microcomputer program listings in FORTRAN and BASIC to compute the longitude of the Sun, the geocentric longitudes and latitudes of Mercury, Venus, Mars, Jupiter, Saturn, Uranus and Neptune with a precision that is always better than $0.01°$ over a period of time up to 12,000 years (Uranus and Neptune 1,200 years). The positions calculated using the procedures detailed in this book will be far more precise than is usually required by astronomers or historians working on the dating of historical events or documents. For the Sun through Mars each coordinate is represented by only one formula and Jupiter through Neptune for time-spans of five years by power series with seven coefficients.

Prices Subject To Change Without Notice